Pink Brain, Blue Brain

Pink Brain, Blue Brain

*How Small Differences Grow
into Troublesome Gaps—
and What We Can Do About It*

Lise Eliot, Ph.D.

MARINER BOOKS HOUGHTON MIFFLIN HARCOURT
Boston New York

First Mariner Books edition 2010

Copyright © 2009 by Lise Eliot

www.hmhco.com

Library of Congress Cataloging-in-Publication Data
Eliot, Lise.
Pink brain, blue brain : how small differences grow into troublesome gaps—
and what we can do about it / Lise Eliot.
p. cm.
Includes bibliographical references and index.
ISBN 978-0-618-39311-4
1. Sex differences. 2. Sex differences (Psychology) 3. Developmental
neurobiology. I. Title.
QP81.5.E45 2009 612.6—dc22 2009014746

ISBN 978-0-547-39459-6 (pbk.)

All illustrations by Michael Prendergast
Book design by Lisa Diercks
Typeset in Whitman

Printed in the United States of America

DOH 10 9 8 7 6 5

Illustration credits appear on page 356.

Contents

Acknowledgments

IF BOOKS ARE LIKE BABIES, this one was an overdue, high-risk delivery. Little did I know when I first conceived a book about sex differences in the brain how complicated and politically sensitive the topic was. Fortunately, I've had many capable midwives to help maneuver through the twists and turns.

Foremost has been Amanda Cook, my wise and probing editor, who's kept me on track and helped wrestle the data into a coherent frame. My fabulous agent, Kim Witherspoon, was always there to hold my hand, along with my dear friend Jenny Cox, who had great insight as both a mother and an editor. I am also grateful to Lisa Glover, Tracy Roe, and Melissa Lotfy at Houghton Mifflin Harcourt, and to Laura Van Dam, who saw the book's promise but who, sadly, I was never able to meet in person.

As for the place of the book's birth, I could not ask for a more supportive environment than Rosalind Franklin University of Medicine and Science. Recently renamed to honor this important female scientist, it is the most strikingly gender-balanced institution I've had the privilege of working at. Thanks to Marina Wolf, Michael Welch, Art Ross, and Michael Sarras for their unparalleled support of scientific scholarship, and to Kevin Robertson for zapping all that scholarship right to my laptop. Many other faculty, staff, and students have buoyed me along the way and I thank them all for making RFUMS such a warm and truly collegial place.

I began this project with a fellowship to the Helen Riaboff Whiteley Center of the University of Washington at Friday Harbor Laboratories and am grateful to Arthur Whiteley and Dennis Willows for creating such a beautiful, inspiring space to work. On the opposite coast, I was able to complete several chapters at the incomparable Marine Biological Laboratory at Woods Hole, Massachusetts, where I thank Colleen Hurter for facilitating all my interlibrary loans and Andrea Walters, Francine Coeytaux, Terry Walters, and David Glanzman for their encouragement and lively dinner discussions.

This book further gestated at the Erikson Institute, another Chicago treasure, where I'm indebted to Linda Gilkerson, Sharon Syc, and many other terrific colleagues who taught me as much about child development as I was able to impart about the brain.

For their continued support of my career, I am grateful to Eric Kandel, Dan Johnston, Margie Ariano, Fred Sierles, Carol Ann Stowe, and Christiana Leonard. I'm also indebted to Susan McGee Bailey for our collaboration on single-sex schooling and to Karen Adolph, Susan Levine, Peter Meijer, Scott Moffatt, Emma Nelson, and Elise Temple for allowing me to reproduce figures from their papers.

Many other scientists have helped by answering questions and sharing insights that don't always appear in their published articles. For their valuable correspondence, I thank Simon Baron-Cohen, Larry Cahill, Sue Carter, Celine Cohen-Bendahan, Lee Cronk, Lisa Donovan, Teresa Farroni, Richard Francis, Jay Giedd, Melvin Grumbach, Janet Hyde, Tatjana Ishunina, Marc Johnson, Sandy Kirkman, Yulia Kovas, Holly Krogh, Mark Lewis, Lynn Liben, Michelle Luciano, Eleanor Maccoby, Dennis McFadden, Helen Neville, Maritza Rivera-Gaxiola, Carolyn Rovee-Collier, Elizabeth Shirtcliff, Courtney Stevens, Andrew Stuart, Shanna Swan, Davida Teller, Claes von Hofsten, and Toni Ziegler. Though we may not agree on everything, I appreciate their open discussions and hope I've done justice to their diligent research.

On the personal side, I could not have completed this work without Veena Bhalala, who befriended and so ably cared for our children after school, and the many wonderful teachers at Dearhaven and the Lake Bluff public schools.

Most of all, I thank my family—my parents, Caryl and Allen, who have always been my bedrock; my brothers, Brian, Dan, and Bruce, and in-laws, Mary Helen and Frosty, who, though far away by land, are always near in my heart; and, especially, my wise and patient husband, Bill, who has never doubted me and has managed, miraculously, to keep the rest of the brood sane along the way. I feel enormously lucky to share this journey with him.

It is this brood of ours that deserves my last and deepest thanks. I can't say that writing this book has made me a better parent, but I do believe it's helped me better appreciate each child's unique mind and personality in ways that go far beyond pink and blue. I love them fiercely and regret all the time and attention this book has taken away from them. Though it's hardly repayment, I dedicate the book to Julia, Sam, and Toby.

Pink Brain, Blue Brain

Introduction

You're finally getting to know the new neighbors. They moved in a week ago, but you've had no chance to chat, which is surely why you didn't notice sooner that the woman is pregnant. *Very* pregnant, by the looks of it.

"How wonderful!" you croon over your common fence. "Do you know if you're having a boy or girl?"

Why is this always the first question we ask when learning about a new baby? The answer is simple: because sex is a *big* deal. Not just the act of it, but the fact of it. Of all the characteristics a child brings into the world, being male or female still has the greatest impact—on future relationships, personality, skills, career, hobbies, health, and even the kind of parent the child is likely to become. That's why 68 percent of expectant parents learn the sex of their child before birth and why you know your neighbor is naive to answer, "We really don't care, as long as the baby is healthy!"

Most American parents hope to have at least one child of each sex. We enjoy the differences between them, even as we worry about their consequences. Will this little boy, now so active and exuberantly affectionate, settle down enough to begin school? Will he form meaningful relationships with his friends and teachers? Will he still express his feelings or, for that matter, communicate with us at all when he grows up?

For parents of girls, the fears run in the opposite direction. Here she is, so confident and full of life. Will she still dig for worms and wonder about the planets when she's in middle school? Will she be assertive enough when she lands her first job out of college? Will it be any easier for her generation to juggle career and family when she grows up?

Boys and girls are different. This fact, obvious to every previous generation, comes as a bewildering revelation to many parents today. Raised in an era of equal rights, we assume—or at least hope—that differences between the sexes are made, not inborn. We mingle comfortably with members of the opposite sex, harangue as easily about sports as cooking, and cheerfully compete in the workplace—all the while pretending the two sexes are more or less the same.

Until we have kids of our own, at which point the differences are impossible to ignore.

Like many parents, I could cite endless examples of the differences between our daughter and two sons: Julia loves shopping, while Sam and Toby can barely be persuaded to try on jeans at the mall. Then there was the evening not so long ago that Julia spent drawing pictures of fairies while Sam and Toby raced around the house having a light-saber battle. Even as a young toddler, Julia would lay all our kitchen towels on the floor and then put a stuffed animal on each one for "nappy time." The only thing that absorbed Sam and Toby as much at that age was seeing how many objects they could jam inside a VCR.

Also, like other parents today, I feel compelled to excuse this gender-typical play with the obligatory "We certainly didn't encourage Julia to play only with girl toys and Sam and Toby to play only with boy toys." On the contrary, many of our kids' building toys—the wooden blocks, Duplos, and Lincoln Logs—were originally purchased for Julia, our oldest child. I try to make a point of praising the boys' nurturing behavior—like when Sam hugs Toby or cuddles his pet gerbil—and never stand in the way of their attempts to help me cook.

Of course, parents are never truly neutral about gender. Regardless of which toys or clothes we buy them, we cannot help but react in different ways to our sons and daughters—if only because of our own long experience of "male" and "female." But still, I had thought our kids would be different. As a neuroscientist and Martha Stewart dropout, I am hardly the typical female role model. My husband, also a scientist, fits the male type in many ways (he can fix almost anything around the house—as long as it doesn't interfere with *Monday Night Football*) but is kinder and gentler than many of the women I know.

And yet, there they are: Julia quietly makes paper flowers or sets up her Playmobil house while Sammy launches cars off his Hot Wheels track or begs me to pitch Wiffle balls to him outside. Even little Toby, with his clear, high-pitched voice, started steering the boy course early, judging from his toddler fascination with trucks, airplanes, balls, and any kind of electrical appliance.

Yes, boys and girls are different. They have different interests, activity levels, sensory thresholds, physical strengths, emotional reactions, relational styles, attention spans, and intellectual aptitudes. The differences are not huge and, in many cases, are far smaller than the gaps that separate adult men and women. Little boys still cry, little girls kick and shove. But boy-girl differences do add up, leading to some of the more alarming statistics that shape the way we think about raising our children.

Here's a stark one: Boys are at greater risk than girls for most of the major learning and developmental disorders—as much as four times more likely to suffer from autism, attention deficit disorder, and dyslexia. Girls, for their part, are at least twice as likely as boys to suffer from depression, anxiety, and eating disorders. Boys are 73 percent more likely to die in accidents and more than twice as likely to be the victims of violent crimes (other than sexual assault). Girls are twice as likely as boys to attempt suicide, but boys are three times likelier to succeed at it.

On the academic side, girls of all ages get better grades than boys. Women now constitute the majority of U.S. college students—a startling 57 percent. And yet males continue to score some twenty-five points higher on the SAT exam and outnumber females four to one in college engineering degrees. In spite of their educational gains, women earn less than eighty cents for every dollar earned by men.

Sex* matters. As much as we may strive to treat them equally, boys

---

* I use the term *sex* instead of *gender* because it is more scientifically correct even if it's less politically so. *Sex* is a biological attribute, defined by chromosomes and anatomic characteristics. It is a binary, either/or trait. *Gender*, by contrast, is a social construct, the sum of all the attributes typically associated with one sex. It is not fixed and binary but a fluid spectrum between masculinity and femininity. For instance, a person wearing a curly white wig would be considered feminine today but wouldn't have been in George Washington's era. While the behavioral traits in this book are properly referred to as *gender-typical,* all of the research has divided subjects according to their biological maleness or femaleness—in other words, by sex—and so this is the term I usually favor.

and girls have different strengths and weaknesses and face very different challenges while growing up. Boys are more vulnerable early in life: they mature more slowly, get sick more often, and are less likely to have mastered the language, self-control, and fine motor skills necessary for a successful start in school. In recent years, as academic expectations have intensified, boys' slower start is stretching into a significant handicap even into the middle-school and high-school years, where they trail girls in graduation rates, academic performance, and extracurricular leadership positions.

Girls pull through the early years more easily than boys, hitting their vulnerable phase around puberty, when their confidence slips, their math and science interests wane, and young womanhood comes to be defined by beauty and submissiveness. Then, after girls navigate the minefield of adolescence, they face even greater challenges out in the real world, where they struggle with the contradictions of ambition and femininity and the conflicting values of the workplace and child rearing.

These differences between the sexes have real consequences and create enormous challenges for parents. How can we support both our sons and daughters, protect them, and still treat them fairly when their needs are so very different?

I study the brain and believe we can't even hope to tackle these issues until we know where the differences come from. What is going on inside boys' and girls' heads that triggers such different interests, emotional reactions, and mental abilities? Are male and female brains fundamentally different from each other? Are boys and girls wired differently from birth?

When I set out to answer these questions, I figured it would be simple: just dig up the studies that compare boys' and girls' brains and link any differences between the two sexes to their emerging verbal, emotional, math, and other skills. In this way, I would be able to paint a clear picture for readers of how the brain develops in pink and in blue.

As a biologist, I know that the different hormones boys and girls are exposed to before birth can exert powerful effects on their later behavior. The receptors for such hormones are present in children's brains from an early age, where presumably they act—along with a handful of sex-specific genes—to shape the neural circuits that eventually underlie these boy-girl differences.

But what I found, after an exhaustive search, was surprisingly little solid evidence of sex differences in children's brains. Sure, there are studies that *do* find differences, but when I looked closely at *all* the data—not just the research that confirms what we already know about boys' and girls' behavior but a truly balanced collection of findings—I had to admit that only two facts have been reliably proven.

One is that boys' brains are larger than girls'; somewhere between 8 and 11 percent larger, depending on the study. This difference is present throughout life and can be a source of humor or defensiveness (depending on your sex), but it's conspicuously similar in magnitude to males' greater height and weight at birth and in adulthood.

The second reliable fact is the difference that shows up around the onset of puberty: girls' brains finish growing about one to two years earlier than boys'. Again, this mirrors the overall sex difference in children's physical growth, since girls also enter puberty a year or two before boys do.

I suppose one could weave all kinds of theories about how having a bigger brain makes boys more active, physically aggressive, and skilled at spatial and mechanical tasks. It's harder to figure out how having a *smaller* brain explains females' verbal and interpersonal advantages. The fact that girls' brain size peaks earlier than boys' does suggest some overall faster program of maturation. But there is precious little evidence to support this from electroencephalography (EEG) or any other measures of actual brain activity.

The reality, judging by current research, is that the brains of boys and girls are more similar than their well-described behavioral differences would indicate. Certainly, there are some data showing subtle sex differences in children's sensory processing, memory and language circuits, frontal-lobe development, and overall neural speed and efficiency. Throughout the book, I will evaluate such findings and explain their implications for children's behavior. But overall, boys' and girls' brains are remarkably alike. Just as boys' and girls' bodies start out more androgynous than they end up in adulthood, their brains appear to be less sexually differentiated than adult men's and women's.

This doesn't mean, however, that neuroscience can't teach us something about sex differences in children. It can teach us plenty. To see that, we must turn to another branch of brain science, a newer area that happens to be where my own research is focused.

It is the study of *plasticity,* an admittedly ugly term used to describe the very beautiful fact that the brain actually changes in response to its own experience. Just as petroleum-based plastics can be molded into endless varieties of grocery bags, milk jugs, pipe fittings, playground equipment, and more, so are our brains fantastically capable of modifying themselves to the jobs at hand. Every physical feature of the human nervous system—the brain cells, or neurons, that transmit information; their axons and dendrites that reach great distances to connect with one other; the tiny synapses that are the actual sites of connection; and the supporting cells, or glia, that keep it all going metabolically—responds to life experiences and is continually remodeled to adapt to them. The brain changes when you learn to walk and talk; the brain changes when you store a new memory; the brain changes when you figure out if you're a boy or a girl; the brain changes when you fall in love or plunge into depression; the brain changes when you become a parent.

Plasticity is the basis of all learning as well as the best hope for recovery after a brain injury. And in childhood, the brain is far more plastic, or malleable, than it is at any later stage of life—wiring itself in large measure according to the experiences in which it is immersed from prenatal life through adolescence.

Simply put, your brain is what you do with it. Every task you spend time on—reading, running, laughing, calculating, debating, watching TV, folding laundry, mowing grass, singing, crying, kissing, and so on—reinforces active brain circuits at the expense of other inactive ones. Learning and practice rewire the human brain, and considering the very different ways boys and girls spend their time while growing up, as well as the special potency of early experience in molding neuronal connections, it would be shocking if the two sexes' brains *didn't* work differently by the time they were adults.

So it's all biology, whether the cause is nature or nurture. Sex differences in behavior *must* be reflected as sex differences in the brain, but the older children are, the less confidently their differences can be ascribed exclusively to genes and hormones. There are, to be sure, a few truly innate differences between the sexes—in maturation rate, sensory processing, activity level, fussiness, and (yes!) play interests—which I will describe in detail in the next few chapters.

However, the male-female differences that have the most impact—cognitive skills, such as speaking, reading, math, and mechanical ability; and interpersonal skills, such as aggression, empathy, risk taking, and competitiveness—are heavily shaped by learning. Yes, they germinate from basic instincts and initial biases in brain function, but each of these traits is massively amplified by the different sorts of practice, role models, and reinforcement that boys and girls are exposed to from birth onward.

Scientists themselves no longer pit nature and nurture against each other as distinct, warring entities but appreciate that they are intricately interwoven. Obviously, boys and girls come into the world with a smattering of different genes and hormones. But actually growing a boy from those XY cells or a girl from XX cells requires constant interaction with the environment, which begins in the prenatal soup and continues through all the dance recitals, baseball games, middle-school science classes, and cafeteria dramas that ceaselessly reinforce our gender-divided society.

Biologists refer to this interaction as *epigenetic*—the environment acting on or through our genes—and every human attribute is shaped in this way. Height, for example, is strongly determined by genes, and yet no matter the genetic potential, a child cannot grow tall if he or she is undernourished. Weight is another example that's strongly biased by genes but ultimately determined by a child's diet, eating habits, and environment (which is now, in many parts of the world, absurdly calorie-laden). Mental traits are even less heritable than height and weight—typically around 50 percent for most measures of intelligence and personality—but this genetic potential too is meaningless outside of the environment in which it develops.

Consider language, our most distinctly human ability. Every healthy child is born with special areas, usually in the brain's left hemisphere, that give him or her the innate capacity to understand and produce speech. But actually molding these areas into a full-fledged linguistic circuit depends on language itself—on a baby hearing literally millions of words in meaningful contexts throughout the first few years of life. We know this because children who are born deaf suffer permanent language deficits if their deprivation lasts beyond the first several years. The areas of the brain that are innately biased for language simply do not develop into this circuit in the absence of normal language experience.

Deafness is the most extreme example, but the power of experience is evident in all children, whose native languages and dialects reflect the wiring of their brains in response to the precise languages they're reared in. That's why a baby can be adopted halfway around the globe from where he or she was born and grow up sounding just like his or her adoptive community. We call this early learning phase the critical period, and it begins at birth and ends around puberty. Adults *can* learn new languages, of course, but it is hard, hard work—nothing like the effortless, instinctive learning that takes place in early childhood—and the results are nearly always less than perfect.

Growing up as a boy or girl is a lot like being immersed in one of two different languages from birth. Boys' and girls' brains are not identical from the start—the few extra genes on a boy's Y chromosome* harness a cascade that clearly influences later behavior and, presumably, underlying brain structure. But fully developing the mental characteristics known as male and female also depends on each child's automatic immersion into either the male or female culture, each one as powerful as the lullabies we sing and the nourishment we provide.

Admittedly, this is not the story parents have been hearing in recent years. If you've read anything about boy-girl differences, you're probably become convinced that scientists have discovered all kinds of disparities in brain structure, function, and neurochemistry—that girls' brains are wired for communication and boys' for aggression; that they have different amounts of serotonin and oxytocin circulating in their heads; that boys do math using the hippocampus while girls use the cerebral cortex; that girls are left-brain dominant while boys are right-brain dominant.

These claims have spread like wildfire, but there are problems with every one. Some are blatantly false, plucked out of thin air because they sound about right. Others are cherry-picked from single studies or extrapolated from rodent research without any effort to critically evaluate all the data, account for conflicting studies, or even state that the results have never been confirmed in humans. And yet such claims are nearly always presented to parents, with great authority, as well-proven and dramatic facts about boys' and girls' brains, with seemingly dire implications.

For now, I want to point out one particularly insidious way in which

---

* The human Y chromosome (by far the smallest of the forty-six chromosomes) contains about sixty genes; compare this to the approximately eight hundred genes on the X chromosome and some twenty-five thousand in the total genome. This means that males and females *share* roughly 99.8 percent of their genes.

neuroscience has been misused: this is the idea that the brain's sex differences—most of which have been demonstrated in adults only—are necessarily innate. Ignoring the fundamental plasticity by which the brain learns anything, several popular authors confuse *brain* with *nature,* promoting the view that differences between the sexes are fixed, hard-wired, and predetermined biological facts.

Psychologist Michael Gurian is perhaps the most prominent of those stressing the nature side as being responsible for boys' and girls' different troubles. Much the same tack is taken by physician Leonard Sax, who trumpets "biologically programmed" differences in his book *Why Gender Matters.* And consider this from psychiatrist Louann Brizendine in her recent book *The Female Brain:*

> The female brain has tremendous unique aptitudes—outstanding verbal agility, the ability to connect deeply in friendship, a nearly psychic capacity to read faces and tone of voice for emotions and states of mind, and the ability to defuse conflict. All of this is hardwired into the brains of women. These are the talents women are born with that many men, frankly, are not.

All I can say is "Thank goodness I'm a woman!"

No question, the tide has turned: sex differences in behavior are widely known and well accepted. We no longer pretend that men and women, boys and girls are fundamentally the same. And somehow—reflecting our current infatuation with genetics, as well as the convenient fact that neither sex appears globally smarter than the other—we have become comfortable with attributing these differences to innate, inborn, intrinsic, or hard-wired causes.

But how did we get from accepting the different strengths and weaknesses of each sex to assuming these differences are fixed by nature?

The answer lies in a fundamental misunderstanding of biology. Yes, men and women are different. Yes, their brains are different. (They pretty much have to be, if you take the modern scientific view that the brain is responsible for all thoughts and feelings.) But—and this is the point lost on most popular interpretations of neuroscience—nearly all of the evidence for sex differences in the brain comes from studies of *adult* men and women. Who's to say that such differences are caused by nature and not by learning—by the thirty or so years of living as a male or female that any research subject invariably carries into the MRI scanner?

Sex differences in the brain are sexy. You can hardly pick up your favorite newsweekly without coming across some reference to this research. It gives an easy explanation for the bewildering Mars/Venus gaps we find so amusing: why men can't multitask and women can't read maps; why men love *The Three Stooges* and women have their chick flicks. So anytime researchers put a bunch of men and women in MRI machines and find some difference between their brains, we say, "Aha!"—as if that explains everything.

But there are some serious problems with the research on sex differences in adults and many more when this research is extended to children. One problem is what statisticians call the file-drawer effect—the fact that a study finding a statistically significant *difference* between men and women is simply more interesting and, therefore, more likely to be published than a study finding *no difference* (leaving the more boring no-difference result to languish in researchers' file drawers).

Another problem, which is also due to the sexy nature of this research, is that such data rarely get verified before they hit the media. Take one claim that has been widely reported in the popular literature: that the corpus callosum, the major fiber bundle that connects the two sides of the brain, is proportionally larger in women than men. This finding was first published in 1982 in the prestigious journal *Science,* even though it was a tiny study based on five female and nine male brains. No matter. TV talk show host Phil Donahue soon proclaimed it the basis for "women's intuition," a remark echoed in both *Time* and *Newsweek* magazines.

What the news media has not reported, however, is the conclusion of an authoritative 1997 review of fifty such studies, which collectively found no significant sex differences in the corpus callosum of adults. Nor have more recent studies detected any difference between the corpus callosum of boys and girls or between male and female fetuses. The point is: real data don't matter if educators and psychologists can find some nugget somewhere that confirms their preexisting theories. So you can still find Michael Gurian and Leonard Sax talking about the corpus callosum in relation to differences between boys' and girls' learning. For the record: the corpus callosum does *not* differ between boys and girls in any meaningful, statistically significant way.

Also for the record: there is a lot of good research on sex differences in the adult brain—research that has been replicated in different labs around the globe and that may have important implications, especially for treating diseases like depression, ADHD, Alzheimer's, schizophrenia, and drug addiction.

But when it comes to differences between boys and girls, and even most psychological gaps between men and women, the fact is that the gaps are much smaller than commonly believed and far from understood at the level of the brain or neurochemistry.

I promise I won't get too technical in this book, but there is one statistic we cannot avoid if we're really going to tackle this thorny topic. It's called, simply, *d,* for "difference value." You calculate it by subtracting the mean score of females on a given test from the mean score of males and then dividing the result by the standard deviation of both groups (which is basically a measure of the overall spread of abilities, or the width of the curve).

The d values range from small positive to small negative numbers: d is positive for traits or skills at which males outscore females, and d is negative for traits or skills at which females outscore males (a convention that may strike females as offensive). Also by convention, differences are considered small when the the d value is around 0.2 (either positive or negative), medium when the d value is around 0.5, and large when the d value is 0.8 or higher.

So how big are the behavioral and psychological differences between males and females? Much smaller, it turns out, than our physical differences are and, notably, quite small compared to the range of performance *within each sex.*

On page 12, for example, are graphs showing the distribution of measures for two traits, one with a d value of 2.6, and one with a d value of 0.35. The curves on the left show the difference between the sexes in adult height, a gap that is considered very large because the six-inch difference between men's average height (5 feet 10 inches) and women's average height (5 feet 4 inches) is considerably greater than the 2.3-inch standard deviation that describes the spread of each curve.*

---

* In a normal-distribution, or symmetrical, bell curve, the mean, or average, is at the top and middle of the curve, and the standard deviation includes about two-thirds of the population, split evenly on either side of the mean.

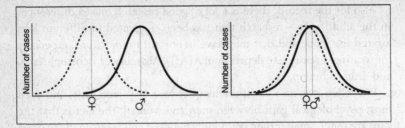

Figure 0.1. Distribution of performance for two traits that differ with d values of 2.6 and 0.35, respectively. Females are represented by the dashed curve, males by the solid curve. Mean score for each sex is shown by the vertical line at the middle of the curve. The graph on the left shows the sex difference in adult height, which is considered very large and for which there is little overlap between men and women. The graph on the right shows the distribution for a sex difference with a d score of 0.35, which is actually on the large side for many psychological differences. Note that the curves overlap extensively. Of the many psychological sex differences that have been repeatedly measured, 77 percent are *smaller* than the difference between the curves on the right.

A very large difference like height is obvious, noticeable, and has predictive value. If I told you that a relative of mine is six feet tall, you'd likely guess that that person is male, and you'd be correct. (My brother and dad are both this height, though come to think of it, I had an aunt who was also six feet tall.) There is not much overlap between the curves of male adult height and female adult height, even though we all somehow manage to drive the same cars and sleep in the same beds.

Now consider the curves on the right, which show a small-to-moderate difference. The d value of 0.35 in this graph is close to what's seen in measurements of sex differences on standardized science-test scores or (if the male-female curves were swapped) in evaluations of verbal fluency (that is, the speed and accuracy of speech). The curves obviously overlap through most (76 percent) of their range, and the difference between average males and average females is small compared to the ability range within each sex. This means that, in spite of our group differences, plenty of women are better than the average man at science, math, and asserting their opinions, while loads of men outperform the average woman at speaking, reading, and deciphering other people's feelings. More important, the wide range within each sex and the substantial overlap between sexes mean that you can never use these data to make predictions about individuals—that Johnny will have trouble learning to read or controlling himself in the classroom, or that Susie will struggle with fractions or is too timid to play hockey.

According to Janet Hyde, a psychologist at the University of Wisconsin who's dedicated her career to quantifying the various differences between the sexes, only a limited number of sex differences even make it out of the small range. Out of 124 extensively analyzed psychological traits that show differences between men and women—from self-esteem and moral reasoning to aggression, throwing ability, sex drive, and all types of academic skill—96 were in the small-difference range, with d values of less than 0.35. Dr. Hyde titled her article "The Gender Similarities Hypothesis" to emphasize the fact that for most psychological traits, sex differences are quite small, and, fundamentally, men and women are more similar than different.

Or as others have succinctly put it: "Men are from North Dakota, women are from South Dakota."

There's no question that small differences can add up, especially at the extreme ends of the curve, where there is much less overlap between the sexes. This distribution helps explain why issues like dyslexia, ADHD, and anxiety disorders affect one or the other sex more profoundly. And at the other end of the curve, it may also explain why so few women make it into the most elite ranks of mathematics and science, an argument former Harvard president Larry Summers famously put forth to explain the limited number of women professors in these fields.

But the problem with focusing on extremes—and the reason Dr. Summers ran into trouble for invoking them—is that it gives a warped sense of the typical differences between the vast majority of boys and girls, and men and women. Sex differences between the *mean* male and female are nowhere near as great as those between the *extremes,* but it's only the extremes that make headlines—an all-girls list of top-ranking students or an all-boys math team or special-education class. These are the noticeable differences that set people off and fuel the idea that children's abilities are limited from the outset by the X or Y chromosomes they inherit.

In other words, focusing on extremes leads to stereotyping. Our big, magnificent human brains love to find categories in the world, and so anytime people see a difference, especially one as emotionally significant as sex, we tend to exaggerate it.

Stereotyping harms both sexes, though women have traditionally been the ones more wary of it. I remember well at the beginning of my own scientific career, during a seminar about the effects of sex hormones on brain cells, some prominent scientists stood up to protest the findings as "dangerous" and "pointless," given their potential for misuse. Just imagine, the argument continued, what would happen if neuroscientists started publishing articles about differences between the brains of blacks and whites.

But the tide has turned dramatically in the last twenty years, ironically because of the new focus on women's intellectual strengths: the verbal and relational skills that Louann Brizendine boasts about and which most people agree balance the spatial and analytical strengths of men. When you put these together, it's clear neither sex beats the other. So what's all the fuss?

The downside of this emphasis on difference equality has been a brand-new wave of stereotyping, fueled by the careless proselytizing of Brizendine, Gurian, Sax, and others. The true, small size of the differences are ignored, replaced in our categorizing minds with large Mars/ Venus chasms, which parents and teachers accept as facts and believe they can do nothing about.

Unlike a generation ago, when parents actually worried about stereotyping their children, the new focus on nature seems to be encouraging parents to indulge sex differences even more avidly. Boys and girls are identified in utero and the nurseries painted to match months before birth. From girls' preschool ballet lessons and makeovers to boys' peewee football, hockey, and baseball leagues, our world is in many ways more gender divided than ever. The more we parents hear about hard-wiring and biological programming, the less we bother tempering our pink or blue fantasies, and start attributing every skill or deficit to innate sex differences. Your son's a late talker? Don't worry, he's a boy. Your daughter is struggling with math? It's okay, she's very artistic.

Even teachers are now preaching the gospel of sex differences, goaded on by bad in-service seminars, by so-called brain-based learning theories, and, perhaps, by the chance to excuse their own lack of success with one or the other gender, usually boys.

And fueling it further is the massive pink-versus-blue (or make that pink-versus-Nintendo-black) marketing of dolls with ever thinner waists and action figures with ever broader shoulders. While some parents still fight valiantly to avoid stereotyping, the larger culture has embraced it with a vengeance. In today's hypermarketed world, what niche is easier to exploit than male or female? We now have magazines, movies, and even entire TV networks dedicated exclusively to men or to women, slicing the culture into ever more gender-segregated pieces. And kids, our most naive consumers, eat it up. They have their own endless TV shows, DVDs, and websites—a round-the-clock indoctrination into the world of brave, cheeky boys and cute, squeaky girls well before they even start kindergarten.

Even today's college students embrace the differences more than the students of a generation ago. For twenty-five years, sociologist Lloyd Lueptow has administered the same survey about sex roles to his classes at the University of Akron. In spite of the dramatic changes in women's achievement over this period, college students today actually perceive *greater* differences between men and women than students did in the 1970s, especially when it comes to the labels "sympathetic," "talkative," "responsible," "friendly," and "affectionate." And although women are inarguably more athletic and ambitious than a generation ago, college-age men grant them only greater "decisiveness" while viewing themselves as more "aggressive," "adventurous," and "self-confident" than male college students rated themselves in the 1970s. So even though objective measures—such as math scores, sports participation, and graduate-school enrollment—show that many differences between the sexes have decreased over the past several decades, society's beliefs about male-female differences have actually become more exaggerated.

We love this stuff. It's fun to be different. It makes romance more exciting and provides endless fodder for late-night comedians. *But there's enormous danger in this exaggeration of sex differences*, first and foremost in the expectations it creates among parents, teachers, and children themselves. Kids rise or fall according to what we believe about them, and the more we dwell on the differences between boys and girls, the likelier such stereotypes are to crystallize into children's self-perceptions and self-fulfilling prophecies.

My goal is to do just the opposite: to present the real magnitude and multiple causes of sex differences in children; to better expose the full range of human potential and the many domains of intelligence that merit cultivating in both sexes.

This book takes a chronological tour of sex differences at each stage of development, from conception through adolescence. At each point,

I'll explain what's known about differences in the brains and behaviors of boys and girls, as well as the relative roles of genes, hormones, and environment in creating those differences. Based on this information, I'll suggest specific ways that parents and teachers can help narrow the more worrisome gaps.

Boys and girls do have different interests, abilities, and personalities, which is part of the fun of having children of both sexes. But what boy or girl, or man or woman, wouldn't be more successful with a fuller deck of cognitive and emotional skills? Studies of gifted teenagers confirm that intelligence and academic excellence are associated more with cross-gender abilities and less with stereotypical gender roles. There can be no doubt that success in our world increasingly requires a mixture of male and female strengths—speaking, reading, writing, math, spatial ability, mechanical dexterity, and physical skills, along with equal measures of empathy and ambition, diplomacy and assertiveness. The earlier we can step in and tweak kids' growing neurons and synapses, the better our chances of raising *both* boys and girls with well-balanced sets of skills.

The task is not simple. Some sex differences are most prominent in early life and then fade as they are trained out of children. Others emerge quite gradually and reach their peak in adulthood as boys and girls grow into their different niches. Nature exerts its pull at all ages, but what is most important is how children *spend their time.* The more similar boys' and girls' activities are, the more similar their brains will be. That's why, as we'll see throughout this book, the cognitive and academic differences between boys and girls tend to be much smaller than their interpersonal and recreational differences. Children are explicitly taught how to read and calculate but not how to take risks or express their emotions. While we can't erase—nor would we want to—all the differences between boys and girls, it's clear that the *size* of the gaps depends on what parents emphasize and how teachers teach.

Nor is the pattern of sex differences as simple as most people assume. Hidden among the obvious differences are some surprising exceptions and a lot of underlying similarities. Yes, we can measure average boy-girl differences in activity levels, empathy, reading, and math scores. But these generalities obscure important subdomains that each sex excels at—such as boys' rich vocabularies, girls' talent for arithmetic calculation, and the unique though sometimes troublesome brands of competition and aggression that members of *each* sex engage in.

So it's time to get *specific* in our discussion of sex differences, to move beyond the broad stereotypes to a more precise understanding of how boys and girls differ. Only then can we determine how such differences emerge and decide what best to do about them.

The real story starts in the womb, the topic of chapter 1, where we'll see how a few genes and hormones kick off the sexual differentiation of both brain and body. We'll also consider ways in which boys and girls are detectably different before birth. Chapter 2 looks at sex differences in newborns, the only age at which the differences can unambiguously be attributed to nature, but also the point when powerful social learning begins. Chapter 3 focuses on the toddler and preschool years, when children themselves first become aware of gender. The differences at this age are in some ways more dramatic than at any other time of life. Chapter 4 compares girls and boys at the beginning of formal schooling, a critical time for forging academic identities. In each of these chapters, we'll explore the true magnitude of boy-girl differences, observe the advantages and disadvantages of each sex's interests and play styles, and find ways to capitalize on the nurture side to help both boys and girls develop more fully.

Our developmental trajectory continues in the second part of the book, though with a special emphasis in each chapter. Chapter 5 focuses on language and literacy skills, a dominant factor in children's early school success and an area of special concern for boys. Chapter 6 looks at math, science, and technical skills, which begin diverging between the sexes somewhat later and remain challenging for many girls. Chapter 7 takes aim at the emotional and interpersonal differences between boys and girls, which are not taught in school but arguably factor in even more importantly than all the academic gaps we monitor so closely. Such differences are generally modest in early childhood but grow more dramatic after puberty, an age when nature (hormones) and nurture (peer environment) become inextricably tangled. I end, in chapter 8, with some perspective on how gender gaps have morphed in recent years, as well as an evaluation of the relative advantages of coed versus single-sex schools for dealing with those gaps.

Boy-girl differences have always been fodder for controversy, but in the last two decades, the topic has taken a sour political turn. It started with a girl crisis, back in the *Reviving Ophelia* days, when people worried about girls' crashing self-esteem and their lack of attention from teachers in school. That movement served a valuable purpose, but all the focus on girls was interpreted by some as a "war on boys" cooked up by feminists. Now we're racked by an epidemic of "boys adrift"—inattentive, unmotivated, and unable to compete with girls academically or to launch themselves beyond their parents' protection.

It's time for a truce. The problem with each crisis is that it has demonized the other sex, pitting boys and girls against each other, as if learning and achievement were zero-sum games. The truth, however, is that neither sex is in serious trouble. Yes, there remain some conspicuous gaps between boys and girls, the causes and remedies of which we'll address throughout this book. But the difference in achievement between the sexes remains much smaller than the gaps in achievement among different racial and economic groups, where we should no doubt be directing more of our energy. Nor have the gender gaps changed precipitously in the last two decades. In fact, both sexes are earning higher grades, graduating from high school at higher rates, and attending college in greater numbers than ever before. Neither sex is sinking into the abyss predicted by each wave of crisis books.

As a mother of both a daughter and sons, I believe we've got to find a better balance. Both sexes have their strengths and vulnerabilities, their easy and troublesome periods while growing up. The reason for studying sex differences is not to tally up who's winning or losing but to learn how to compensate for them early on, while children's brains are still at their most malleable. How can we help boys express their feelings, learn to read and write better, and feel at home in the school classroom? How can we help girls stay confident in math, learn how to read a map, and embrace technology and competition?

A better understanding of sex differences and neural plasticity can help us raise better children—teach us how to make the most of their strengths and help forestall their weaknesses. Piecing together the different influences at each stage is the only way to truly understand them, equalize opportunity between the sexes, and, ultimately, bring out the best in every child.

1

Pink and Blue in the Womb

YOUR PERIOD IS ONLY one day late, but you can't wait any longer. You do the test, waking up to pee on a stick at 6 A.M. Then you wait, smiling nervously while your husband stares at the result window that's shaking subtly in your hand. Finally, some faint lines begin to emerge. You flash back and forth between the tester and the instruction diagrams. . . . It's looking good. Yes, it's definitely looking like a positive. Your eyes meet. It's going to happen! You're going to have another baby!

"Oh, please, let it be a boy this time," your husband blurts out; it's still too early in the day for him to keep his deepest wish in check. You share his hope, your two daughters having satisfied all your frilly-dress dreams.

Home pregnancy tests are amazing, but they can't yet tell the sex of a baby. That's partly because the earlier it is in pregnancy, the less difference there is between the sexes. Boys and girls are identical for the first six weeks* of development. While many expectant parents begin fanta-

---

* The sixth week of fetal development is actually the eighth week of pregnancy according to the dating system used by physicians and midwives. Of course, development begins at conception, but clinicians traditionally time a pregnancy from the first day of the last menstrual period, an unambiguous date that typically falls two weeks before ovulation and conception. So there is a two-week discrepancy between the true developmental or *conceptual* age of a fetus and the *menstrual* age; the latter is what medical practitioners use when referring to the forty-week pregnancy. Actual gestation time is therefore thirty-eight weeks.

sizing about one or the other sex from the very beginning of pregnancy, the embryo itself appears conspicuously uncommitted.

Sexual differentiation begins about midway through the first trimester but isn't obvious by ultrasound until the end of that trimester, at the absolute earliest. Fetuses take their time before presenting themselves as clearly male or female on the outside. Inside their brains, sexual differentiation is slower still.

Nonetheless, differences in the brain and mind do take root before birth. You can't see it in an ultrasound scan or hear it in the fetal heartbeat, but boys' and girls' brains are influenced in the womb by their different genes and hormones. We know this from their different medical risks and especially from studies of children who for some reason were exposed to abnormal amount of levels of sex hormones before birth. For eons, old wives have claimed that there are ways to tell the difference between male and female pregnancies, and there is some reason to think this may be possible, even if the methods are not yet rigorously proven.

In this chapter, we'll explore the fascinating process of sexual development before birth. How do the sex chromosomes work? When are the sex hormones, including testosterone and estrogen, first produced? Are there any differences in the behavior or abilities of boy and girl fetuses? Is there any truth to the folklore about differences between male versus female pregnancies? And most important, what is the impact of these first nine months on the later differences between boys and girls?

The Moment of Chance

Development may be an intricate dance of nature and nurture, but here's one instance when nature unambiguously sets the ball rolling.* A baby's sex is decided at the moment of conception, when one sperm, carrying either an X or a Y chromosome from dad, penetrates one egg, carrying one of mom's two X chromosomes. If an X-carrying sperm wins the race, the merger produces an XX genotype, and the fertilized egg, or zygote,

---

* Sex determination is not as cut-and-dried in all animals as it is in humans. Some lizards and turtles, along with all species of crocodiles, have no sex chromosomes; they become males or females depending on the temperature at which their eggs are incubated. Other species, especially certain fish, can change sex as adults on the basis of which sex has the greatest chance for reproductive success in that particular environment.

will grow into a girl. If it is a Y sperm that outswims its five hundred million competitors, the baby will be a boy.

You can't see the difference between male and female zygotes, even under the highest-power microscope. Doctors performing in vitro fertilization (IVF) screen tens of thousands of fertilized eggs each year, but separating male from female is a complicated, somewhat risky procedure involving the removal and testing of one cell from an eight-cell embryo. Known as preimplantation genetic diagnosis (PGD), this procedure can be used to identify male embryos for mothers known to be carriers of sex-linked recessive genetic disorders, such as hemophilia, Duchenne's muscular dystrophy, and fragile X syndrome (which produces mental retardation in affected boys). Such disorders have a 50 percent chance of afflicting a carrier mother's son, who inherits his only copy of the X chromosome from one of his mother's two X chromosomes.

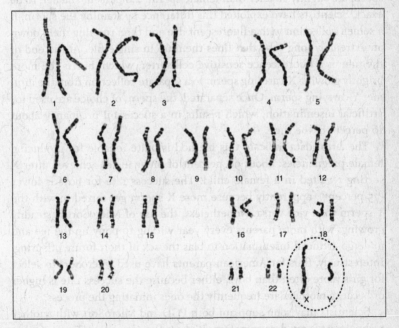

Figure 1.1. Male human chromosomes magnified about two thousand times, with the sex chromosomes circled. Of the forty-six total chromosomes, the only difference between the sexes is the presence of the small Y chromosome in boys. Girls have two copies of the X chromosome but no Y chromosome.

More recently, however, some doctors have extended the use of PGD for the purpose of "family balancing"—permitting parents who already have one or more children of one sex to select a fetus of the other sex. It's an expensive procedure and controversial, since it involves the creation and disposal of unwanted embryos. Nonetheless, an increasing number of parents are using PGD to guarantee that their next baby will be the "right" sex.

For parents who don't require in vitro fertilization, there is another scientifically based method to select their baby's sex. Called MicroSort, it was developed in 1998 and, like PGD, is performed in just a few reproductive centers in the United States, for a hefty fee. In this case, the sorting takes place before fertilization and is based on a subtle difference between Y-carrying and X-carrying sperm. Because the Y chromosome is only about one-third the size of the X chromosome, male sperm are slightly smaller than female sperm: 2.8 percent smaller, to be exact. Scientists have exploited this difference by staining the sperm in a semen collection with a fluorescent dye and then running them down an extremely long tube that lines them up in single file. At the end of the tube is a fluorescence-sensitive cell sorter, which diverts the more brightly glowing X-carrying sperm to a separate collection from the dimmer Y-carrying sperm. Once separated, the sperm of choice are used for artificial insemination, which results in a successful pregnancy about 10 percent of the time.

The latest data indicate this method is quite reliable for producing female pregnancies: about 90 percent of pregnancies achieved after X sorting resulted in a female child. The success rate for boys is lower (75 percent), apparently because more X sperm get mixed in with the Y sperm than vice versa. Nonetheless, the use of MicroSort is steadily growing, with more parents every year willing to pony up the fee and undergo artificial insemination to bias the sex of their future offspring. Interestingly, thus far American parents have used MicroSort to select for girls more often than boys, either because the success rate is higher or because mothers are frequently the ones initiating the process.

Scientists may soon supplant both PGD and MicroSort with another technology for sex determination. It's based on the recent discovery that small amounts of fetal DNA actually cross the placenta and enter a pregnant mother's bloodstream. Using probes for specific DNA sequences located on the Y chromosome, researchers have shown that they can screen a woman's blood as early as seven weeks into pregnancy and de-termine whether she is carrying a male fetus.* Already, a slew of genetic entrepreneurs have hung out their Internet shingles promising "guar-anteed" sex determination for just a few hundred bucks and a couple of drops of blood collected at home. Such testing is not subject to regula-tion by the Food and Drug Administration, so there is no proof of claims of "as high as 99.9 percent accuracy." One company, Acu-Gen of Lowell, Massachusetts, is even facing a class-action lawsuit brought by parents who were given the wrong results using Acu-Gen's Baby Gender Mentor testing service—usually, parents were told they were having a boy when in fact the mother delivered a girl. But as long as they stay legal, such services will inevitably become widespread. The opportunity to find out a baby's sex so early and in such a private, inexpensive, and noninvasive way is simply too alluring to avoid exploitation in this Wild West era of genetic testing.

Sex selection: high-tech, low-tech, and the ethics of choosing babies by gender

MicroSort and PGD are only the latest in a long history of methods par-ents have tried to bias the sex of their future children. Most have been pure folklore, no more scientific than relying on the direction of the wind (Aristotle's theory) or on which side of the bed the husband hangs his pants (right for a boy, left for a girl, according to medieval legend). However, things turned more empirical in the 1960s and 1970s, when scientists began characterizing differences between X- and Y-containing sperm. One of these scientists was obstetrician Landrum Shettles, who detailed his theory in the paperback guide *How to Choose the Sex of Your Baby*. Republished as recently as 2006, this widely read book claims a success rate of 75 percent for girls and 80 percent for boys. Though origi-nally advising a combination of measures (specific intercourse positions and precoital douches to favor male or female conceptions), Dr. Shettles eventually settled on the *timing* of intercourse as the key factor. Male sperm, he argued, are not only smaller but also faster and shorter lived, so intercourse closer to the moment of ovulation (egg release) is claimed to result in more male conceptions, while intercourse a few days before

---

* DNA from female fetuses is also detectable in a mother's blood, but it's much more difficult to distinguish from the mother's own DNA.

ovulation is said to produce more females. (Sperm can live up to six days inside a woman's reproductive tract, but an egg lives less than a day following ovulation, so the overall window for fertilization spans from about six days before to one day after ovulation.)

The theory makes sense, except for the fact that male sperm are not the quicker, wilier critters Dr. Shettles imagined them to be. Research has exploded since the advent of in vitro fertilization, but other than Micro-Sort, no reliable way to separate X-carrying from Y-carrying sperm has been identified. Still, there are plenty of parents who swear by Shettles's system, and it's charming to imagine all the dog-eared copies of his book hastily tossed off the bed by parents passionately trying to conceive either a boy or girl. Of course, it's easy to collect glowing testimonials when the chance of conceiving the desired gender is already 50 percent, but the theory was debunked by a large 1995 study that found that the timing of conception has no bearing on whether a boy or girl is conceived.

Parents do care passionately about the sex of their children, which raises all sorts of ethical issues as modern technology makes it increasingly possible to control this fate. Already, the ratio of boys to girls has become dramatically skewed in several Asian countries, where millions of parents have illegally aborted unwanted female fetuses after ultrasound sex determination. Several nations, including Canada and most of Europe, have responded by banning sex selection used for nonmedical reasons (that is, beyond detecting sex-linked disorders such as hemophilia). The U.S. government does not appear poised to do this, although the American Society for Reproductive Medicine, the organization that oversees such technologies, declared in 2001 "that initiating IVF and PGD solely to create gender variety in a family should at this time be discouraged." But this policy is already being flouted by several high-profile IVF clinics that are unable to resist the multimillion-dollar revenue.

The situation is bound to get even worse with this newest wave of blood testing. According to data from the 2000 U.S. Census, parents who identified themselves as Chinese, Korean, or Indian showed an excess of second- and third-born boys. For parents' whose first children had been girls, the ratio for second-born children was 117 boys for every 100 girls; if the parents' first two children had been girls, the ratio was a striking 151 boys for every 100 girls. (For white Americans, the ratio was 105 boys to 100 girls, regardless of older siblings' gender.) The inescapable conclusion is that these nationalities are selectively aborting female fetuses—right here on U.S. soil, and even before the advent of at-home sex-determining blood tests.

The debate about sex selection is certain to heat up as the technology comes within the reach of more and more parents. Thus far, fertility doctors have barely blinked each time ethicists have challenged their increasing liberties over human reproduction. The issue is more critical in Asia, where feminists realize that the greatest hope for squelching sex selection is not through its legal restriction but through improving the status and equality of women. Nonetheless, it seems clear that sex-selection technology is here to stay unless doctors themselves reject the idea that, as one fertility specialist puts it, "being a boy or girl is a medical handicap."

How Boys Are Made

Of course, fertilization is just the first step in making a boy or a girl. What happens next depends on one tiny gene, which is located on the short arm of the tiny Y chromosome. As scientists only recently discovered, it doesn't take the whole Y chromosome to make a boy, merely this one microscopic stretch of DNA, known as SRY, scientific shorthand for "sex-determining region of the Y chromosome." SRY was discovered by British researchers studying a rare population: men with two X and no Y chromosomes. These XX males look, act, and consider themselves to be men, all because the single SRY gene jumped aboard one of the X chromosomes sometime during sperm formation. SRY's function has been confirmed based on the opposite, and equally rare, group of XY females: individuals who each have both an X and a Y chromosome but whose SRY gene has been deleted or mutated. In spite of their male genotype, children with this mutation look and feel like normal girls, at least until puberty, when their unformed ovaries prove incapable of supporting menstruation and breast development.

SRY is first activated about five weeks after conception within the genital ridge, the primordial, unisex gonad that has the potential to develop into either ovaries or testes. By six weeks, SRY has initiated its most important act: molding this generic gland into the male testes. SRY codes for a type of protein known as a transcription factor, which binds to the DNA in a cell's nucleus and turns on or turns off genes that generate the structure of cells and tissues. Although the details of its actions are still being worked out, SRY is clearly the master switch in this sequence, flipping on the full cascade of male development in normal XY embryos.

Once SRY finishes its job, the testes take over the most critical work of male differentiation. They produce testosterone, of course, but first they secrete a factor known as anti-Mullerian hormone (AMH), which is essential for linking up the correct plumbing for male urogenital function. Testosterone and AMH are hormones, which means they are released by glands (in this case, the testes) and flow through the blood to affect many other tissues, including the urogenital system, muscles, bone, and, of course, the brain.

The gonads are not the only reproductive organ that passes through a brief unisex phase. Each of us begins life with two complete sets of would-be reproductive tracts connected to these undifferentiated gonad. One set is the Wolffian ducts, named for an eighteenth-century German anatomist, which can morph into the vas deferens, seminal vesicles, and other internal parts of the male reproductive tract. The other is the Mullerian ducts (named for another German scientist), which can become the oviducts (fallopian tubes) and uterus in females. In males, the testes begin secreting AMH just six weeks after conception. As its name describes, AMH causes degeneration of the Mullerian ducts, obliterating the baby's potential female plumbing. By ten weeks after conception, the testes have begun producing significant quantities of testosterone, which transforms the Wolffian ducts into the various sperm- and semen-carrying tubes of the male reproductive tract. Females retain their Mullerian ducts (because they have no testes to secrete AMH) but lose their Wolffian set due to their lack of testosterone.

All of this sounds quick, and sexual differentiation does get an early start in the developing embryo. But finishing up the process takes several more months. In girls, the uterus and vagina aren't fully formed until about twenty weeks past conception. Externally, boys and girls are indistinguishable until at least twelve weeks past conception. Both the penis and the clitoris evolve from a common structure, the genital tubercle, which grows larger under the influence of testosterone, smaller if testosterone is absent. Still, these structures can remain hard to distinguish under ultrasound, even at five months' gestation. (That's why ultrasound technicians typically issue a disclaimer at this stage about not being able to guarantee they can tell you the sex of the fetus.) Slowest of all steps in genital development is the descent of the testicles, which form within the body cavity but move outside to the scrotum during the latter half of gestation. About 4 percent of full-term baby boys are born with undescended testicles, but the number is much higher—up to 33 percent—for boys born prematurely. If the testicles fail to descend by twelve months of age, which happens in about one out of every hundred boys, a simple surgical procedure can correct the problem.

Like the Wolffian ducts', the development of most male reproductive organs is critically controlled by testosterone. The testes secrete their first shots of testosterone just six weeks after conception. This hormone peaks in the male fetus between fourteen and sixteen weeks of development (sixteen to eighteen weeks of pregnancy) at a level some eight times higher than in females, and then it gradually declines until twenty-four weeks, when its level is merely twice that of females'. Testosterone is responsible for expanding the penis out of the unisex genital tubercle, for fusing the urethral folds along the midline (which forms the shaft of the penis), and for creating the scrotal sac.

The importance of testosterone before birth is highlighted by individuals who can't respond to it. A genetic male who has a normal Y chromosome but lacks the receptors for circulating testosterone looks like a normal *female*, because the hormone can't exert all these critical actions. This condition is known as androgen insensitivity syndrome, or AIS, because the receptors that bind testosterone also bind other male hormones, which are called androgens. AIS is rare, occurring in about three of every one hundred thousand children, and it often remains undetected until puberty, when the child everyone had assumed was a girl fails to begin menstruating. In spite of their normal female external genitalia, AIS individuals do not form a uterus because their Mullerian ducts have degenerated under the influence of AMH (which does not need an androgen receptor). Lacking both a uterus and ovaries, an individual with complete AIS is not capable of bearing children.

More interesting is the fact that males with AIS regard themselves as unequivocally female. They play like regular girls in childhood, develop breasts at puberty,* are sexually attracted to men, typically get married, and often become mothers through adoption. About their only distinguishing feature is height, which tends to fall in the normal male range.† But while individuals with AIS prove the great potency of prenatal tes-

---

* Breast development in AIS is fueled by the pubertal surge of testosterone, which certain cells can convert to estrogen.

† A quick Internet search will lead you to some statuesque celebrities suspected of being AIS males, though I'd rather not invade their privacy here.

tosterone on bodily development, they are less helpful for the under-standing of psychological sex differences, since they both appear to be and are treated like girls from the moment of birth.

How Girls Are Made

So far we've been focusing on male development and have seen how the presence of at least one specific gene (SRY) and two hormones (AMH and testosterone) are required to make a male body. If any of these in-nate signals is missing, the fetus ends up female (more or less). Does this mean, as we were taught in high-school biology, that females are the default sex—the generic form that will develop in the absence of any sex-specifying signals?

Not exactly. While genetic males lacking SRY or testosterone recep-tors generally do look more female than male on the outside, they are invariably infertile and don't go through normal female pubertal devel-opment. What's more, scientists have recently discovered another gene, called DAX1, that resides on the X chromosome and plays an active role in producing the ovaries (analogous to SRY's role in producing testes). Males, of course, have one X chromosome, but the action of their DAX1 gene is overruled by SRY. However, if the DAX1 gene is duplicated, as it is in some rare XY individuals, then these two copies of DAX1 override the one copy of SRY. Ovaries are produced, and these genetic males ap-pear female.

Nor are testosterone and AMH the only hormones that influence sexual differentiation. If you've been wondering, What about ovarian hormones? animal studies suggest they also play a role in brain develop-ment. However, estrogen cannot exert an effect until *after birth*, because the fetus is largely shielded from it by a unique substance called alpha-fetoprotein that is present in the blood before birth only. In fact, both boy and girl fetuses are exposed to high levels of estrogen (made by the placenta), but this blood-borne estrogen cannot enter the fetal brain be-cause it's tightly coupled to alpha-fetoprotein.*

Another reason to reject the idea of girls as the default sex comes from a condition known as Turner's syndrome. About one in four thousand

---

* Alpha-fetoprotein, or AFP, is the same factor that is measured in the maternal blood test that screens for neural-tube defects and Down syndrome.

babies is born with this disorder (though many more fetuses with the syndrome are spontaneously miscarried). Turner's syndrome involves the loss of part or all of one of the X chromosomes, producing what's known as an XO genotype. All affected individuals appear to be girls, but they lack functional ovaries, are infertile, and do not enter puberty naturally. They tend to be short and broad-chested, and they frequently suffer from certain cardiovascular and kidney problems. Girls with Turner's syndrome can be treated during childhood with growth hormones and androgens to increase their height, and with estrogen at the normal age of puberty to promote the development of breasts and other female characteristics. But their physical deficits tell us that it takes more to make a female than just a missing Y chromosome. In fact, the differentiation of male and female bodies is turning out to be much more complicated than researchers ever suspected, involving a number of male- and female-specific genes along with various sex-specific hormones and the several different receptors they bind to.

The same is likely true for the sexual differentiation of babies' brains. Women with Turner's syndrome exhibit lifelong problems with attention and spatial reasoning, which suggests an important role for the second X chromosome or for early postnatal estrogen in female brain development. At this point, however, most of our understanding centers on prenatal testosterone and the fascinating, if sometimes troublesome, ways this hormone appears to shape boys' and girls' later behavioral differences.

Prenatal Testosterone and the Brain

Compared to babies' reproductive organs, their brains develop incredibly slowly. Midway through gestation, when ultrasound images are already revealing a fetus's gender on the bottom end, the top end can barely coordinate a hiccup. The brain is by all measures the slowest to mature of all the organs, so little of what makes boys and girls different from each other can be fully attributed to prenatal development. At the same time, the fetal brain is not immune to the surges in gene expression and hormone secretion that so impressively sculpt the reproductive organs. Of these various biological influences, researchers know the most about testosterone, that most notorious of steroid hormones mothers love to blame when catching their sons racing around the house or wrestling too close to the coffee table.

Few parents, however, realize how early in the process testosterone acts. The surge begins just six weeks after conception and finishes before the end of the second trimester. By birth, there is little difference in boys' and girls' testosterone levels, although boys experience another, smaller surge shortly after birth that lasts through the first six or so months of life. Nonetheless, the brief four-month window of testosterone exposure before birth is enough to masculinize male babies down between the legs and—to some degree—up in their developing brains.

Most of our knowledge about prenatal testosterone comes from lab rats and doesn't necessarily translate to human boys. Still, researchers have uncovered many fascinating effects in these uncuddly albinos, which, like all creatures, become a lot more interesting when the subject turns to sex.

Compared to human infants, rats are much less developed at birth, equivalent to a human fetus after six months in the womb. This means that male rats at birth are still in the midst of their testosterone surge, which extends from the last four days of gestation through about ten days after birth, a window of development known as the perinatal period. Researchers have exploited this difference to perform all sorts of hormonal manipulations on rats, the easiest of which is (sorry, guys) castration.

Not surprisingly, male rats that had their testes removed at birth show little interest in mating with females as adults. They are also less aggressive than intact adult males, a difference that first appears in the play-fighting of pups. Like boys, young male rats normally engage in a good deal of rough-and-tumble play, but male rats castrated within the first six days of birth play more like females. However, if the procedure is delayed three weeks, to the age when play-fighting normally peaks, castration has no effect. Similar results have been found for other behaviors that differ between the sexes, such as the preference for salty foods (higher in female rats) and spatial-learning ability (higher in males). In other words, it is not the testosterone circulating in young males' blood that drives their aggressive play and other behaviors but the testosterone that washed through them at the very beginning of life.

Much the same effects have been proven through experiments on female rats. When a female pup is injected with testosterone on the fifth day of life, she becomes permanently sterile: her ovarian hormones never begin cycling, she never ovulates, and she will avoid mating with males. Early testosterone exposure also triggers more masculine-type play in females—romping and tackling their littermates in a more aggressive style than their uninjected sisters—along with malelike salt preferences and maze-learning skills. Remarkably, these behavioral reversals do not take place if female pups are shot up with testosterone beyond ten days of age. Like the genitals, the rat's brain is uniquely open to sexual differentiation during just a brief period in early development.

As fascinating as these findings are, scientists have yet to unravel the brain mechanisms by which early testosterone affects rats' later behavior. The strongest clue comes from the hypothalamus, a tiny but potent brain area that coordinates reproductive and other instinctive behaviors. Within the hypothalamus, there is an even tinier zone, called the preoptic area, that is five times larger in male rats than in females and that researchers have named the sexually dimorphic nucleus of the preoptic area, mercifully abbreviated to SDN-POA. In rats, this nucleus is critically shaped by the perinatal testosterone surge. A male rat castrated shortly after birth ends up with a small (female-size) SDN-POA, while a female given a shot of testosterone during the first week of life—but not in adulthood—ends up with a large (male-size) SDN-POA. There's also another hypothalamic nucleus, known by the abbreviation AVPV, that is larger in females and that shrinks in males under the influence of prenatal testosterone.

In rats, testosterone receptors appear as early as seven days before birth and are located throughout the brain, though they are most heavily represented in the deeper, emotion-controlling structures that make up the limbic system. The hypothalamus is part of the limbic system, but so is the amygdala (which controls aggression and other emotional experiences), the hippocampus (the memory center of the brain), and even portions of the cerebral cortex (home of the intellect and personality). There is some evidence that prenatal testosterone shapes the amygdala and hippocampus in rats, but so far, there is little understanding of how behaviors like rough-and-tumble play are linked to specific sex differences in rats' brains.

Boys Are Not Rats

Rats are easy to study. They have a short gestation and life span, so researchers can quickly see the effects of hormonal manipulation during pregnancy. But we need to be careful about extrapolating this large body of rodent data to human behavior. For one thing, human infants are much more mature at birth than rats are. Because of our long nine-month gestation time, the critical period for testosterone action on the human brain takes place exclusively *before* birth, whereas in rats it spans both pre- and postnatal phases. Research on monkeys, whose growth, development, and even social life is a lot more similar to humans', confirms this. For the most part, sex hormones exert very little effect after birth on either male or female monkeys' behavior.

By the same token, the effects of hormone manipulations before birth in monkeys are also less dramatic than in rodents. For instance, young female monkeys, like rats, engage in much less rough-and-tumble play than their brothers, but exposing females to high levels of testosterone prenatally does *not* make them start pouncing on their peers. Nor does prenatal testosterone lessen females' interest in babies or increase their tendency to mount other monkeys, two other traits that differ dramatically between young male and female monkeys. Two other behaviors—infant vocalizations and spatial navigation—were indeed affected when researchers manipulated monkeys' prenatal testosterone levels. But generally speaking, prenatal testosterone does not affect monkeys' behavior as strongly as it does rats'.

Nor, for that matter, are sex differences in the hypothalamus as dramatic in primates as they are in rats. Studying our own human brains, four different groups of researchers have described male-female differences in the preoptic area of the hypothalamus, but they still cannot agree on which tiny zone corresponds to the rat SDN-POA. What's more, the sex differences they *have* found in human brains are much more modest than the five-fold difference in size between male and female rats' SDN-POA. Even in monkeys, which are easier to study than humans, sex differences in the hypothalamus have not been clearly linked to sexual behavior. For instance, one recent study of Japanese macaque monkeys found a tiny zone of the preoptic hypothalamus that was two to three times larger in males than in females, but there was no difference in the size of this region between "lesbian" and "straight" female macaques, contrary to the findings in rats.

Experiments of nature

So what *has* been proven about the effects of prenatal testosterone on children? The bigger the brain, the less instinctive the behavior, and the more the brain's abilities are influenced by learning. That's why prenatal hormones have less impact on monkeys than on rats, and why if we really want to know how hormones shape boy-girl differences, we have to study human children.

This is easier said than done. For all their freakish portrayal in Hollywood movies, scientists are not going around castrating boys or injecting girls with testosterone to figure out how sex hormones affect their behavior. However, owing to some rare medical conditions, a handful of children have been raised as the opposite sex of what their chromosomes (or prenatal hormone exposure) would have dictated; these individuals have helped researchers sort out the roles of nature and nurture in gender development.

One such condition is androgen insensitivity syndrome, which, as we've already seen, produces a complete reversal of gender identity and attributes in genetic boys. These children lack the receptors for testosterone and other androgens, and so, unable to respond to male hormones, genetic males with AIS look like normal girls at birth and are invariably raised as females. They have no issue with their female identity and grow up to be typical heterosexual (though infertile) women. Their strongly female psychological profile demonstrates that the presence of male genes does not seem to masculinize the brain in the absence of functional testosterone receptors.*

But while AIS clearly demonstrates the importance of prenatal testosterone on the body, it is not very helpful in untangling the roles of nature and nurture in psychological development, since such individuals lack *both* prenatal testosterone responsiveness and a normal male upbringing. There are, however, other cases in which testosterone and rearing are at odds with each other—examples that give us a better understanding of the extent of the hormone's influence.

The most famous such case was David Reimer, the "boy who was raised as a girl," whose life was chronicled in a fascinating book by journalist John Colapinto. This tragically true story—of an infant boy whose penis was destroyed in a circumcision accident and who was then raised,

---

* This is an important point, since recent research on rodents and birds has found that SRY and perhaps other Y chromosome genes can directly alter males' brains and behavior without the intervention of testosterone. However, the behavioral profile of AIS individuals suggests such direct effects of Y chromosome genes are minor in humans.

most unhappily, as a girl—does indeed support the premise of Colapinto's title: *As Nature Made Him.*

Born in 1965, David was eight months old when he suffered his botched circumcision. His distraught parents sought various medical opinions and were finally persuaded to raise him as a girl, a switch that took many months to accomplish and was never truly successful. When David was seventeen months old, they cut his hair, started clothing him in dresses, and changed his name to Brenda. At twenty-two months, he underwent surgery to remove his testicles and to fashion a rudimentary vagina.

But the switch came too late. David never fit in as a girl. The child preferred toy guns to dolls, liked to play dress-up in men's clothing, and even insisted on urinating in a standing position. School was a struggle, peers were cruel, and Brenda/David became prone to aggressive outbursts. Finally, at age fourteen, Brenda/David was told the truth about his medical history, and he immediately reverted to a male identity. Eventually, David married a woman and became the adoptive father to her three children, but he struggled with his past; tragically, at age thirty-eight, he took his own life.

David Reimer's story makes the poignant argument that gender behavior and gender identity are hard-wired from an early age. There was little feminine about him, in spite of the way his case was reported to the medical community at the time.* Nonetheless, the lessons from his plight are not entirely clear-cut. David was actually quite old—nearly two—by the time he was truly reassigned to a female role. As we'll see in the next chapters, babies much younger than this already know a great deal about the difference between male and female, already prefer gender-appropriate toys, and are often already consciously aware of their own sex. Another crucial variable—and the reason his case became so

---

* David and his twin brother became famous in the medical annals as the John/Joan twins, code-named by Dr. John Money, the Johns Hopkins University psychologist who had advised David's parents to raise him as a girl and who published and lectured widely about the twins' development. But as depicted in Colapinto's book, Dr. Money was neither an objective scientist nor a compassionate clinician. Whether because of wishful observation or deliberate fabrication, Dr. Money portrayed David's conversion to a girl as a smashing success; the reality was quite different. According to Colapinto, Money's refusal to admit the failure of his "experiment" contributed mightily to the prolonged anguish of David and his family.

celebrated—was the fact that David was an identical twin. In addition to his own early life as a boy, David could see his male identity mirrored every moment in his twin brother, Brian, a perfect clone with whom he was tightly bonded.

As it turns out, David Reimer was not the only unfortunate boy to lose his penis during circumcision. Another, less famous case was reported in 1998. It had a very different outcome, probably because the accident occurred when the child was just two months old. By seven months, this boy had been reassigned, both socially and surgically, to the female gender. Now an adult, she unambiguously regards herself as a woman. Though the patient has had sexual relationships with both men and women, she never cross-dressed and always preferred to play with other girls.

In all, researchers have studied several dozen children who were genetically boys but who, for a variety of medical reasons,* were raised as girls, and the answer is nowhere as clear-cut as David Reimer's case. In one 2005 review by Columbia University psychologist Heinz Meyer-Bahlburg, only seventeen of seventy-seven such individuals had chosen to revert to the male gender. While many of these girls exhibited some signs of masculinity and often reported being more sexually attracted to women than men, the majority were nonetheless clear in regarding themselves as women. As Dr. Meyer-Bahlburg concluded: "These data do not support a theory of full biological determination of gender identity development by prenatal hormones and/or genetic factors, and one must conclude that gender assignment and the concomitant social factors have a major influence on gender outcome."

So while these cases clearly demonstrate that prenatal testosterone is important for shaping play behavior and sexual preference, they also prove that it's not omnipotent. Rearing also matters in establishing a child's sense of gender identity and in the many other behaviors that accompany this potent piece of self-knowledge.

---

* Only a few cases are due to penile ablation, like David Reimer's. Most such intersex cases result from congenital abnormalities affecting the development of male external genitalia. The decision about which sex to raise such children is extremely difficult. While most of these genetic boys raised as girls are content with this assignment, some are not. On the other hand, none of those raised as boys chose to revert to a female identity, so the preferred way of raising such children seems to be male, if the urogenital system can be adequately reconstructed through surgery.

Another "Experiment": Girls Bathed in Androgens

Very similar conclusions have been reached through studies of another group of children, in this case, genetic girls who have been reared in a manner opposite their prenatal hormone exposure. In a rare genetic disorder known as congenital adrenal hyperplasia (CAH), individuals produce very high levels of androgens, including testosterone, beginning early in gestation.* The imbalance has few developmental effects on boys, who already release high levels of androgens from their testes. In girls, however, the high levels of androgens have a strong masculinizing effect: the clitoris enlarges into a small penis, and the labia may be partially fused at the midline, creating a scrotumlike structure. The good news is that their internal reproductive organs are unscathed,† so girls with CAH are fertile and can bear children, provided their external genitalia have been surgically feminized, as is typically done in infancy.

The disorder is usually identified at birth and can be controlled with ongoing hormone supplements. But this treatment does not fully reverse the effect of prenatal androgens on girls' developing brains, which is a bit like what happens in rats and monkeys. Girls with CAH behave more boylike than their unaffected sisters, the optimal control group in many studies. As children, they are more aggressive, more physically active, more fond of rough play, and spend more time playing with boys than do typical girls. The difference is most pronounced in their preference for traditional boy toys, such as trucks, balls, and building blocks, and their relative lack of interest in dolls and—usually the biggest attention-grabber for young girls—real human infants. As adolescents, girls with CAH continue to be more interested in male pastimes, such as cars and sports, and to express greater interest in pursuing traditionally male careers, such as engineering and airline piloting.‡

When it comes to gender identity and sexuality, however, the picture

---

* CAH occurs in about one in sixteen thousand births and is caused by the loss of a single enzyme, 21-hydroxylase, that functions in the adrenal glands to produce the stress hormone cortisol. Lacking this enzyme, patients with CAH produce an excess of other adrenal hormones, including androgens.
† Like all girls, a female with CAH does not produce anti-Mullerian hormone, so her Mullerian ducts develop into normal ovaries, fallopian tubes, and a uterus.
‡ Girls with CAH are not, however, reliably gender shifted in their cognitive skills, such as verbal and spatial reasoning, a point we'll return to in later chapters.

is foggier. In spite of their tomboyish behavior, CAH girls are happy to be girls. As children and as adults, few females with CAH regard themselves as males living in female bodies. Rather, they have a definite sense of female identity, are comfortable in their role as girls and women, and only rarely choose to switch to a male identity as adults. Their sexual orientation, however, is modestly shifted. Though most marry and carry out normal heterosexual lives, CAH females admit to more lesbian and bisexual attraction and report less sexual attraction to men than typical women do.

Overall, CAH females demonstrate the potency of prenatal androgens in shaping some, but not all, aspects of masculine behavior. Although critics maintain that the more masculine behavior of CAH girls is a result of socialization—that their parents, aware of their daughters' ambiguous genitalia, permit more boylike play than they would in typical daughters—most scientists now accept that it is the hormones, not the parents, that shift CAH girls' behavior, interests, and sexuality in a somewhat masculine direction. (Careful studies have also helped demonstrate that parents of CAH girls do not encourage boylike play in them any more than they do in the CAH girls' normal sisters. On the contrary, one might expect these parents to be *more* motivated to bring out the feminine side of their CAH daughters' behavior, to help solidify their gender identity.)

But while most CAH girls are comfortable with their gender identity, there has been a movement in recent years to question the routine female assignment that doctors traditionally counsel at birth. CAH is the leading cause of intersex gender, which is the preferred term for individuals born with ambiguous genitalia. Some intersex individuals have become quite vocal about rejecting the surgery and other aspects of gender assignment imposed on them at birth, arguing that gender identification should be a matter of individual choice, made later in childhood or adolescence. (Such surgery often destroys sexual sensitivity, and patients typically go through several rounds of genital reconstruction during childhood.) The problem, of course, is that parents would have an impossible time raising a unisex child. No matter how open-minded the parents may be, our society demands that everyone be labeled either male or female, if for no other reason than figuring out which bathroom to use at school. And so, current medical practice dictates that, based on their reliable female identity and predominantly heterosexual orientation, genetic females with CAH should be raised as girls, though with every effort to minimize surgical damage.

Opposite-sex twins

CAH girls are not the only females exposed to virilizing hormones before birth. As it turns out, any female who shares a womb with a male fetus will be bathed in androgens, though not in the dramatic amounts of CAH. In some animals, a female born with a male twin is altered enough by male hormones to end up sterile. These intersex females, known as freemartins, occur in some 90 percent of cows born with a male twin. The other 10 percent escape this fate because they shared fewer placental blood vessels and, therefore, less testosterone and other hormones with their male twins.

Female rats are also known to be affected by the prenatal testosterone of their brothers. Mixed-sex litters are the norm among rats, which typically gestate in litters of ten to twelve, lined up like peas in a pod along the two extended uterine horns located on each side of the dam's body. Because of this arrangement, some females are exposed to more testosterone than others; it depends on the number of brothers they grow next to. Researchers have found that female rats and mice that gestate between two males are slightly more masculine than those gestating between one male and one female. Their genitals look a bit more masculine; they engage in more rough-and-tumble play as pups; they are more aggressive; they roam over a wider territory as adults; and they are more likely to exhibit male sexual behaviors, such as mounting and urine-marking their territory. These females, in turn, are slightly more masculine than females that gestated between two sisters (and vice versa for males gestating between brothers or sisters).

Importantly, most female rats end up being heterosexual—that is, they mate and rear pups—regardless of their nearest neighbors in the womb. But females exposed to higher levels of prenatal testosterone are less preferred by males and end up raising fewer offspring than females exposed to lower levels of prenatal testosterone.

So what about human females? Are girls with male twins or triplets exposed to higher-than-normal levels of testosterone, and if so, does it affect their personalities, cognitive abilities, or sexuality? The issue is increasingly relevant, considering the large number of children now born as multiples through in vitro fertilization and other assisted-reproduction technologies.

Parents of boy-girl twins are often struck by the great *differences* between their children. One of my colleagues, a professor of biochemistry and the father of twin toddlers, a boy and a girl, couldn't get over their different interest in toys, "even though," he quickly adds, "my wife and I are poster children for gender-neutral rearing." But while Marc sees these differences and can't help but conclude that the Y chromosome is responsible, researchers have been more focused on the *similarities* between boy-girl twin pairs. They've asked the question raised by the rodent studies: are girls whose twins are brothers more masculine than girls whose twins are sisters?

The answer is: not much. While there is some fascinating evidence for the slight masculinization of certain anatomical and physiological traits, most research has been unable to identify reliable differences in the behavior and mental skills of girls with twin brothers compared with those with twin sisters.

According to one 1999 study, girls with male twins have larger (more male-size) teeth than girls with female twins. Another study, conducted by Celina Cohen-Bendahan and her colleagues in the Netherlands, found that girls with male twins are more lateralized in their auditory processing. That is, normally, when two different words are played simultaneously into each ear, boys can recognize the word in the right ear better than the word in the left, while girls can detect both words equally. However, girls with twin brothers are more lateralized than girls with twin sisters, suggesting an influence of prenatal testosterone.

Two other studies have described subtle differences in hearing between girls with male twins and girls with female twins. One was published in 1993 by University of Texas psychologist Dennis McFadden. He was investigating some mysterious sounds known as otoacoustic emissions (OAEs), tiny echoes that our ears actually produce in response to incoming sounds and which are now used in hospitals to screen newborns for congenital hearing loss. OAEs are well known to differ between the sexes. From birth through adulthood, OAEs are both stronger and more frequent in females than in males. But as McFadden discovered, girls with male twins actually exhibit OAEs in the male range. This shift, McFadden suggested, could be caused by prenatal testosterone transfer. He bolstered this argument with a 1998 finding that lesbian and bisexual women also have more male-range OAEs than heterosexual women do. (Male homosexuals have OAEs within the normal male range.)

Needless to say, McFadden attracted a lot of media attention with the finding about gay women. But his theory—that prenatal androgen exposure shapes both OAEs and the sexual preference of women—is not well supported by other studies. While androgens do slightly shift the sexual preference of girls with CAH, they are not the cause of homosexuality in most lesbian and bisexual women. What's more, girls with twin brothers are no more inclined toward lesbian or bisexual orientation than any other girls.

Which brings us to the many studies of behavior and mental skills in girls with male twins. It's true that a handful have reported a shift in the masculine direction—finding that girls with male twins are more prone to aggression and risk taking or are better at spatial skills than girls with female twins. However, the bulk of such research has found no significant difference: girls with male twins show just as much preference for girl toys, appear just as feminine, and score no higher on math and other male-type cognitive skills than girls with female twins.

The most convincing of these twin studies tracked literally thousands of women—every female twin born in Finland from 1983 through 1987. When compared with women with female twins, women with male twins were not found to differ in their age of puberty, their interests in feminine or masculine activities, or their subsequent number of children—differences you'd expect to result from prenatal testosterone transfer.*

So while there may be some modest physiological effects of prenatal exposure to testosterone (or some other male factor) in girls with male twins, it doesn't seem sufficient to alter girls' behavior or cognitive skills. What's more, the few studies that do find a shift in such girls' risk taking or aggressiveness cannot rule out the very real possibility that girls with boy twins act or think a little more like boys because of the time they spend with their twin brothers *after* birth.

---

* Recently, a study based on a historical population—Finnish birth records from the eighteenth and nineteenth centuries—found that women with male twins did indeed have fewer children than women with female twins, suggesting prenatal testosterone reduced their fertility. However, the lack of difference in the number of offspring in the much larger study of modern Finns, as well as in a recent study of Dutch, Australian, and American twins, indicates that any effect on fertility is insignificant in the context of contemporary reproduction.

In fact, just such an effect was shown by researchers Brenda Henderson and Sheri Berenbaum at my own university. In comparing girls with twin brothers with girls with twin sisters, Henderson and Berenbaum found no difference in toy interest and only a very modest shift in sports participation. But they also included an important control group—nontwin girls with older brothers. Such girls, it turns out, *are* likelier than other girls to play with boys' toys and to enjoy sports, and they're much less inclined to play with girls' toys, as shown in Figure 1.2.

Older brothers of girls, unlike twin brothers of girls, do not share their prenatal testosterone with their sisters, and yet apparently they encourage an even stronger shift toward toy trucks, balls, and sports than the twin brothers do. Other research confirms this influence of older siblings, who, much more than younger siblings, shape the gender behavior of their brothers and sisters. Especially when one considers the research on opposite-sex twins, this study makes clear that any masculine behav-

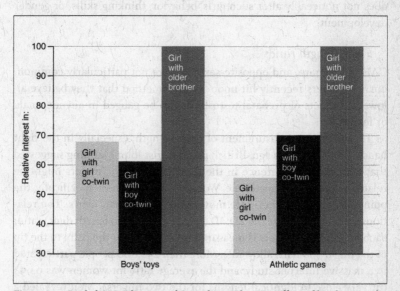

Figure 1.2. Study by Henderson and Berenbaum showing effect of brothers on the play interests of girls between three and eight years old. For data on the left, girls were given a choice of toys. Compared to girls with a twin sibling, singleton girls with older brothers showed a marked preference for toy trucks, cars, and Lincoln Logs and less interest in dolls and toy dishes. The data on the right are based on mothers' ratings of their daughters' enjoyment of various types of games, including sports.

ior—such as athleticism or spatial skill—in a girl with a male twin could be due to nurture, the shared playing, bonding, studying, and general time spent with an identical-age brother, rather than to prenatal testosterone transfer.

In sum, girls with male twins are not appreciably more masculine than other girls, nor are boys with female twins more feminine than other boys. Prenatal testosterone (or some other male factor) may induce some subtle physical differences, but it is not strong enough to override the typical socialization of girls as girls and boys as boys. Or, to put it another way, the lack of effect shows that personality and cognitive traits are much less potently shaped by nature than traits like tooth size and otoacoustic emissions are. Further research is needed to determine exactly how much testosterone boys pass on to their twin sisters in the womb, but for now it seems that the slight testosterone elevation (which is unlike the higher levels of androgens experienced by girls with CAH) does not noticeably alter such girls' behavior, thinking skills, or gender development.

Finger-length ratios

CAH is very rare, and opposite-sex twins are not particularly common, but researchers recently hit upon another method that they believe allows the effects of prenatal testosterone to be gauged in anyone, male or female.

The method is measurement of finger length, especially in the right hand. About ten years ago, British psychologist John Manning suggested that a subtle sex difference in the lengths of specific fingers might be related to prenatal testosterone. Women's fingers—more specifically, the pointer and ring fingers—are more symmetrical than men's. The relationship is described as the 2D:4D (for second digit to fourth digit) ratio; each of these two fingers is measured from the fold at the palm to the tip of the finger (no nails allowed!). More than 250,000 people participated in a massive Internet study, and the average ratio for women was 0.994 (or very close to identical lengths for the two fingers). Men revealed a lower ratio, meaning their ring fingers were longer than their pointer fingers, with a ratio averaging 0.984. So there's about a 1 percent difference in 2D:4D ratio between women and men. (See Figure 1.3.)

Though tiny, this difference between the sexes is very reliable. It is found in children and even in fetuses as early as nine weeks of gestation, which is about two weeks after testosterone production begins in male embryos. And that's the reason everyone is so excited about it: there is some, admittedly weak, evidence that small variations in the 2D:4D ratio are shaped by the amount of testosterone babies are exposed to in utero. In one study by Manning and colleagues, 2D:4D ratios in a group of thirty-three boys and girls were found to correlate with prenatal testosterone and estrogen levels measured from amniocentesis samples that had been taken during their mothers' pregnancies. Unfortunately, these researchers did not report whether hormone levels were related to digit ratios *within* the groups of boys or girls alone, which is considered the more crucial test for proving an effect of prenatal testosterone. Other evidence that 2D:4D ratio is shaped by prenatal testosterone comes from three studies of girls with CAH, two of which found these girls had smaller (more masculine) digit ratios compared to girls without excess hormone exposure. Similarly, two out of three studies of girls with male twins found their digit ratios to be significantly smaller than other girls', although in this case, the third study, which failed to find a difference, was larger and more convincing than the other two.

While the ratio's link to prenatal testosterone is pretty weak, this hasn't discouraged an enormous number of researchers from measuring 2D:4D ratios and trying to relate them to seemingly every possible behavioral sex difference: spatial skills, literacy, ADHD, dominance, fertility, athletic ability, aggression, delinquency, thrill-seeking, and—the most frequently studied—sexual preference. To cut to the chase, finger ratios are not clearly linked to female homosexuality but may be linked to male homosexuality. We'll revisit digit ratios and some of these other sex differences in later chapters, but the bottom line is that for some traits, digit ratios do correlate with behavioral sex differences. However, such relationships tend to be quite small and aren't nearly potent enough to predict, for instance, that your son will be gay or your daughter will be a gifted athlete.

In sum, these several lines of research on prenatal testosterone—studies of girls with CAH, of males with androgen insensitivity, of finger-length ratios, and (to a lesser extent) of opposite-sex twins—generally support the idea that it contributes to later behavioral sex differences. However, prenatal testosterone is hardly the only factor. As Manning's own research team puts it, prenatal sex hormones exert only "a modest predis-

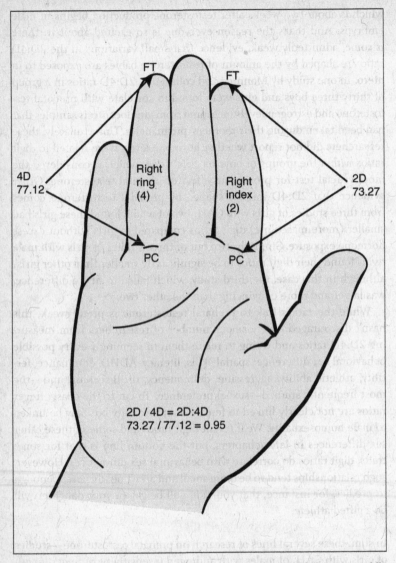

Figure 1.3. How to measure digit ratio: Using a metric ruler held along the middle of the palm side of the finger, measure the distance between palmar crease (PC) and fingertip (FT) for both ring (4D) and pointer (2D) fingers. Divide 2D by 4D to obtain the digit ratio. Women's ratio tends to be very close to 1.0, while men's is slightly lower, meaning their pointer fingers are shorter than their ring fingers.

posing influence on human development" that "merely bias, rather than determine" the different behaviors we see in our sons and daughters.

How Do Boys and Girls Differ Before Birth?

Sarah is pregnant with her second child. Her first, now the curly-headed three-year-old Maggie, seemed pretty quiet as a fetus. Sure, she kicked and squirmed a bit—Sarah would have been very worried if she hadn't—but that movement was nothing compared with what Sarah is experiencing now. Punch, wiggle, kick-kick-kick: it's like some kind of wrestling match is going on in there! Sarah and her husband chose not to learn their baby's sex before birth, but all this action, given her quiet pregnancy with Maggie, has Sarah convinced that their second baby is a boy.

Considering its sway over boys' later dispositions, prenatal testosterone might be expected to reveal itself in boys' behavior before birth. Boys are unquestionably more active than girls after birth, a difference that grows during early childhood. But despite Sarah's assumption, there is no reliable sex difference in fetal movements. Now that prenatal ultrasound imaging has become widespread, researchers have conducted many studies carefully comparing movement in male and female fetuses. While a few do report slightly greater activity in males, overall, most have found no sex differences in the frequency or intensity of fetal movements. Boys and girls pull off roughly equal numbers of punches, kicks, and rolls before birth. The fetal heartbeat, another important indicator of babies' health and development in the womb, also matures comparably in males and females.

Nonetheless, some differences—beyond the obvious genital ones—are apparent in utero. Boys grow more quickly than girls from early on in gestation. Doctors who perform in vitro fertilization can often guess whether an embryo will be male or female based solely on the number of cell divisions that have taken place in a given number of hours since fertilization; male cells have a higher metabolic rate, which accelerates their early growth and cleavage. Evolution seems to have favored this faster growth so that male embryos can get through the critical period of testes differentiation before their mothers' estrogen, which rises steadily during early pregnancy, interferes with the development of their sex organs. As a result of their faster growth, boys are larger, heavier, and physically sturdier than girls at birth, with thicker skulls and, yes, bigger brains.

While boys' bodies get bigger faster, girls' bodies mature faster physiologically, adding up to a clear advantage for females by the end of gestation. By most measures, girls are better able than boys to handle the challenge of life outside the womb, while boys are more vulnerable to various illnesses, cognitive and behavioral problems, and even death during the prenatal and early postnatal period.

This difference is especially pronounced in the developing respiratory system. Babies born prematurely can have difficulty breathing, and the risk for serious respiratory problems is considerably higher in male preterm babies than in female preterms. Girls' lungs are about one week more mature than boys' during the later fetal period, a seemingly small difference that can have great consequences for survival following premature delivery. Overall, the risk of newborn death due to respiratory problems is a striking 50 percent higher for boys than for girls.

The difference also shows up in certain measures of neurological function. Girl fetuses respond earlier than boys to external stimuli, according to one study in which researchers applied a vibrating electric toothbrush to mothers' abdomens to study the onset of fetal habituation. By twenty-six weeks of gestation, most female fetuses responded to this vibration with some kind of movement—kicking, punching, twisting, or turning—while it took the boys two additional weeks to reach this response rate. Females also habituated, or ceased responding, to the repeated vibratory stimuli about two weeks earlier than males did.

One other prenatal sex difference is a bit more amusing: female fetuses move their mouths more than males do, according to two recent studies. Beginning at around sixteen weeks of gestation, girls were found to wiggle their tongues and open and close their jaws more often than boys, a difference that grew larger over the subsequent month of prenatal development. While this difference seems to set the stage for females' more talkative futures, it is probably more a reflection of their overall earlier maturation. Mouth movements are important for survival after birth—they allow a baby to suck and swallow, and thereby gain nourishment. In fact, preterm girls are known to suck harder and more frequently than boys of the same age, a difference that may contribute to their faster growth and healthier development as compared with premature boys.

So while they may not be obvious to parents, some sex differences do emerge before birth. Generally speaking, they are subtle—nothing like the kind of pink-versus-blue behaviors we see later in childhood—but they do help set the course for different developmental pathways of boys and girls.

Male vulnerability

Men are bigger, stronger, and, in most physical ways, tougher than women, so one of the more surprising facts about prenatal development is that male babies are far more vulnerable than females. The difference is truly astounding when you realize just how many boys are lost in early development. Although boys and girls are born in nearly even numbers (the precise ratio in the United States for babies born at term is 105 boys for every 100 girls), the ratio at conception is far higher, with some estimates ranging up to 170 males for every 100 females conceived. We don't know why Y-carrying sperm are more successful at fertilization; they are not any faster than X-carrying sperm, but their smaller size may confer some other advantage, such as getting through tight spaces.

In spite of their numerical advantage at conception, male embryos fare much worse during the rest of development. When a pregnant woman suffers a miscarriage, it's about 30 percent more likely that the baby was a boy. Boys are also about 7 percent more likely to be born prematurely. Preterm birth is never easy, but of those babies born three or more weeks early, boys are less likely than girls to survive, and those boys who do survive are much likelier than girls to suffer long-lasting behavioral or neurological impairments, including cerebral palsy. This difference in vulnerability is so pronounced that of all the influences on preterm babies' health and survival, the greatest risk factor is male sex. This statistic, often cited to parents of premature daughters, is probably carefully avoided when neonatologists discuss prognoses with the parents of premature sons.

Even boys who make it to full term remain at greater risk than girls after birth. Overall, the U.S. infant mortality rate is some 22 percent higher for boys than girls, with similar ratios reported in Sweden and Japan. On a happier note, this sex difference appears to be declining, especially among babies born prematurely, thanks to many advances in their medical treatment.

All of these factors together explain how the larger number of males at conception gets winnowed down to a fairly even sex ratio by birth. But even after males are born, their vulnerability remains a dominant theme in early development. As we'll see in the next two chapters, boys are more at risk for a huge number of physical and mental problems, making them in many ways the more challenging of the two sexes to rear, at least during early childhood.

Male vulnerability is not unique to humans. In other species, the ratio of male to female offspring is known to be skewed by how fit a mother is. That is, healthy female deer, caribou, geese, quail, and other species end up raising more male offspring than do mothers in poorer condition, a contrast named the Trivers-Willard effect, after the two evolutionary biologists who figured it out. Many human societies also conform to the Trivers-Willard rule, revealing a higher ratio of boys to girls among families who are socially and economically better off. In the United States, the son of a married mother with some college education will have a nearly 1 percent higher chance of surviving to age one than the son of an unmarried mother who did not complete high school. Because males are conceived in higher numbers but are more vulnerable at every step of early development, this extra Darwinian process results in more boys surviving in advantaged families than in poorer or otherwise stressed families.

Why are males—not just human babies, but males of many animal species—so much more vulnerable than girls in early life? One hypothesis is that male fetuses are literally rejected by their mothers' immune systems. Male cells express a protein known as the H-Y antigen, another product of the Y chromosome. This protein, which causes tissue rejection when organs are transplanted from men to women, could trigger a slight immune reaction by the mothers, leading to subtle damage of male fetuses and thus to their greater susceptibility to infections, malnutrition, drug exposure, and any other kind of prenatal insult. Such damage has never been proven, but it remains one theory for males' greater vulnerability in early life.

Another factor is certainly the slower maturation of male fetuses, especially of their respiratory and immune systems. Testosterone and other androgens are known to suppress immune function in certain circumstances, which may make boys more vulnerable to infection during early postnatal life. Boys also have a higher metabolic rate and higher activity levels, so they need more calories to thrive. In sum, at the end of nine months of gestation, boys are both more needy and less ready to enter the world than girls are. While most boys weather this stormy period without difficulty, their greater vulnerability does tip the balance for some, threatening their health, mental development, and very survival.

Are the Old Wives' Tales True?

Considering the several differences between boys and girls before birth—in vulnerability, growth rate, hormone exposure, and certain behaviors—you might expect a pregnant woman to be able to tell whether she's carrying a boy or a girl, even without modern technology. Of course, most parents today *do* opt to learn their baby's sex before birth, usually in a midgestation ultrasound exam or an amniocentesis or by the newer, albeit less reliable, mail-order blood-testing kits. But what about all the old-fashioned ways of divining a baby's sex before birth? Is there any truth in the old wives' tales about differences between male and female pregnancies?

Sex prediction is an ancient art. About 2,400 years ago, the Greek physician Hippocrates claimed that women pregnant with boys had healthier skin tones than women pregnant with girls. In modern times, we have Internet folklore, with dozens of methods that can supposedly be used to distinguish male and female pregnancies. Some of these, such as a difference in fetal heartbeat or in fetal position within the mother's abdomen, have been clearly refuted by science. As we just saw, boy fetuses do not have a slower heart rate than girls, in spite of their slightly larger body size. Nor do women pregnant with boys tend to carry them lower and more in front (like a basketball) than women pregnant with girls, who are erroneously said to sit higher and spread more sideways across the abdomen. One recent study found that women pregnant with boys consume about 10 percent more calories during the second trimester than mothers carrying girls, but it seems unlikely that most women could consciously detect such a difference.

Then there are some completely unbiological theories, such as the way a woman's wedding ring swings when she hangs it from a thread above her belly. (Circular motion supposedly indicates a girl; side-to-side motion says it's a boy.)

There is, however, one difference between male and female pregnancies that *has* held up well to scientific scrutiny. Severe morning sickness, known medically as hyperemesis gravidarum, is more common among women pregnant with girls than with boys. This finding, first reported by a Swedish research team in 1999, has been replicated several times. The original study tracked all Swedish births between 1987 and 1995. Of these, 51 percent of babies were male. However, among the women with the most severe pregnancy nausea (those bad enough they had to be hospitalized during the first trimester), only 44 percent of babies were boys. A similar study carried out in the UK found a 46 percent male birth rate among women with hyperemesis gravidarum in the first trimester as compared to 51 percent male births overall.

It's not a huge difference, nor is it known if having milder nausea predicts fetal sex. So it's hard to say whether a mother like Sarah could compare her first and second pregnancies and, noting she is less nauseated this time around, predict that her second child is a boy. Nonetheless, we do know one hormone that contributes to the nausea of early pregnancy: human chorionic gonadotropin (hCG), the detection of which in urine serves as the basis for home pregnancy tests. HCG levels rise dramatically in early pregnancy, and they are higher in women with hyperemesis gravidarum than in those with normal pregnancies. What's more, hCG levels are higher in women pregnant with girls than in those pregnant with boys. So it's likely that this hormone does make moms a little sicker, on average, when they're carrying girls than when they're carrying boys. Hippocrates didn't know what hCG was, much less how to measure it, but considering how pale most people get when they're feeling nauseated, perhaps he was on to something with the lack of facial color in mothers pregnant with girls.

In fact, one recent study adds to the list of scientifically proven differences between male and female pregnancies, and it may be related to hCG levels. Psychologists at Simon Fraser University in British Columbia made the surprising finding that women pregnant with boys actually scored higher on certain cognitive tests, especially short-term memory and spatial tasks such as calculation and mental rotation. Beginning at twelve weeks of pregnancy, women carrying boys were found to score some 30 percent higher on these tests than women carrying girls. (None of the women in this study knew the sex of the fetus.) What's more, their advantage held throughout the pregnancy and even several months after their babies were born.

It's a startling result, and it must be replicated before we put too much stock in it. But if it's true, it suggests that either there's something about carrying boys that's good for mothers' brains, or there's something about carrying girls that's bad for them. I'm sorry to say that the study's authors, Claire Vanston and Neil Watson, favor the latter view, speculating that the higher hCG levels in women carrying girls may suppress their thinking skills. It's not such a stretch—if you're feeling sick, chances are you're not going to perform as well on tricky mental tests. But it is remarkable that the effect lasted so long—well after pregnancy had ended and hCG levels had returned to baseline levels in the mothers of both boys and girls.

The sunnier view is that fetal factors actually *enhance* cognitive skills in women pregnant with boys. There is, as I'll return to later in this book, some weak evidence that testosterone promotes male-type cognitive skills, including mental rotation. Since male fetuses start producing androgens as early as six weeks after conception, it's possible that testosterone and other male factors enter a mother's circulation and influence her cognitive skills—in this case, for the better—during pregnancy and for several months thereafter. But considering the small size of Vanston and Watson's study, as well as the generally weak link between testosterone and cognitive skills, this idea remains highly speculative.

Meanwhile, it's worth exploring another notion about the mommy brain and prenatal sex prediction: whether a woman can use her famous intuition to detect the sex of her unborn child. Lots of women profess to having a feeling about whether they're carrying a girl or a boy even before the ultrasound pictures come through. Is there anything to these gut impressions? Can a mother somehow sense her baby's gender, perhaps through some subconscious mind-body cues that scientists have yet to pin down?

Researchers at Johns Hopkins University tested this possibility by interviewing about a hundred women who had chosen not to learn the baby's sex. (They had intended to study a larger group, but such mothers are not easy to find these days.) On average, women's intuitions were not very accurate. However, when the researchers divided these women by the number of years of higher education they had completed, it turned out that women who had completed at least one year of college were considerably more accurate at predicting the baby's sex than women without any higher education. The more educated women guessed correctly 71 percent of the time, while the less educated group was right only

43 percent of the time. The authors of the study admit it could be a fluke, some statistical anomaly resulting from the small number of women they were able to recruit. But interestingly, when these educated women were asked how they knew the baby's sex, the ones who had been correct were the ones who had relied more on emotional reasons—that it was "just a feeling," or a dream about the baby's sex—than on objective methods such as fetal activity level or comparison to a previous pregnancy.

Neuroscientists are increasingly appreciating the power of emotion over supposedly rational thinking. The body signals the brain in countless ways—through sensory nerves, hormone release, metabolic changes, and immune responses. Most of the time, we are unaware of these cues, but they undoubtedly sway our thoughts, memories, planning, and decision-making. A pregnant woman's brain may receive an extra jolt from that second tiny body with its own physical and chemical life. Perhaps, then, there really is something to a mother's intuition, and it starts cluing her in, even during pregnancy, about the mysteries of her new child, including his or her emerging gender.

My husband and I were in that odd group that chose not to learn the sex of any of our babies before birth. With our first child, we convinced ourselves we didn't care, nor can I remember feeling any particular intuition about her. With the second, I was pretty darn curious and can now confess that I inadvertently saw the XY chromosomes while peeking at the amniocentesis report in my medical chart. By the third, I was downright obsessed with our baby's sex. Still, we managed to avoid the answer in my ultrasound and amniocentesis reports. Instead, I waited, monitoring every little hiccup and queasy bout for hints about our family's final makeup.

As it turned out, I was wrong. I had Toby pegged as a girl, probably reflecting wishful thinking more than anything. (I'd hoped Julia would have the sister I'd always wanted.) But in hindsight, and after three pregnancies, I believe there was one clue, which leads me to suggest a new wives' tale about differences between male and female pregnancies.

It has to do with sex drive. As we've seen, male fetuses produce testosterone, beginning about six weeks after conception. Some years ago, scientists confirmed that women carrying male fetuses have higher levels of testosterone circulating in their bloodstreams than women carrying female fetuses, especially during the second half of gestation. Apparently, some of the testosterone pumped out by boy fetuses crosses the placenta and makes it into their mothers' blood. The difference is reliable enough that researchers in the 1970s proposed using maternal blood levels of testosterone as a method for prenatal sex determination. (They would have persevered if ultrasound imaging hadn't taken off shortly thereafter.)

It may interest you to know that testosterone acts as a kind of aphrodisiac in women. Women with low sex drives, as sometimes happens in those who have undergone hysterectomies, are often treated with testosterone, which has been shown to increase both their interest and pleasure in sex.* Testosterone is produced in females by the ovaries and adrenal glands, but the levels are obviously much lower than in males. Could the elevated levels of testosterone in women pregnant with boys cause them to have higher libidos than women pregnant with girls?

Inspired by my own male pregnancies, I've scanned the scientific literature for evidence of such an effect. So far, I've found nothing. Either researchers haven't uncovered any particular difference in sex drive between women carrying boys and those carrying girls, or they haven't bothered to ask. Until they do, I'll continue to spread my own new wives' tale. It may be just my overactive imagination, but whatever the reason, I confess that I really enjoyed being pregnant with our boys.

Sex and the Prenatal Universe

Newborns are hardly the most sexual beings around, but their male- or femaleness has already begun to be shaped by the nine months of development before birth. For most babies, this development proceeds normally, according to the male or female plan. Equipped with a Y chromosome, a boy's SRY gene kicks off the development of the testes, which then send testosterone surging to the brain, where it presumably (though the evidence is still largely lacking) begins shaping circuits for later male behavior. Or if the baby has two X chromosomes, she makes ovaries, which produce hormones with less sway, at least prenatally, over brain development. The X and Y chromosomes may also act in more direct ways to masculinize or feminize babies' brains before birth, although scientists are just beginning to piece together this part of the puzzle.

Clearly, nature sets the ball rolling, biasing boys and girls toward

---

* There are also some risks of testosterone treatment, ranging from the growth of facial hair to liver damage.

different interests and behavioral styles even before they're born. But prenatal testosterone is not everything. What happens next, when these small, immature brains meet our inexorably gender-divided culture, is also crucial. From their faint beginnings, sex differences become quickly magnified as babies enter a world that sees them as, above all, either boys or girls.

2

Under the Pink or Blue Blankie

WE HAVE A NEW BABY in the family. He's so small and sweet that everyone keeps squabbling over who gets to play with him and cuddle him at night.

Okay, he's a kitten, but our affectionate urges are much the same. And just like a human baby's, his gender is not much of an issue at this point. While full-grown male and female cats have very different personalities, little of that is evident in kittens. (Then again, little Snookie isn't going to experience much life as a tom, since we plan to neuter him at six months of age.)

Similarly, when it comes to boys and girls, few differences are obvious in early infancy. Scientists have certainly looked for them, reasoning that the earlier they can detect sex differences, the likelier they're due to purely innate factors and not simply to moms and dads nurturing their tiny sons and daughters differently. Sure enough, when scientists look closely and study large numbers of newborns, they have been able to document a few significant differences between infant boys and girls. But generally speaking, the differences are few and far between, nothing like what we see later in childhood.

In this chapter, I'll explain what's known about the similarities and differences in boys' and girls' bodies, brains, skills, and maturation in the first year of life. Some differences are present at birth; others turn up along with the many new skills infants acquire during the exciting first year. Hormonal surges, while not as dramatic as before birth, may also play a role in the growing differentiation of boys and girls during infancy.

But parents are another big part of this equation. In spite of our best efforts, moms and dads don't treat sons and daughters exactly the same. The two sexes start out a little bit different, and we, with our own life-times of cultural experience, react differently to them from their earliest days. In this chapter, we'll explore what psychologists have discovered about the different ways parents interact with boys and girls, not all of which are conducive to raising strong, well-rounded children. I'll con-clude with suggestions of what parents can do from the very beginning to promote the emotional and cognitive development of their infant sons and daughters.

Boys and Girls at Birth

Like kittens, newborn babies are all pretty much the same. Don't get me wrong; I find newborns amazing—almost magical in their tiny perfec-tion. But aside from marveling at their microscopic toes and startlingly strong grips, parents don't have a whole lot to "ooh" and "ahh" about. They sleep (blessedly), feed, cry, and sometimes make amazing eye con-tact. But there is little that jumps out to reveal their future personalities, interests, talents, or even endearing quirks.

Still, there are a few consistent differences between the sexes, and these probably do influence the way boys and girls get started in life. The most obvious is size: at birth, boys are larger than girls, on average about four to five ounces heavier and half an inch longer. This difference helps explain why boys have a more difficult time being born. Women labor an average of twenty-four minutes longer and are more likely to use an-algesic medications when delivering boys. Electronic monitors are now used to track the baby's heartbeat throughout the birthing process, and boys are found to undergo fetal distress more often than girls do during this dicey time. This means that boys are about 50 percent more likely to end up being delivered by C-section, an operation that, while often lifesaving, is associated with greater risks than vaginal delivery for both mother and baby. At birth, boys are about 30 percent more likely than girls to receive a low Apgar score, which is a measure of a newborn's color, breathing, and cardiovascular function that indicates how well the baby fared during the birthing process and how the baby is adjusting to life outside the womb. Newborn boys also score a bit lower than girls do on the Brazelton scale, which is a more thorough assessment that measures many aspects of neurological maturity.

All of this should sound a bit familiar by now. As we saw in the last chapter, male vulnerability—both physical and emotional—is a common theme in early development. From conception through at least four years of age, boys have a rougher time fighting off infections, adapting to new environments, and generally overcoming the many hurdles that threaten early growth. Some of this vulnerability has to do with their larger size, but it's also a reflection of more specific physiological differences, such as boys' higher metabolic rates and less mature lung function at birth.

Happily, my boys fared fine during childbirth, but I learned that larger infants are indeed harder to deliver. Both Sam and Toby weighed in at close to nine and a half pounds at birth; compare that to their eight-pound sister. I'd say it's at least twice as painful to push out that extra pound and a half of baby!

With bigger bodies come bigger heads. Boys' heads are about one centimeter (four-tenths of an inch) larger than girls' at birth and about one and a half centimeters (six-tenths of an inch) larger by four years of age. Pediatricians still use head circumference as a quick check for how well a child's brain is growing: wrap a tape measure around the head and you can get a rough idea of how big the brain is. And just as they use different growth charts for boys' and girls' heights and weights, pediatricians use different graphs to track head size for boys and girls.

Parents of boys tend to get very excited about this fact: a bigger head indicates a bigger brain, so this must mean that boys are smarter, right? In fact, head circumference is a pretty inaccurate way of estimating brain size. Some people have more hair, thicker skulls, and so on, all of which tends to distort brain-size estimates. With the advent of MRI, we now have much better ways of measuring brain size, and in the last few years, scientists have been able to track sex differences in brain volume as children grow. One large study of children between four and eighteen years old confirms that boys' brains are about 9 percent larger than girls'. This is the same difference found between men and women. So males really do have larger brains, from birth onward, though it is not clear how this relates to the different mental abilities of the two sexes. (Boys also have larger kidneys than girls—even during the prepubertal age, when boys' and girls' heights and weights are indistinguishable—but no one seems to be arguing that this gives them better urinary function.)

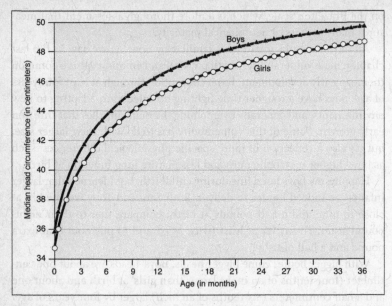

Figure 2.1. Median (50th percentile) head size of boys and girls from birth to age three. Data are from the Centers for Disease Control.

What boys have over girls in head size, girls appear to make up for in maturity. Girls develop faster, beginning from midgestation and ending with their earlier entry into puberty. At birth, their skeletal system is some four to six weeks more mature than boys'. Greater maturity is said to be the reason why infant girls handle labor, delivery, and early infancy better than boys and why in early life they are generally less vulnerable to most illnesses, including asthma and infections of the ear, respiratory system, and gastrointestinal tract.

When it comes to brain maturity, however, the evidence for girls' faster development is surprisingly weak. The easiest method for measuring brain activity, electroencephalography (EEG), has not demonstrated any global maturity advantage for girls. While one Swedish EEG study did find more mature cortical function in girls at birth, another found little sex difference at three months but a female maturity advantage emerging at six months. And, surprisingly, several studies of school-age children have reported more mature EEG patterns in *boys*, with girls not catching up until adolescence.

So it's not accurate to say that girls' brains are globally more mature than boys'. A likelier scenario is that *certain aspects* of brain function mature earlier in girls and, likewise, certain aspects mature earlier in boys. As infants, boys are typically a bit fussier and more irritable than girls, suggesting that at birth, their nervous systems simply aren't as ready to tackle the transition from their quiet, protected wombs to a noisy, complex world. But the truth is that there is no simple neurological explanation for why boys are at greater risk for the many developmental disorders involving the brain.

Sensory Differences

Do boys and girls differ in their basic sensory abilities at birth? Babies learn primarily through their senses, so any differences here could, in principle, launch the two sexes down very different developmental roads. The physician and author Leonard Sax has built his movement for single-sex education based in large measure on the idea that girls and boys see and hear differently from birth. But a close look at the research on sensory differences in newborns reveals that they are small and of little relevance to children's learning.

Consider tactile ability. In 1974 psychologists Eleanor Maccoby and Carol Jacklin reviewed all the existing studies of newborn touch sensitivity in their landmark tome *The Psychology of Sex Differences*. Newborn girls were found to be more sensitive than boys in five out of thirteen different experiments, while there was no sex difference found in the other eight experiments. Not exactly overwhelming evidence for an innate sex difference in the response to touch, nor has more recent research revealed any dramatic differences between newborn boys and girls.

The findings are similar for pain. Although studies of adults generally find that women have a lower *tolerance* for pain than men, studies of pain *threshold*—the lowest intensity at which a stimulus is perceived as painful—have not revealed any reliable differences. Pain perception is notoriously subjective, strongly influenced by both prior experience and an individual's own sense of "toughness." So it's important to analyze responses to this sensation in newborns, who are blissfully free of such cultural expectations.

In one study of two-day-old infants, girls made more dramatic facial responses than boys did during the heel stick used for routine newborn blood screening. However, there was no difference between boys and girls on another measure of pain that takes into account not only facial response but also crying, arousal, limb position, and breathing patterns. So either girls are more sensitive to pain than boys or they're simply more likely to express it on their faces. The latter conclusion seems to be the right one, based on recent research by Swedish scientists, who devised a fancy, noninvasive way of gauging brain activity by using infrared light to measure cerebral blood flow right through the skull. Surprisingly, this team found that it was the *boys* who exhibited greater cortical activation during the heel-stick procedure, suggesting they feel pain at least as well as the girls.

Misperceptions about boys' pain sensitivity have important consequences. Many newborn boys still undergo circumcision without any anesthetic, even though physiological measures tell us it is *very* painful: their heart rates accelerate and the stress hormone cortisol rushes through their bloodstream. The American Academy of Pediatrics appreciates this fact and recommends that all boys undergoing circumcision receive either a topical anesthetic or, preferably, an injection of a numbing agent prior to the procedure. But while young obstetricians, family practice doctors, and pediatricians are currently being trained to do this, many older doctors continue to circumcise boys without any pain relief. Some 65 percent of boys in the United States are circumcised at birth (versus about 15 percent worldwide). This fact raises a crucial issue when considering any reported difference between the sexes in the behavior of newborn boys and girls: whether differences in sensory responsiveness, fussiness, social interest, and so on are truly innate or are a consequence of boys' suffering from their recent circumcisions.

Surprisingly little of the research has controlled for this important variable, though there is one study—of olfactory perception—that included a comparison of circumcised and uncircumcised newborn boys. Psychologists at Vanderbilt University introduced one-day-old babies, both boys and girls, to a pleasant odor, either ginger or cherry. An hour after that, the baby was presented with both odors, each one soaked into a gauze pad and placed on either side of the bassinet. The girls in this study showed a consistent preference for the familiar odor—they would turn their heads toward the cherry scent, for example, if that was the one they had smelled an hour earlier. By contrast, boys turned their heads to the right, no matter which scent was on that side of the bassinet. (Most babies, male and female, do show a right-head-turning bias, which is thought to predict later hand preference.) This study hints that newborn girls can smell or remember scents better than newborn boys can, whether or not those boys are circumcised. Similarly, another study of two-week-old infants found that girls preferentially turned their heads toward a breast pad that had previously been worn by a lactating woman. Remarkably enough (considering their later interest), boys showed no such preference for breast-milk odor over the odor of a clean nursing pad. So it seems that newborn boys, like males of all ages, have a less sensitive sense of smell than girls, although once again, recent measurements of newborns' cerebral blood flow revealed no sex difference in frontal-lobe activity when boys and girls were presented with strong odors.

Now let's consider the sense of hearing. While scientists have documented all kinds of auditory differences between men and women—some favoring one sex, some the other—these are generally too slight to have any impact on real-world hearing abilities. On the one hand, women's threshold for hearing is about 3 decibels lower (that is, more sensitive) than men's; while 3 decibels corresponds to a doubling of sound pressure level (because decibels are measured on a logarithmic scale), it is nonetheless a trivial difference in the real world, where most sounds are well above threshold and range across a scale of some 130 decibels. On the other hand, men are better at localizing sounds in space, a difference that is explained mostly by their larger heads.* Auditory differences between the sexes have also been measured using EEGs, where a particular brain-stem response to sound stimuli, known as wave V, is both larger and faster in women.

So what about babies? Are auditory differences present at birth? Researchers have indeed found that newborn girls and adult women both have the same faster wave V response, but the effect is small, with a d value of 0.26, meaning the average girl's response is faster than about 60 percent of boys'. What's more, another study found that the threshold for this response—the lowest level of sound that triggers wave V—was actually lower, or more sensitive, in newborn *boys*, just the opposite of what you'd expect from either the sex difference in adults or the EEG latency data.

There is, however, one unambiguous auditory difference between

---

* The brain calculates sound-origin location by comparing the difference in time at which a particular sound arrives at each of the two ears. Because men have larger heads, their ears are farther apart, increasing their ability to resolve slight timing differences between the ears.

girls and boys from birth. This shows up in the otoacoustic emission test introduced in the last chapter; it screens for hearing deficits, and most newborns now receive it in the hospital within the first day or two of life. OAEs are sounds generated by the tiny hair cells of the inner ear. You can't hear them with your naked ear, but sensitive earphones can pick them up and give a wonderfully accurate view of babies' auditory function, well before the age when children can actually tell you what they hear.

Many studies have now confirmed that from birth to adulthood, OAEs are slightly larger in females than in males. A notable exception is the group of girls who have twin brothers; their OAEs fall in the typical male range (suggesting some influence on the developing auditory system by prenatal testosterone or another male factor). But while OAEs are a great way of testing the integrity of a baby's inner ear organ, they do not translate directly to hearing ability. In fact, girls' approximately 1 decibel louder OAEs can be explained by the shorter length of the female cochlea, the inner ear's snail-shell-like coil in which sounds are converted to electrical signals. What's more, the sex difference is so small (with a d value of about 0.15), it doesn't even warrant separate OAE standards for boys and girls when screening for hearing deficits.

It's therefore disingenuous to suggest, as Leonard Sax does in his book *Why Gender Matters*, that "these built-in gender differences in hearing have real consequences." Based on the small differences in auditory thresholds and wave V latency, Sax claims that girls hear shouting when their dads speak to them in a normal tone of voice, and he advises teachers to seat the boys in the front of a classroom (where the teacher's voice is loudest) and the girls in the back. Or better yet, Sax advocates, send them to separate schools, where teachers can whisper to the girls and yell at the boys.

But if you really want to test the built-in hearing abilities of boys and girls, you need to test babies, using real-world stimuli and behavioral measures that reveal how infants are actually perceiving this cochlear and brain-stem processing. Happily, this was done more than thirty years ago, in data reviewed by Maccoby and Jacklin (but notably overlooked by Leonard Sax). Maccoby and Jacklin examined the existing research on newborns' hearing ability and found that in only one of six studies did girls respond more overtly than boys to auditory stimuli. The results were no different for older infants, leading them to conclude: "The bulk of the evidence over the period from birth to 13 months shows that the sexes are highly similar in their attentiveness to auditory stimulation."

Maccoby and Jacklin's conclusions were the same for vision, the one sense in adults that is generally sharper in men than women. In nine different studies, involving nearly four hundred newborns, no differences were detected between boys' and girls' attention to checkerboards, colored lights, swinging objects, faces, geometric designs, and other stimuli. In another fifty-four experiments (published in thirty-three papers) involving babies between one and twelve months of age, girls were more visually attentive in nine experiments, boys in a different nine, and there were no sex differences in the remaining thirty-six experiments.

So the common belief that baby boys are more visual than girls is untrue. If anything, girls' visual systems develop a few weeks earlier than boys', based on EEG measurements and a handful of other recent studies. One of the more intriguing found that girls' stereovision emerges about three weeks earlier, on average, than boys'. Stereovision is the ability to fuse the images from each eye into a three-dimensional picture. It's critical for depth perception but not present in babies at birth. However, this 3-D perception pops up quite rapidly between three and five months of age, turning babies into highly visual creatures.

This particular ability is known to depend on the cerebral cortex, so the difference in onset between the sexes suggests that a girl's visual cortex matures about three weeks ahead of a boy's. There's even a hint that testosterone may contribute to this difference, based on a preliminary study that found a correlation between the hormone's level in baby boys' blood and their ability to fuse images.

Put all of this information together, and it's clear that at birth, boys and girls do not differ dramatically in their perceptual abilities. Although girls, like adult women, appear more sensitive to tactile and olfactory cues, the differences are quite small, and the more dominant senses, hearing and vision, do not differ meaningfully between young boys and girls. So it's a mistake to generalize, as many do, that boys learn through their eyes and girls through their ears. In fact, research by Carolyn Rovee-Collier and her colleagues at Rutgers University using various kinds of stimuli has uncovered no evidence for sex differences in infants' learning or memory. As we've seen, girls are actually more visual than boys in early infancy, and boys clearly learn a tremendous amount through verbal and other auditory input. Normal-developing boys and girls are *both* eminently capable of learning through either sensory channel and indeed should be encouraged to exercise all senses, especially in infancy, when every mental advance has such a cumulative effect on later learning.

Motor Skills

Baby Jack is working so hard to lift his head and look around. Only six weeks old and used to sleeping on his back, Jackie starts fussing when his mom puts him face-down on his colorful play mat for some "tummy time." To turn his head, he first has to lift it, not an easy feat for the little eleven-pounder, whose weight resides disproportionately in that well-protected cranium. But he's getting close. Finally! He manages to raise his chin an inch above the mat and turn his head to the opposite side; then he collapses, exhausted, but with a brand-new view of the world around him.

Motor skills are the most obvious proof of a baby's development. Compared to their more private sensory skills, babies' motor abilities are right out there for all to see (and parents to obsess about). Whether it's a newborn working on his head control or a one-year-old trying to take her first steps, parents pay more attention to their babies' motor skills than to any other aspect of their development.

So how much do boys and girls differ in their motor development during infancy? Surprisingly little. You might expect girls to take the lead here, considering their greater sensory and physiological maturity. Girls are sometimes described as ahead in fine motor control, the ability to move their fingers in a purposeful way, though there is no sex difference in onset of the pincer grasp, the ability to hold a small object, such as a Cheerio, between the thumb and forefinger. Some studies give boys a slight advantage in other skills, such as crawling and grip strength. But for the most part, there are very few sex differences in motor development. So few, in fact, that pediatricians don't bother with separate charts for boys' and girls' motor milestones, in contrast to their separate charts for physical growth. Even the two most widely used checklists of infant development, the Denver Developmental Screening Test and the Bayley Scales of Infant Development, have not revealed any sex differences in motor achievement. On average, boys and girls master the same skills, in the same sequence, at the same ages after birth: rolling over in the fourth month, sitting in the seventh month, and walking at right around twelve months. My own children, though hardly a scientific sample, confirm this last statistic: Julia first walked at twelve and a half months, roughly in between Toby (twelve months) and Sam (fourteen months).

One difference that does show up at birth concerns symmetry of movement. When a baby is born, the pediatrician tests a number of reflexes that are good indicators of a baby's overall health and development. And one feature that clinicians look for in these reflexes is symmetry: both sides of the body responding equally to the same stimulus. No child or adult is perfectly symmetrical. In fact, there are hints of later hand preference in the bias that most babies, male and female, show for turning their heads to the right side. But boys are more asymmetric than girls in certain reflexes involving the legs and feet, including the plantar grasp (gripping with the toes, one of several primitive newborn reflexes apparently left over from our ape ancestors). Such asymmetry, or lateralization (a difference between the two sides of the body), is also more prevalent in adult males, so this finding supports the idea that the male nervous system is innately more lateralized.

But beyond these minor differences, the striking fact about motor development in infancy is that it does *not* differ between boys and girls. Considering that girls are perhaps a month or so more mature in other functions, one way to think about this similarity is that boys actually have a motor advantage. That is, boys manage to hold their own in motor development despite their generally greater immaturity.

Size may be one factor keeping boys on par with girls. The fact that boys are bigger and heavier may help stabilize them better for weight-bearing skills such as standing and walking. Another advantage comes in boys' higher activity levels. Though this difference becomes more pronounced later in childhood, it begins in infancy, when male infants are found to kick, punch, and, eventually, move around the house more than female infants. Like all athletic feats, motor milestones depend heavily on practice. Babies don't just lie around waiting to roll over. They spend weeks pushing up, strengthening their arm and neck muscles, and learning to balance those big heads. Boys' greater activity level may mean that they give themselves more practice from an earlier age, overcoming their slower maturation to keep pace with girls' advancing movement abilities.

But parents are another important factor in boys' and girls' comparative motor development. Research shows that parents often place a higher premium on their sons' physical development than their daughters'. The difference becomes quite obvious by Little League age, but it's evident even in infancy. In one study, moms and their eleven-month-old babies were brought into the lab to teach the babies a new motor skill: crawling down a carpeted slope. In the first part of the experiment, the mothers were shown the sloped walkway, which was rigged with a hinge to adjust the angle of the slope. By pushing a button, a mom could change the angle to the maximum slope she judged her baby capable of crawling down. In the second part of the experiment, the babies were tested to see how steep a slope they were willing and able to crawl down regardless of their mothers' judgments.

The results of this study were eye-opening. Boys and girls did not differ in the steepness of the slope they were able to crawl down. Girls attempted and successfully descended slopes ranging in angle from 10 degrees to

Figure 2.2. Setup for infant-crawling experiment. Mothers estimated the slope their babies would tolerate crawling down, and then boys and girls were tested on their ability and willingness to crawl down different slopes.

46 degrees, while boys attempted slopes between 12 and 38 degrees and successfully navigated those up to 30 degrees. So this study revealed no sex difference; if anything, infant girls had greater courage than boys for crawling down steep slopes. Mothers, however, had assumed just the opposite. While the mothers of boys were within 1 degree of accuracy in predicting their sons' slope-crawling abilities, mothers of girls *underestimated* their daughters' abilities by an average of 9 degrees.

Are mothers therefore the culprits in limiting girls' athletic prowess? This study does suggest that parents, aware of physical differences in older boys and girls, set up different motor expectations for their sons and daughters, even in infancy. So here is a place where parents can step in and challenge their daughters physically rather than hold them back. Moms who sign up their sons for infant gym classes—"because he's so active"—must be careful not to overlook their daughters, who may benefit even more. We may not be able to flip men's and women's Olympic records, but there is certainly more we can do to increase girls' physical confidence and fitness.

Language

"Girls talk earlier than boys." If there is one mantra most people can recite about sex differences in infants, this is it. And it is accurate, though like most gender traits, the difference is small and there are many exceptions to this rule. (Our own Sam was the earliest of our three babies to talk—and the chattiest—even compared with his sister.) The finding is present in rhesus monkeys as well; their female infants vocalize more and become fluent at an earlier age than their male infants, a difference that's known to be influenced by prenatal testosterone. So there may indeed be something innate about the sex difference in age of vocalization, though it's a long way from a few instinctive coos and cries to a fully fluent vocabulary. Much more important, for parents of sons *and* parents of daughters, is to appreciate how remarkably early language learning begins and how a baby's language exposure, from even the first days of life, crucially influences his or her eventual verbal ability.

Most babies manage to say their first real words* close to their first birthday. The typical girl, however, will tend to blurt out a *hi* or a *baba*

---

* Sadly, researchers don't count *mama* or *dada* as valid first words.

(for "bottle") about a month earlier than the typical boy will. As parents, we pay a lot of attention to these first words. We write them down in the baby books, call the grandparents, and have lots of fun finally conversing, albeit in single words, with our babies. But first words are not the true beginning of communication. Babies begin much, much earlier to inform us of how they're feeling—by crying, squealing, making eye contact, and pointing and other gestures. What's more, babies perceive and understand language a long time before they're able to make themselves understood. As we now know, language learning begins at birth, if not before, and sex differences in language, which are among the earliest of the cognitive differences between the sexes, have been found as early as researchers can study them.

Seven-pound Ellie lies comfortably sleeping in a reclining infant seat, oblivious to the measuring devices adhered painlessly to her fuzzy head. She's cooperating beautifully for a research study in the hospital where she was born a mere twenty-one hours ago. As she rests, a loudspeaker located a few feet above her plays simple repeated speech sounds, *ba* and *ga*, in a varied order. Ellie is sleeping peacefully, but nonetheless, her ears take in all the sounds, convert them to electrical signals, and send them up to her brain, where the sensitive electrodes record some rather remarkable findings.

Newborns can tell the difference between speech and other forms of sound. Already, their little left hemispheres devote more effort to processing speech than their right hemispheres do. Other sorts of sounds, like bells or barks, do not get such preferential neural treatment, which tells us that even before birth, the brain is specialized to perceive language.

However, girls' and boys' brains do not handle language identically. When one-day-old Jacob is given the same test at the hospital, his sleeping brain produces a different pattern of brain waves. Researchers do not agree on exactly how boys and girls differ at this young stage: one early study said that baby girls' brains are more lateralized, or left-biased, for processing language, a seemingly more mature pattern since most adults process language more with their left hemispheres. However, this frequently cited study has been disputed, while a very recent study of four-week-old babies found exactly the opposite result—stronger left-hemisphere activation in boys. In fact, studies of newborn auditory function indicate that boys, not girls, are more lateralized, producing stronger otoacoustic emissions in the right ear and also showing a lower threshold, or higher sensitivity, for perceiving sounds with the right ear than the left. Because most auditory information crosses the brain on its way up from the ears, a right-ear advantage for perceiving sounds corresponds to a left-brain advantage. Adult males may be modestly more left-hemisphere dominant for language than women (see chapter 5). So this greater lateralization in newborn boys' hearing suggests that they do not get as even an exposure to language as girls do, perhaps paving the way for later differences in verbal and written language skills.

Infancy is, by definition, a preverbal phase in development. (The Latin word *infans* means "without language.") And yet, we now know that babies learn a tremendous amount about language during the first year of life. Most of their growth is in the realm of understanding, or receptive, language. Although it's rarely obvious to parents, babies manage to recognize familiar speech sounds, words, and even grammatical patterns, all within the first year of life. Speech therapists appreciate the importance of receptive language to overall communicative development. While parents worry most if their toddler is late to talk, clinicians focus more on receptive-language skills to determine whether a child is language delayed.

In one large study of early language development, researchers at San Diego State University documented slight, but significant, differences between boys' and girls' receptive-language development. Female infants are about a month ahead of boys in the number of words they understand. The typical nine-month-old girl understands about fifty words, such as *no, dog, bath, bottle,* and the like. The typical boy reaches this vocabulary size at ten months of age (see Figure 2.3).

A similar difference turns up in babies' early gesturing, a big part of communicative development. Many scientists believe speech evolved from gesture—that our progenitors "spoke" with their hands before using their vocal tracts.* Babies too gesture before they speak, a discovery that inspired all the popular sign-with-your-baby books and videos that teach parents how to use gesture to encourage their infants' communicative development.

---

* This manual origin of speech might explain how language came to be localized in the brain's left hemisphere, which controls movement of the (usually) dominant right hand.

Girls lead boys in the number of gestures they produce during later infancy. On average, they begin pointing, waving bye-bye, and raising their arms to be picked up a few weeks earlier than boys. Again, their advantage is rather small; in one large Swedish study, eighteen-month-old girls were found to produce just 5 percent more gestures than boys. However, some of the gestures babies make are already sex-typed: female babies between eight and sixteen months old are more likely than boys are to imitate parenting behaviors such as hugging or rocking a doll; boys in this age range pretend they are reading, driving a car, or pounding with a hammer.

After gestures come words, babies' first true expressive language. Girls maintain their small lead in expressive language throughout the toddler period, producing an average of three hundred different words by twenty-two months old, while boys get to this mark at twenty-three or twenty-four months.

After age two, children truly begin to talk—combining words into little sentences such as *Mommy home. More milk. Let's go park.* Here again, girls take the lead: by two and a half years old, girls are stringing together about eight words at a time, boys about six. From their first emergence, girls' sentences are longer and more complex than boys', a difference that holds throughout the preschool period.

These differences are reliable but really quite small. Compared to the overall range of language ability in young children, the average difference between girls and boys is tiny, accounting for just 1 to 2 percent of this total variance. Consequently, you can find many boys who are more verbal than the average girl, and lots of girls who are less verbal than the average boy. In practical terms, the very small magnitude of this gender difference means that parents should be just as concerned about a son's slow language development as a daughter's. Being male is no excuse for talking late (in spite of what many pediatricians tell parents).

There is ample reason for parents to pay close attention to their children's language development. Whereas language ability is only subtly affected by a child's gender, it is strongly affected by his or her environment. Study after study has now confirmed this startlingly simple equation: Language in = Language out. From the moment of birth, the amount and quality of language addressed to babies directly influences the quantity and quality of the babies' own verbal ability. Researchers have documented this relationship through painstaking studies: visiting

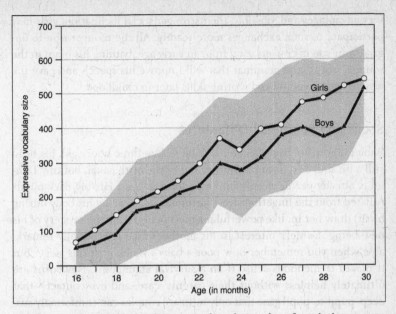

Figure 2.3. Girls lead boys by about a month in the number of words they can say, but the difference between sexes is much smaller than the normal range for all toddlers, two-thirds of whom fall within the shaded region. Data are based on more than eighteen hundred children as reported by Larry Fenson and colleagues.

families' homes and counting literally every word spoken from parent to child. By analyzing hours and hours of audio recordings, these studies were able to show significant correlations between the amount of language addressed to young children and those children's later vocabularies and other skills, including reading ability. The input doesn't have to be all talk; songs, stories, poems, and nursery rhymes are as good as or better than conversation. But it does have to be live: TV, DVDs, and recorded music simply don't provide the kind of feedback and encouragement children need to best develop language skills.

In other words, while boys may be at a slight disadvantage when it comes to language development, there is enormous opportunity to improve the verbal skills of every child, male or female. All it takes is talk: conversing, reading, singing, and generally narrating the events of the day can provide *both* boys and girls the optimal experience to hone the brain circuits that underlie language. Given their slight verbal and auditory advantages and, possibly, their stronger social inclination, girls may participate in such exchanges more readily. All the more reason to engage your son in conversation from an early age, bathing his brain in the sounds, words, and grammar that will improve his speech and pave the way to better reading and writing skills later in childhood.

Social and Emotional Differences

Babies are instinctively social. Here he is, just three hours old, his head still a bit squished from his journey out of the birth canal, but tiny Daniel is already gazing deeply into his mother's eyes. His big dark pupils (dilated from the lingering stress hormones released by his body during birth) draw her in, like powerful magnets penetrating the mystery of his new being. Daniel's interest in his mother's face is especially remarkable when you remember how poor a baby's vision is at this stage. But it makes tremendous sense from a survival standpoint. Newborns are ultimately helpless without their parents' care, and eye contact—that deep, pupil-to-pupil gaze that only intimate couples can comfortably sustain—is probably the quickest way to get their dads and moms to fall in love with them.

Considering the importance of this early eye contact, it is not surprising that newborn boys and girls engage in it equally. Although one 1979 study found that girls held eye contact longer than boys, a larger 2004 study found no sex difference in the newborn period. Both studies were carried out blind—that is, the experimenters were deliberately kept unaware of each infant's sex, and all the pink-or-blue cues, such as balloons and gender-coded diapers, were carefully removed from the hospital rooms.

But wait a minute. Aren't girls supposed to be innately more people oriented than boys are? This idea has gained a lot of currency in the last few years, but it is based largely on a single study. Maybe you heard about it: it was carried out by a team of psychologists led by the University of Cambridge's Simon Baron-Cohen (not to be confused with his cousin Sacha Baron-Cohen, of *Borat* fame). The experimenters presented newborn babies with two visual stimuli: a live human face followed by a colorful mobile with a little ball hanging down from it (or vice versa). One hundred and two newborn babies were tested in their hospital rooms the day after birth. And guess what: the boys spent more time looking at the mobile, while the girls looked longer at the live human face. The difference was not large: boys spent an average of 52 percent of their time looking at the mobile and 46 percent looking at the face, while girls split their time 41 percent and 49 percent between the mobile and the face. (Girls must have been looking elsewhere for the remaining 10 percent of the time.) But it was statistically significant, which somehow emboldened the researchers to state in their peer-reviewed article that the results "demonstrate beyond a reasonable doubt that these differences are, in part, biological in origin."*

Since it was published, in the year 2000, this experiment has been hailed by seemingly every commentator on sex differences: Leonard Sax, Louann Brizendine, Steven Pinker and his sister Susan Pinker, among others. It's been taken as proof that girls are pre-wired for social interaction and boys are pre-wired for objects and mechanical interests and its results extrapolated to explain later gaps in language, empathy, math, engineering—pretty much every social and cognitive sex difference ever described.

But there are some problems with this study that should urge a little more restraint. The most serious, to my mind, is the fact that the person whose face the babies were looking at, lead author Jennifer Connellan, was not always blind to the sex of the babies at whom she was smiling. As we've seen, researchers studying newborns have to be very careful to remove (or have others hide) all gender cues in the hospital room so that they do not unwittingly bias babies' behavior; for instance, by unintentionally making more eye contact with girl babies than boy babies. Another problem is that neither Baron-Cohen's group nor any other researcher has replicated the results of the study, even some nine years after its original publication.

And the results really need to be replicated, because as Elizabeth Spelke, a Harvard psychologist who has studied infant cognition for decades, has observed, it stands in opposition to many similar studies that have looked for but failed to find evidence for sex differences in object and motion perception. In 1996, for instance, Canadian researchers found that at eleven weeks of age, *both* boys and girls gazed longer at a

---

* Once again, the claim that any behavior is *biological* is hardly earthshattering (because all behavior, both learned and instinctive, is controlled by our biological brains), but the authors seem to be using this term to mean "innate" or "hard-wired," which is how it has been construed by several commentators.

toy (a colorful mobile with small bells) than at their own mothers' live faces. Another study found that *boys,* not girls, paid greater attention to faces than to geometric displays. In fact, there was a much greater flurry of research in this area some years ago, from which Maccoby and Jacklin concluded that "there is no evidence that girls are more interested in social, boys in non-social stimuli."

So I'm not convinced that, as Louann Brizendine asserts, "girls, not boys, come out wired for mutual gazing." Nor is there any evidence, as she also states in *The Female Brain,* that boys' "testosterone surge in utero . . . shrinks the centers for communication, observation, and processing of emotion." If anything, existing data suggest testosterone *grows* the one structure most strongly associated with the recognition of faces and emotional expression—the amygdala—though this finding is based on research in rats and monkeys.

The amygdala is part of the limbic system, a complex network of conscious and subconscious brain zones that controls our rich social and emotional lives. But while girls do not necessarily pop out of the womb more socially wired than boys, there is another sex difference that hints at an innate difference in the limbic system.

The difference is fussiness. In contrast to their more stoic adult selves, infant males are actually the more emotional of the two sexes. As we've seen, boys are more irritable, more easily distressed, and harder to soothe than girl babies. They startle, cry, and grimace more often, at least during the newborn period, and take longer than girls to establish a stable sleeping pattern. Once again, these differences have mostly been described by examiners who weren't blind to the sexes of the babies, and some studies have failed to detect any sex differences in newborn irritability or consolability. But overall, a difference in emotional fragility appears more reliable than any social sex difference in newborns, and one study even found a possible physiological basis for it: compared to girls, newborn boys produced a greater surge in the stress hormone cortisol when they were disturbed by a surprising stimulus or an unwanted social advance. The release of cortisol is ultimately governed by the hypothalamus, the one part of the limbic brain most clearly shaped by prenatal testosterone. Since irritability and stress reactivity are also greater in preterm than in full-term neonates, this sex difference supports the idea that a boy's hypothalamic-endocrine system, along with the autonomic nervous system that it governs, are less mature than a girl's at birth.

Parents may not be aware of this sex difference in fussiness, but it almost certainly influences the way we interact with our newborns, helping to shape some of the more obvious social-emotional differences that emerge later in infancy. Baby boys are generally not as "easy" as baby girls (though there are, as always, many exceptions to this generalization). I remember well the exhaustion I felt when Sam came along. Julia had been such a good baby, such a breeze to care for, but Sammy would fuss every few minutes. It seemed like I was always nursing him, holding him, or carrying him around in a sling or backpack. (Toby was somewhere between the two in fussiness.)

And it's not just me: Rutgers University researchers Jeanette Haviland and Carol Malatesta found that parents (particularly mothers) do work harder to manage the emotions of their infant sons. While it's easy to respond more positively to girls—since they are already a bit happier—parents may respond more negatively to boys, shushing them or ignoring their cranky bouts in an effort to dampen them. While their goal is to protect boys from their own emotional overload, the actual result may be to send a negative message to boys about their expressiveness, deterring further social interactions and contributing to the later suppression of emotion that is more characteristic of boys' development.

Whatever the cause, boys and girls do start diverging in their social responses over the next few months of life. The 2004 study that found no sex difference in newborns' eye contact found a very different pattern when the experiment was repeated with the same infants four months later. Boys had increased their eye contact only slightly, but girls had increased it more than four-fold! A similar increase in mother-daughter gazing was found during the third month of life in a 2002 Italian study that tracked boys' and girls' face-to-face communication from birth. In fact, baby boys begin to do something quite different over this period: they avert their gaze. Gaze aversion is the tendency everyone has to look away from an oncoming person when walking down the street or a grocery aisle, thus avoiding any awkward social engagement. Both sexes do it, but males of all ages resort to gaze aversion more frequently than females do.

Between three and six months of age, girls and boys also begin diverging in the variety of facial expressions they display. Although both sexes spend roughly equal amounts of time looking happy, sad, and angry, girls in this age range show the raised eyebrows, relaxed mouth, and wiggly tongue that express *interest* more often than boys do. And, while Haviland and Malatesta found encouraging evidence that mothers are almost equally attentive to the emotions of baby boys and baby girls, their careful analysis did reveal a conspicuous lack of response to boys' expressions of pain, while another study found that mothers of girls tend to ignore their expressions of anger.

So while differences between the sexes in sociability and emotional expression are not obvious at birth, they grow significantly during early infancy. The differences probably originate in boys' greater fussiness and immaturity, but they are amplified by the way that parents respond to these differences, as well as our preconceived notions of what boys and girls are like. By ignoring boys' expressions of pain, mothers may be trying to "toughen up" their sons, while by ignoring girls' expressions of anger, mothers may be attempting to dampen their assertiveness. The fact that the sex difference in eye contact is not present at birth but emerges after a few months supports the idea that it is shaped, at least to some degree, by parents' different styles of interaction with boys and girls.

Are Baby Girls More Empathetic?

In 1971, psychologist Marvin Simner wanted to figure out if newborn babies were capable of empathy. So he set up shop in a quiet room next to the neonatal nursery at the Lying-in Hospital in Providence, Rhode Island. There he tested individual babies to see how each responded to a simple social stimulus—a brief audio clip of another newborn crying. Simner discovered that babies do indeed cry in response to one another, but he also observed that after hearing the tape, female newborns cried longer than males. Statistically, the sex difference was quite marginal, but the fact that girls outcried the boys in four different repetitions of the test, as well as in two later studies by other authors (where the effects were also not statistically significant), has been hailed as proving that girls are innately more empathetic than boys.

Now, it's a long stretch from this imitative crying to conscious recognition and sharing of another person's emotional experience. Nonetheless, a sex difference in empathy is one of the more reliable findings in adults: women outperform men at accurately deciphering whether another person is displaying anger, fear, interest, and other emotions. There are some important caveats* to this rule, and the difference is not large, with a d value of about 0.40. But it raises the issue, as psychologist Baron-Cohen puts it, of whether "the female brain is predominantly hard-wired for empathy" and, conversely, whether boys have some innate deficiency in this area that renders them less fit for more interpersonally sensitive careers such as nurse, doctor, teacher, minister, social worker, and—oops!—psychologist.

Happily, there are actual data addressing this topic, based on about two dozen studies of babies' abilities to recognize or discriminate facial expression in others. As reviewed by Emory University psychologist Erin McClure, female infants are better than males at deciphering people's facial expressions, with a difference score of 0.26—smaller than the difference in adults but, surprisingly, *larger* than the difference during later childhood, when it shrinks to a modest 0.16. McClure speculates that girls are indeed more capable of detecting others' emotions in infancy, but their advantage is mostly a matter of neurological maturation. Just as their sensory and language skills are slightly ahead of boys' at birth, so girls' brains may be sufficiently more mature in infancy to enhance their awareness of other people and their facial expressions. With time, and experience with other people, the gap closes, and boys and girls are not so different in their sensitivity to others' feelings during the rest of childhood.

It's true, then, that girls are more empathetic than boys. But the difference is small and clearly shaped by learning. It is not the fixed, black-and-white distinction that many people believe. Consider this, from Louann Brizendine:

> Anyone who has raised boys and girls or watched them grow up can
> see that they develop differently, especially that baby girls will connect
> emotionally in ways that baby boys don't.

---

* The size of the sex difference in empathy depends on how it's tested. Women are likelier than men to say they feel sorry or guilty about another person's experience, but the difference is much smaller when empathy is tested using more objective measures, such as identifying the emotion on another person's face. What's more, the ability to recognize others' emotions also depends on who is looking at whom. Men are better at detecting emotion on men's faces than on women's, while women perform about equally for men's and women's expressions (see chapter 7).

In fact, the d value of 0.26 in infants translates to about 60 percent of boys falling below the average girl in ability to detect emotion in other people. This means that 40 percent of baby boys are actually *more* attuned to other people's expressions than the average girl, and the proportion is even larger for older boys. Hardly a lack of emotional connection.

So Brizendine's statement is not only wrong, it's downright subversive. Imagine parents of a newborn boy *expecting* that they won't be able to bond with their son! Boys won't have a chance. As we've seen, parents already react differently to the emotional expressions of their very young sons and daughters. Telling them that their boys are not even equipped for emotional connection is only going to exacerbate this bias and further suppress boys' social-emotional development.

If anything, the small sex difference in a newborn's emotional awareness appears to be more a matter of timing than a fixed and permanent deficit. Neuroscientists have learned that the processing of facial expressions depends on a particular zone in the temporal lobe, known in monkeys as area TE. There is evidence that area TE develops more slowly in males than in females, and that this delay is due to testosterone. So, like other infant sex differences, girls' greater sensitivity to facial expressions may be a simple matter of a few extra weeks of neurological maturation—hardly enough to warrant an adjustment in parenting philosophy.

Emotional learning is a two-way street. Boys and girls enter the world with slight differences in social and emotional styles, but in reacting to these differences, parents end up training boys and girls in different ways. Girls' stronger social bias is highly reinforcing to parents and so becomes easily strengthened, while boys, who are less mature and consequently slightly less social and more fussy at birth, compel parents to a more cautious style of interaction, reinforcing boys' already weaker orientation toward other people.

The good news is that the small sex difference in emotionality and social responsiveness is not reflected in the depth of the bond between babies and their parents. Infant-parent attachment has been extensively studied over the past half a century, and there are no differences between boys and girls in the quality of attachment to either their mothers or fathers. So while boys and girls do begin differing in social style even early in infancy, this doesn't affect their fundamental connections to their parents or the sense of security the bond provides.

But parents aren't the only ones who interact with infants, and there is some research suggesting that boys may not fare quite as well as girls when they are cared for by other adults—either by babysitters or in daycare centers. Although most studies of daycare find few overall effects—either positive or negative—on children's development, there is some evidence that boys receive less positive caregiving in such settings, and even that infant daycare may impair boys' attachment to their parents. It has no effect on girls' attachment and may even benefit girls' cognitive development when compared to being home with their mothers. Considering that boys are both fussier and somewhat less social, they simply may not charm their caregivers as much as girls do, leading to less-positive interactions. While parents may be willing to go the extra distance with their sons, working harder to comfort and engage them, paid caregivers may not be so willing, or, given a choice between the girls and boys in the room, they may simply direct more of their attention in the pink direction . . . all of which could have a negative impact on boys' emotional and behavioral development.

Is Autism a Male Trait?

While the difference in empathy between boys and girls is relatively small, one glaring sex difference has pressed the idea that boys are innately less attuned to others than girls are: the high ratio of boys to girls in autism. This devastating disorder was not even identified until the mid-twentieth century but is now one of the most prevalent syndromes of childhood, diagnosed in 3.4 of every 1,000 children. And of these children, nearly 80 percent are boys.

Autistic disorders range in severity but share a core deficit: a lack of social awareness or understanding of other people's feelings and motives. Autism may be diagnosed as early as age two or, more typically, in the fourth year of life, although there may be some signs in the first year, such as a lack of shared gaze and joyful expressions. Children with autism (or its milder form, Asperger syndrome) fit Brizendine's stereotype of males being unable to connect with other people. Autistic children have difficulty communicating, making eye contact, and, especially, understanding that other people have thoughts and emotions different from their own. In other words, they lack empathy, and so the fact that three or four boys are diagnosed with autism to every girl with the disorder has suggested that boys are more vulnerable because their brains innately lack the circuitry for empathy.*

The strongest advocate of this view is Simon Baron-Cohen, who actually describes autism as the consequence of an "extreme male brain." His theory is based on both the empathy and communication deficits of autism as well as the tendency for people with autism to demonstrate highly restricted interests — such as the compulsive calculating by Dustin Hoffman's character in the film *Rain Man*. Baron-Cohen and his colleagues have constructed numerous questionnaires that probe these two dimensions — empathizer versus systematizer.† The questionnaires look like any battle-of-the-sexes game, and, not surprisingly, women do score higher on the empathizer scale and men higher on the systematizer scale. One problem, however, is that these scales rely on self-report instead of more objective measures of facial perception or analytic ability; in fact, men and women did not score significantly differently from each other on one of Baron-Cohen's tests that required subjects to identify a person's emotion based on a photograph that showed only the eye region of the face.

But the theory goes even further, suggesting that the same factor that makes a brain male is also what *causes* autism. Baron-Cohen and his colleague Rebecca Knickmeyer target the usual suspect: prenatal testosterone. In an ambitious research project, they measured testosterone in amniocentesis samples taken from dozens of women during midpregnancy. They then followed the babies to four years of age, when they tested the children for two autistic features, reduced empathy and restricted interests.

Their team has now published many articles on this project that have impressive-sounding titles, but the results are not so convincing when you read between the lines. While they have found that prenatal testosterone levels correlate with many measures — eye contact in infancy, vocabulary size in the second year, empathy and restricted interests in young children — these results often did not hold up when males and

---

* Then again, some experts believe that autism and Asperger syndrome may be underdiagnosed in girls, whose better language and mimicry abilities may mask the same underlying social deficit.
† The somewhat awkward term *systematizer* is supposed to encapsulate one's tendency to focus on objects and how they are ordered or organized. A more accurate name for what this survey probes is *analytical* tendency.

females were analyzed separately. Obviously, boys are exposed to higher testosterone levels in utero than girls are, and boys, as we've seen, show less eye contact, smaller vocabularies, lower empathy scores, and more restricted interests. So, if you lump boys and girls together, you're likely to find some correlation between testosterone and these various behaviors. (You'd also find a correlation between fetal testosterone and who's wearing pink or blue Pampers, but that doesn't mean the relationship is causal.) The real issue—and the way most studies of prenatal testosterone are conducted—is if the relationship between fetal testosterone and empathy is significant *within* either sex. For instance, if autism is caused by extreme testosterone exposure, then you would expect to find that boys with the highest prenatal testosterone levels are the ones who end up being diagnosed with the disorder, while boys with lower prenatal testosterone exposure would be diagnosed much less often.

This is not the case,* nor are autistic boys hypermale in any other sense, such as aggressiveness or body structure. These researchers have also yet to show a relationship between testosterone levels and empathy or systematizing measures within the population of girls alone. As we've seen from studies of girls with CAH (the genetic disorder that exposes them to high levels of male hormones), prenatal testosterone has been linked to certain masculine behaviors, such as a preference for trucks and balls, as well as to higher rates of lesbianism and bisexuality. However, girls with CAH are not any more likely to be diagnosed with autism than girls with normal prenatal testosterone exposure.

A last piece of Baron-Cohen's extreme-male-brain theory concerns the brain itself. Recent research has identified several differences between the autistic and the normal brain, especially in areas known to be involved in social awareness and interaction, such as the amygdala. But the most definitive finding is the somewhat surprising *overgrowth* of the brain in the first year of life in children who are later diagnosed with autism. In other words, the brain is larger in autism (at least during the first year of life), and boys have larger brains than girls. So this observation is said to support the extreme-male-brain theory of autism.

But there is no evidence that prenatal testosterone is responsible for

---

* Very recently, Baron-Cohen's group did report a correlation between fetal testosterone level and one measure of "autistic traits" that was significant within separate populations of boys and girls. However, none of these children, who were tested between six and ten years of age, actually suffers from autism.

the brain overgrowth in autism. For one thing, this overgrowth does not begin until at least a month or two *after* birth and is most dramatic between six and fourteen months of age. As we've seen, boys' head size is consistently larger than girls' from birth onward. If prenatal testosterone were causing the overgrowth, you would expect to find the extra brain volume present in autistic children at birth. In addition, girls with CAH do not have larger brains than normal girls, in spite of their exposure to excess testosterone and other androgens before birth.

So although prenatal testosterone may shape some of the social differences between boys and girls, it's a stretch to say that it causes autism. What, then, is responsible for this heartbreaking disorder?

The answer remains an agonizing mystery, especially for parents whose children suffer from it. Genes are the biggest factor, with heritability of autism estimated at between 60 and 90 percent. However, there is no single gene that causes autism, and the disorder appears to involve the interaction of many different genes. Nor is autism clearly linked to the X chromosome, unlike other disorders that occur primarily in boys (such as Duchenne's muscular dystrophy and colorblindness).* Environmental influences, ranging from diet to TV exposure to the much-maligned vaccination program, have also been proposed. But none of these has been scientifically validated, nor do any purport to explain why autistic disorders are so much more common in boys.

There is, however, one thing we do know about autism, and that is that the earlier children are diagnosed and treated, the better their prognoses. The best treatment involves intense social interaction, as much as forty hours per week with a one-on-one therapist who encourages speech, eye contact, and other aspects of communication while motivating the child through play. It's an expensive proposition but well worth the effort for children who are not too severely affected and who begin such therapy early, preferably by age two. The reason for starting as young as possible is to wire in social and communicative circuits before the brain is taken over by the maladaptive repetitive behaviors and restricted interests that characterize older autistic children. In fact, new evidence suggests that a small percentage of children can even "recover" from autism if they're provided with this optimal early treatment.

---

* An exception is the disorder known as fragile X syndrome, which is due to a mutation on the X chromosome and which is responsible for about 7 percent of autism cases.

So whether or not autism is caused by an extreme male brain, the treatment is the same as the solution to the gender gaps in language and empathy: early social immersion. Live engagement with responsive, sensitive caregivers is by far the most important experience any baby can have, and so a greater focus on this area by parents of boys—and especially those with a family risk for autism—is unquestionably beneficial.

How Parents Stereotype Their Baby Boys and Girls

There's another problem with this concept of autism as a symptom of maleness. Like any focus on extremes, it reinforces the same stereotype about boys that parents should be combating. Most boys are not autistic, and whatever the small differences in eye contact between boys and girls, boys' interpersonal skills will, like their verbal abilities, ultimately be acquired through learning. Telling parents that all boys are a little autistic by nature seems unlikely to promote this emotional tutelage.

Consider Steven, a thirty-four-year-old electrician who's thrilled to be a father. When his daughter, Meredith, was born, he couldn't wait to get home from work and hold her all evening, giving his wife a break during her all-too-brief maternity leave. Meredith would sleep peacefully on his chest, her warm little body emitting tiny sounds and the occasional little yelp to say it was time to move, or nurse, or have her diaper changed. Now they've got another baby, ten-month-old Kyle, and Steven interacts with him quite differently. Though he held Kyle plenty as a young infant, Steven can't get enough of tossing Kyle in the air and rolling balls across the floor to him, hoping it will translate into a passion for baseball or soccer in a few years.

Both types of interaction are great, but why should his daughter be treated in one way and his son in such a different way?

Boys and girls do differ subtly at birth, distinctions that contribute to the different ways parents interact with them. But moms and dads must also admit our own parts in fostering the social and emotional differences between boys and girls. Parents treat baby boys and girls differently, not only because they *are* different but also because of our preconceived notions of what it means to be male or female. Inevitably, we stereotype our babies, even when they're fresh out of the womb. Few of us are aware of this fact, and even those who are conscious of it find it difficult to avoid.

In one classic study, parents were asked to rate their newborns on various attributes. Parents of girls tended to describe them as prettier, softer, more delicate, weaker, finer-featured, and less attentive than parents of boys described their infants, and fathers tended to more extreme characterizations than mothers. Of course, boys *are* larger than girls, so there is some basis for at least a few of these distinctions. However, some fascinating cross-dressing experiments have proven that the differences are less real than they appear in the eye of the beholder. Researchers have taken two approaches to address this issue: they disguise the babies' sex by dressing them in gender-neutral clothing; or they deliberately mislead adult subjects by calling a baby girl Jonathan or a boy Marie. In one such study, girls were described as angry or distressed by adults who thought they were boys more often than they were by adults who knew their true sex. Similarly, boys mislabeled as girls were more often seen as joyful or interested than they were when their sex was accurately revealed. Dozens of other gender-disguise studies confirm that people judge babies differently based on what sex they believe the babies are and regardless of the real sex under the diapers. In addition to rating babies' expressions and physical appearances differently, adults tend to choose different toys for each sex (footballs and hammers for babies they believe are boys, dolls and hairbrushes for those they think are girls) and to engage with them differently—interacting in more physical, expansive ways with boys and more nuanced, verbal ways with girls.

So for all our interest in raising strong girls and sensitive boys, parents still push gender distinctions from an early age. We find out our babies' sex as early as possible, so we can decorate their rooms in the appropriate hues and pick out the right birth announcements. We also receive different birth congratulations: according to one study, the cards parents receive for boys are likely to depict an active infant with balls or cars or sports equipment more suitable for an older child. Cards for baby girls tend to paint a more passive child, using adjectives such as *sweet* and *beautiful* and showing toys like rattles or mobiles that are more appropriate for a young child. So regardless of what parents profess to wish for them, our culture still values boys' strength and girls' appearance more than the other way around (except, perhaps, in Lake Wobegon).

The good news is that parents today are differentiating less between their baby boys and girls than they did a generation ago. For instance, a 1997 study of young women's responses to gender-disguised infants found little difference in the style of speech or amount they spoke to "girls" or "boys." The only significant difference was in their tendency to comment more on the physical activity of infants they thought were boys. A study published in 1995 found that parents perceived fewer differences between newborn boys and girls than parents in the 1970s had. Fathers, in particular, are shedding their stronger biases of a generation ago as they play an increasingly greater role in early infant care and discover that the stereotypes don't necessarily hold for their own children.

Parenting styles change quickly. Many of us would never dream of treating our children the way we were treated or of engaging in certain practices, such as spanking, that were once considered essential tools in child rearing.

Gender expectations are part of that change. While some parents maintain the traditional stereotypes and expectations for their children, many more are keeping an open mind, hoping to nurture both the masculine and feminine sides of their sons and daughters. This can only be to the children's benefit.

But parenting can take children only so far. When it comes to gender stereotypes, children are much less forgiving than adults in enforcing expectations for male-versus-female behavior. Why they're so tough on one another is anybody's guess, but there can be little doubt that their knowledge of gender stereotypes begins very early, and that their parents' examples are a big source of their learning.

What Do Babies Know About Gender?

If you thought infants were too young to be aware of gender differences, think again. Infants of all ages prefer the higher-pitched female voice to the deeper male voice, showing that they can tell the difference between male and female speech even shortly after birth. By three or four months of age, both boys and girls prefer to look at women's faces, even unfamiliar faces taken from a Land's End catalog. However, this holds true only when the mother is the primary caregiver. While most of the babies in this British study showed a preference for adult female faces, the small number of babies reared by stay-at-home fathers showed a preference for adult male faces.

These studies tell us that long before they consciously understand the concept of gender, babies begin to form their own stereotypes of male and female. Gender is just one of many categories babies figure out early in life. Recent research has proven that infants also get the concepts of color, quantity, and animate/inanimate years before they have the awareness or words to express such generalizations. And they are particularly adept at conceptualizing opposites, with male/female perhaps the second dichotomy they grasp, preceded only by big/little.

In the case of gender, this unconscious understanding has a powerful influence on babies' development and self-concept. Babies learn an incredible amount through mimicry, and girls and boys begin to emulate the emotional expression, visual attention, and physical mannerisms of one or the other parent long before they consciously understand the male/female categories they are conforming to.

Gender is the first, most obvious, and often most significant attribute each of us shows the world. And infants understand much more about gender than we ever used to give them credit for. It is not long before this implicit understanding starts shaping their own behavior, actions, and emotional style, and boys and girls begin growing into the sex roles that they see modeled all around them.

In sum, social influences go both ways in molding boys and girls into their typical gender roles. Parents stereotype their babies based on a lifetime of experience with people of both sexes. And babies, in spite of their brief experience in the world, also begin to stereotype their parents, forming the concepts of male and female that will shape their behaviors and emerging gender identities. Socialization is obviously not the only factor making boys and girls different, but it is one that we can try to exert a little more deliberate control over, difficult though that might be.

Hormones and the Mini-Puberty of Infancy

Anna and Paul are both babies in the infant class at a gorgeous new child-care center. Anna is quite a talker, already pointing and babbling up a storm to anyone who comes her way. Paul is quieter but an insatiable climber, trying his hardest to reach the summit of every piece of furniture in the room. Ms. Kimberly, the lead teacher in the class, has seen all types in her many years of caring for babies, but she can't help remarking to Paul's parents, who notice the startling differences between the two babies: "It must be the hormones!"

In fact, by the time babies are babbling and climbing, it's already a bit late to be sorting nature from nurture. Still, when people see the kind of extreme differences between babies like Paul and Anna, many assume that the answer must lie in hormones or other innate factors. As we've seen, testosterone exposure before birth does indeed shape certain features of children's later behavior. But what about the sex hormones that circulate in babies' bodies after birth? Their levels also differ between boys and girls, though only for the first couple of months outside the womb. Thereafter, babies' gonads settle down, and there are virtually no differences between boys' and girls' levels of testosterone, estrogen, and other sex-related hormones until the onset of puberty.

It is, nonetheless, startling that newborn infants undergo what endocrinologists refer to as a "mini-puberty" during the first few postnatal months. Boys experience a surge of testosterone, which peaks within the first month or two and then subsides to its low childhood level by four to six months of age. Girls undergo an estrogen rise within a similar time frame. Both hormonal flurries are activated in the usual way—by pituitary hormones, which are suddenly released from the newborn's brain by a burst of activity in the hypothalamus. During pregnancy, mom's high hormone levels suppress activity in the fetal hypothalamus and pituitary gland, but soon after the umbilical cord is cut, these brain areas let loose, and the pituitary gland begins pumping out the stimulating hormones (known as gonadotropins) that travel through the bloodstream to turn on, if just for a month or two, the ovaries or testes.

Boys' postnatal testosterone surge promotes growth of the penis and scrotum. It is not known whether girls' estrogen spurt influences female genital development, although one study did find a correlation between newborn girls' estrogen levels and their amount of breast tissue at birth. Nonetheless, the fact of this mini-puberty in both sexes suggests neonatal hormones might also influence babies' brains. Just as the surge of testosterone before birth apparently organizes the brain for later male-type behaviors, so this neonatal surge might create a critical period for launching brain development down the boy or the girl trajectory, shaping, as some researchers speculate, everything from children's play styles in preschool to their sexual orientation as adults.

Evidence for such influence in rats is clear, as we saw in the last chapter. Suppressing testosterone in newborn male rats profoundly alters their style of play and even their later sexual proclivity. Female rats

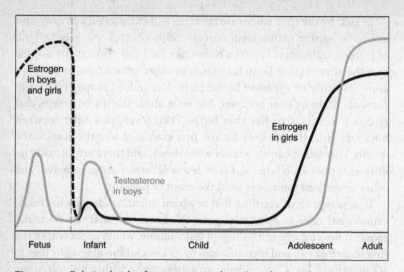

Figure 2.4. Relative levels of testosterone in boys (gray line) and estrogen in girls (black line) at each stage of development. The moderate peaks in both hormones during infancy define the mini-puberty of the neonatal period. During the fetal period, both boys *and* girls are exposed to their mothers' high levels of estrogen (dashed line), but this does not enter the brain or affect mental development because it is bound up by alpha-fetoprotein, a blood protein present during the fetal period only. The vertical line denotes birth.

treated with testosterone shortly after birth show the opposite change, a fondness for rough-and-tumble play. But because rats are born at a much more immature stage than humans are—corresponding to around the end of a human baby's second trimester of gestation—these effects are more relevant to the testosterone surge that human boys experience *before* birth. In rats, there is no distinction between males' pre- and postnatal testosterone surges. It is one continuous explosion whose effects on the developing brain happen to peak right around the time of birth. Human boys, by contrast, experience separate pre- and postnatal testosterone surges: the first begins around seven weeks after conception and subsides by about twenty-four weeks of prenatal development; the second, smaller rise occurs during the first and second months after birth.

This human pattern is better modeled in infant rhesus monkeys, which gestate for about five and a half months and are born at a developmental stage more similar to human babies than are rat pups. Like baby boys, male rhesus monkeys undergo separate pre- and postnatal testosterone surges, allowing scientists to investigate the different effects of hormone exposure during these two developmental periods.

At first glance, such studies suggest that the postnatal testosterone surge is not all that important in primates. As scientists at the University of Cambridge found, blocking testosterone after birth does not diminish male monkeys' fondness for rough-and-tumble play or affect their mother-son interactions in any noticeable way. Similarly, these researchers observed no change in the play style or social interactions among young female monkeys that had received testosterone injections as newborns.

However, other research has uncovered more subtle effects of postnatal testosterone on monkeys' cognitive abilities. Corinne Hagger and Jocelyne Bachevalier at the National Institute of Mental Health found that young male monkeys learn more slowly than females do on a particular test of object discrimination—exploring several pairs of objects and then remembering which one of each pair conceals a tasty banana pellet. For normal young monkeys, it takes males about twice as long as females to remember which object in each pair will get them the treat. But Hagger and Bachevalier found that male monkeys deprived of testosterone (by neonatal castration) learned just as quickly as females. And females injected with a testosterone derivative learned more slowly than usual, at virtually the same rate as normal males. So this study, like others we've seen, hints that postnatal testosterone somehow slows brain development and may be one reason why boys' minds mature more gradually than girls'.

Other research on monkeys, by David Mann and his colleagues at the Morehouse School of Medicine, suggests that the effects of neonatal testosterone are more evident at puberty than in infancy. They found that blocking testosterone at birth delayed the onset of puberty in male monkeys, which typically occurs around three years of age. It also led to reduced testicular size and sex hormone levels once the animals had reached adulthood. Males with reduced neonatal testosterone underwent less bone growth, leading to a shorter stature and lower bone mineral density in adulthood. Neonatal testosterone even appears necessary for normal immune function in males, according to this research. Taken together, the findings of Mann and his colleagues indicate that the brief elevation in neonatal testosterone has significant consequences for male physical and sexual growth, even though the effects are not apparent until puberty.

Does the neonatal testosterone surge similarly shape human boys' development? We don't really know, but there is evidence that boys with undescended testicles—a not uncommon medical condition known as cryptorchidism—experience lower than normal levels of testosterone during the first few months of life. This condition, easily treated by surgery in infancy, is not associated with any deficits in later growth or in pubertal development. However, men who suffered from cryptorchidism as boys often do experience fertility problems and tend to be less sexually active than other males. Whether the neonatal testosterone surge affects the brain, sexual orientation, or other mental functions is still unknown, although the little bit of information I described above—finding correlations between high neonatal testosterone levels and delayed visual or language-processing abilities—suggests it may have some impact on brain development.

Clearly, sex hormones are flowing in early life. Though there is little evidence of it in newborns' behavior, boys and girls are undergoing their separate mini-puberties, which may have lasting consequences for how their bodies and brains develop further. Some researchers propose that neonatal surges of testosterone, estrogen, and other less familiar hormones may even constitute another critical period that influences, for better or worse, outcomes as diverse as physical stature, the timing of puberty, cancer risk, cognitive skills, and sexual orientation. While fascinating, such possibilities are purely speculative at this point and await future research to sort out.

Are Plastic Bottles and Soy Formula Safe?

In spite of our limited knowledge about these early sex-hormone surges, researchers are actively studying how such hormones may run into interference from our increasingly contaminated environment. Attention has been focused on a slew of endocrine-disrupting chemicals: molecules that either mimic or block the actions of estrogen and testosterone. They include many of the usual suspects: pesticides, such as DDT and the fungus-killing agent vinclozolin; industrial chemicals, such as dioxin and PCBs; and, more recently, two startlingly ubiquitous plastic additives, bisphenol A (BPA) and phthalates (also known as phthalate esters). Bisphenol A is a chemical with weak estrogen-like effects that is used in hard polycarbonate plastics (including some, but not all, products with the 7 recycling symbol) as well as in dental sealants and the lining of many food cans. Phthalates are found in soft, flexible plastics, including PVC (polyvinyl chloride, which is used for, among many other purposes, medical tubing and IV drip bags), and in many personal care products, such as hairspray, lotion, fragrances, and nail polish. What's more, endocrine disruption is known to result from phytoestrogens, a large category of natural plant-based compounds. These are generally less potent than the industrial chemicals but act in much the same way and have come under special scrutiny because of their widespread use in soy infant formulas.

The evidence linking such substances to developmental (or other health) problems is largely indirect. Most have been identified based on their effects on rodents. Certain pesticides and phthalates are known to disrupt testosterone's action and can lead to reproductive malformations in male rats that are exposed during the critical prenatal period of sexual differentiation. Two such defects, cryptorchidism and hypospadia (a malformation of the penis in which the urethral opening is located in the shaft of the penis instead of on the tip), have become increasingly common over the last few decades, raising worries that endocrine-disrupting chemicals are responsible. Another concern is a rather alarming worldwide decline in men's sperm count—as much as 50 percent between 1930 and the present—which some researchers have suggested is also attributable to estrogen-disrupting chemicals. Still other fears revolve around the apparent decrease in age of girls' puberty onset, which is easy to imagine being caused by environmental estrogens (though much harder to prove).

These are worrisome issues, but it's important to point out that the risks vary for different endocrine disruptors and that there is little data showing harm in humans, as opposed to rats or mice. The evidence is strongest for agricultural pesticides, which one American study found associated with lower sperm count in male farm workers with higher exposures (especially to the herbicide alachlor and the insecticide diazinon), while a Spanish study found higher rates of cryptorchidism and hypospadia in baby boys whose mothers were heavily exposed to pesticides such as DDT and lindane. However, other studies have failed to find similar effects, and the evidence is weaker for a relationship between breast cancer and DDT exposure, or a connection between the mysteriously decreasing age of puberty in girls and PCB exposure.

Studies of the phthalates and bisphenol A are even newer and less definitive. Both agents have been shown to produce certain reproductive and behavioral disturbances in rats, even at low doses, but the evidence of harm in humans is sparse. Still, there's enough concern, as scientists from the U.S. National Toxicology Program voiced in congressional testimony, to state that bisphenol A "cannot be dismissed" as possibly altering human development. The concerns are greater for phthalates, especially for baby boys and pregnant women undergoing medical treatments that expose them to a particular phthalate, DEHP, which is used in intravenous tubing and is known to interfere with testosterone action. In 2005, the European Union banned the sale of infant toys, feeding nipples, and other baby products that contained DEHP, as well as five other phthalates, but the U.S. government has yet to follow suit. It's simply too early to know whether these plastic additives accumulate in babies' bodies in high enough levels to cause harm; we don't even know how these different chemicals are capable of altering development of the brain or reproductive system. More research is urgently needed.

Compared to the attention paid to these plastics, there's much less focus on soy and other natural estrogens, even though babies' exposure to such chemicals may be much higher. Half a century ago, farmers in western Australia noticed that many of their sheep became infertile after grazing on a certain type of clover. Few people eat clover (though it's increasingly available in herbal supplements), but comparable levels of isoflavones, a major class of phytoestrogen, are found in soybeans, which are consumed by billions worldwide.

No one disputes that the isoflavones in soy and other plant foods act as weak estrogens. Indeed, many postmenopausal women purposefully increase their intake of soy, which is marketed as a natural estrogen, in the hopes of avoiding some of the negative side effects of pharmaceutical hormone replacement. Compared to the endocrine-disrupting pesticides, soy isoflavones are less potent, are cleared from the body much more rapidly, and, presumably, are far safer.

Still, infants reared on soy formula are exposed to much higher levels of isoflavones than adults. For one thing, their bodies are tiny, and formula is their only source of nourishment. So infants end up consuming much more isoflavone per pound of body weight than adults do. Scientists have measured these levels and found that babies reared on soy formula are exposed to isoflavone levels about ten times higher than the dose known to disrupt women's menstrual cycles. Research on rats has found various adverse effects of exposure to soy phytoestrogens during prenatal and early postnatal development, ranging from altered reproductive development to changes in immune function to the masculinization of the female brain and behavior. Less research has been done on primates, but one study of male infant monkeys reared on soy formula found that it suppressed their testosterone levels by more than half. Animal research, then, does raise some concerns about soy's possible interference with normal estrogen and testosterone action in early development.

In the United States, about one-fourth of all the infant formula sold is soy based. In spite of this large exposure, very little research has been conducted on its effects. One study, published in 2001, found little evidence of harm: adults who had been fed soy formula from early infancy showed no difference in growth, pubertal development, or overall health when compared to adults who had been reared on cow's-milk-based formula. Women reared on soy, however, reported slightly longer periods and greater discomfort during menstruation than women raised on cow's-milk formula reported.

This study is encouraging, but most scientists agree that more research must be done before soy formula can definitively be declared safe. In 2003, the British Dietetic Association advised against the feeding of soy formula to infants under six months of age. Unlike in the United States, where soy formula is widely available, in Europe it can only be obtained with a doctor's prescription. Even in Asian countries, where soy-rich diets are more traditional and thought to contribute to lower rates of cancer and cardiovascular disease in adults, babies are generally not exposed to soy-containing foods. Most are exclusively breastfed until at least six months of age, beyond the period of normal estrogen and testosterone surges. In other words, while the estrogenic properties of soy may be health promoting in adults, they should be viewed with caution in very young infants. As always with infant feeding, breast is best.

Summing Up: Tips for Raising Boys and Girls in Infancy

Peering through the window into a hospital's newborn nursery, you'll find little evidence of the difference between boys and girls. Sure, they may be swaddled in either pink or blue blankies or be sporting similarly color-coded caps or bracelets, but without these artificial clues, few people could tell with certainty whether a given baby was male or female.

Still, as we've seen throughout this chapter, boys and girls do differ in subtle ways from early in infancy. Boys are a bit larger but paradoxically more vulnerable than girls. They are also fussier, harder to soothe, and—by three or four months of age—less socially attuned. Their senses of touch, smell, and hearing are a bit less acute, and their language, memory, and fine motor skills also lag during the first year of life. Where boys do excel is in gross motor development—sitting, standing, and walking at the same ages as girls in spite of their overall slower maturation.

All of which leads to the conclusion that boys need a little extra attention as babies. Perhaps, if we are aware of their slower development and specific vulnerabilities, we can help offset some of boys' slower maturation and greater risk for emotional and communicative problems. At the same time, we cannot ignore the special issues posed by girl babies. While girls are often easier to care for and more socially aware, they do not get as much encouragement as boys for their physical development and emotional independence, two concerns that become more pronounced later in childhood but that can be addressed from the first days of life.

What follows is a list of tips that can benefit all babies, though some are better suited to one or the other sex. Brain development is a cumulative process: each change builds on all those that came before, but there's no doubt that babies' brains are far more malleable in early life than at any later age. So the earlier parents are aware of the special needs of boys and girls, the better the chance that all babies will reach their full potential.

• **Talk to your babies, especially boys.** Of all the purported tricks for raising smarter children, this is the only one that has been scientifically proven: the amount of language directly addressed to a child in the first few years of life significantly influences his or her vocabulary size, reading ability, and writing skills for many years to come.

Parents of boys should err on the side of talkativeness. Use every interaction as a chance to communicate: narrate your activities ("Now I'm going to change your diaper"; "Let's put your coat on so we can go outside!"), sing songs, and introduce your baby to word play such as rhymes and alliteration. Baby talk, or "parentese," is an especially effective style of communication, as research has shown that it exaggerates and emphasizes the differences among speech sounds. What doesn't work, however, are baby videos and DVDs (such as *Baby Einstein*). As one recent study found, an hour per day of such viewing between eight and sixteen months of age was associated with a 17 percentile drop in vocabulary development. Another study found that babies were able to learn foreign speech sounds (in this case, Mandarin) only when they heard the language spoken by a live, interactive storyteller, and not when they viewed the same storytelling sessions on TV.

• **Listen too!** Communication is a two-way street. Babies learn as much about language from their own attempts at vocalization as they do from our chatter. Most infants begin cooing, or vocalizing using protracted vowel sounds, somewhere around three months of age. They start adding consonant sounds, progressing to true babbling, around five or six months of age. By the end of the first year, there are probably several words hidden amid their *babababas* and *mamamamas*, though few parents pay close enough attention to notice them.

So here's another chance to promote your child's verbal development, especially in boys: Stop, listen, and respond to his vocalizations. In other words, talk *to* your baby, but don't talk *over* him. Babies don't coo or babble much without an audience. They do it to communicate, and you can increase and improve their verbal output by responding, commenting on, and, especially, imitating your baby's budding vocalizations. Luckily, babies are guileless enough to appreciate imitation as the sincerest form of flattery. Returning your baby's *ooooo*s and *gagagaga*s with more of the same not only shows him that he is communicating effectively but also gives him important auditory feedback about what he sounds like to others.

Research shows that babies whose parents respond to their babbling and other vocalizations verbalize more than babies whose parents are less attentive. What's more, by noting every new vowel or consonant sound your child can produce, you stand a better chance of detecting those first words when they begin slipping in among his other babblings, somewhere toward his first birthday. Speech is an intricate mental and motor skill, so anything you can do to encourage your baby's vocal practice is going to help, especially in boys, who have a greater propensity than girls for later speech problems.

• **Books for babies, especially for boys.** Of all the ways to talk to babies, reading aloud is the best. Reading is the single most effective way for adults to increase their own vocabularies, and guess what: it works the same way for children. Books take a baby beyond his immediate environment, introducing him to animals, places, characters, and events he is otherwise unlikely to encounter. Studies show that parents engage in their most thorough language instruction while reading to their young children, by emphasizing new words and pointing to the pictures that create concrete meaning in babies' minds. Books also inspire parents to ask their babies questions, coaxing along speech and vocal practice. Such dialogic reading, in which parents use picture books as steppingstones for more elaborate two-way conversations, has been shown to substantially accelerate young toddlers' verbal development.

Reading together is the ultimate quality time: cozy, educational, and a wonderful bonding experience. Of course, babies of both sexes should be treated to this experience on a daily basis, but it may be especially important for boys, many of whom could use the extra dose of language and emotional enrichment. Even if your baby's only interest in books involves grabbing them and chewing, this stage shall pass, and you will have extended your baby's "taste" for books into the larger world of literary enrichment.

• **Watch out for ear infections, especially in boys.** Babies need to hear to learn language, but young children are highly susceptible to infections of the middle ear, a secondary consequence of colds and allergies that can leave their ears filled with fluid, often for weeks on end. Such fluid muffles babies' hearing, obscuring the subtle differences among speech sounds that are usually mastered in the first year of life.

Boys are more susceptible to ear infections than girls are, so this can be another factor working against them during the critical phase of early language development. Fortunately, ear infections are both preventable and treatable. Secondhand smoke, formula feeding, and large-group daycare are all known to increase the risk of ear infections, and all can be modified. Another no-no is putting your baby to sleep with a bottle, which can permit milk to contaminate the middle ear.

If your child does come down with a cold or other illness, it's important to have his ears checked periodically for several weeks after he's sick. Antibiotics are not necessary for every ear infection. (They shorten the duration of infection in only one out of seven children.) But what is important is monitoring how long a baby's ears remain filled with fluid. Pediatricians often request that parents return to have their baby's ears checked after an infection appears to have cleared up, but many don't bother, assuming that if the fever is gone, the ears must be fine. This is not true.

Any child who suffers from persistent or repeated infections should have a thorough hearing test. If there is trouble in both ears, he or she may be a candidate for tympanostomy tubes—tiny grommets that allow the middle ear to drain to the outer ear, restoring normal hearing. Boys get these surgically implanted tubes more often than girls, reflecting their greater propensity for middle ear infections. However, ear tubes do have certain risks, such as scarring of the eardrum and slight hearing loss later in childhood. Tubes should therefore be inserted only as an absolute last resort—that is, after several months of careful monitoring to see if the baby outgrows his middle ear troubles.

• **Stop parking your baby.** These days, we have so many devices for holding and carrying babies that young children are not getting enough physical exercise or opportunity to challenge their developing balance and postural abilities. Think about it: between their car seats, strollers, bouncy seats, high chairs, ExerSaucers, and electric swings, babies are rarely free to stretch their limbs or try to support their own bodies. All these baby holders are not unlike the recliner seats that lead to adult paunches: neither adult nor infant recliner seats require people to use their stomach muscles or even to support their own necks. Many infant seats even come with cup holders, helping to create the ultimate junior couch potatoes.

This parking trend is bad for girls and boys alike, though for somewhat different reasons. While all children these days need more exercise, the many variations of infant recliner seats reinforce girls' less active tendency. Recall that girls do not lag behind boys in the attainment of most gross motor skills in infancy. However, differences in speed and strength do emerge during the preschool years, probably reflecting the greater amount of movement and motor practice that boys engage in from early infancy. Any change that gets girl babies moving and encourages their physical independence will therefore be beneficial for their later fitness, gross motor abilities, and perhaps even spatial skills.

Boys' motor development is probably also suffering from being parked in various infant holders. However, an additional issue for boys is the social isolation these seats enforce. Carted around like an extra handbag, a baby in a car seat simply doesn't get the same amount of attention he would receive if he were being held in his parents' arms or in a sling or front carrier. Holding a baby brings him close to a your voice, warmth, touch, and comforting, familiar smell. A baby who is merely carted from car seat to stroller and back again at each stop on an errand list gets about as much inter-action as a sack of potatoes (the non-couch variety).

Car seats are for cars, not for holding a baby at the grocery store, doctor's office, or older child's basketball game. Stroller riding should come after, not before, a toddler has set out for a walk on foot. And ExerSaucers, well, these have little to recommend them. The dan-ger of parking is very high when a baby can stand upright with a tray full of colorful toys (and, often, tasty treats) to entertain her. If you need a safe place to put your child while you're cooking dinner or folding laundry, try a doorway jumper. This harness clamps onto a standard doorframe and at least offers the older infant a chance to flex his legs, get some exercise, and master a new motor skill.

I suggest you limit the number of such seats to the bare mini-mum: a car seat (but only for car rides), a high chair (but only for meals), and a stroller (but only for longer walks).

• **Responsive, sensitive caregiving.** Close physical contact is just one piece of the larger equation for positive parenting. By staying closer to babies—holding them, carrying them, and even giving them daily massages (which has all sorts of benefits for both body and mind)—parents are likelier to be aware of their needs. Re-search consistently shows that caregivers who are more sensitive, responsive, and highly involved with their kids do the best job of building healthy relationships and promoting their children's secu-rity and self-confidence.

This advice sounds simple enough: you can't spoil a baby by responding promptly and warmly to him or her. However, the exact prescription for sensitive, responsive caregiving may differ for girls and boys, or for babies of either sex with different temperaments.

Simply put, boys are often more needy as infants than girls are. They are less physically mature and take longer to develop the self-calming skills, such as hand-sucking or pulling into a tightly tucked posture, that help them compensate when overwhelmed. Parents may need to step in sooner with a boy, picking him up, changing his position, or giving him a soothing ring to grasp and suck on. Here is where stereotypes can get in the way. In the general spirit of "toughening them up," parents may let their baby boys fuss and squirm longer, or they may resort to artificial stimuli—videos, electronic swings, and elaborate toy bars—to entertain them without helping them to discover their own self-calming skills.

Girls, on the other hand, can sometimes be too easy. Quiet, complacent babies may not get as much attention as fussier types, and they may actually suffer from a lack of the stimulation and interaction needed to fully develop their motor and cognitive skills. While there are plenty of exceptions, girls fall into this category more often than boys. So here is another way that close physical contact can pay off: by keeping your baby nearby, you are likelier to interact more—talking to her, pointing out interesting things in the environment, and generally integrating the baby into your daily life, making the most of her waking hours.

These issues are every bit as important—if not more so—for babies in child care. When one caregiver is responsible for several infants, the temptation to stereotype and respond to boys and girls in a one-size-fits-all way may be especially high. Parents should be on the lookout for sensitive, responsive caregivers who do not favor either sex and who appreciate the importance of communicating and connecting with both boys and girls.

• **Girls need challenge, even in infancy.** On a related theme, baby girls can benefit from a greater diversity of activities. Again, this pertains to the quieter, more compliant, and more observant girls—by no means all female babies—who may not be getting as much stimulation and chances to explore as their fussier age mates. Parents often take an if-it-ain't-broke-don't-fix-it approach, but doing this may shortchange some babies, especially girls.

Girls need liberating! Out of those bouncy seats and onto the floor, where they can try pushing up, rolling over, and crawling. Out of their high chairs and into the kitchen, where they can crawl or cruise along the cupboards, rummage through pots and pans, and stack sippy cups. Out of their strollers and onto the jungle gyms, where they can test their growing physical strength. Sure, most babies get to go to the park, but boys, because of their more active nature, may demand it more, and so end up getting more physical exertion, muscular practice, and spatial experience than some of the quieter girls.

Because of baby girls' smaller size, and possibly because of lingering stereotypes, parents tend to be more cautious with infant girls, permitting them less freedom to explore and to push their physical limits. But later on, girls begin falling behind boys in their physicality and spatial skills, and it's clear that girls could benefit from greater physical challenges and earlier opportunities to explore.

So unharness your daughter, help her stretch her muscles, and encourage her physical exploration. You don't need to sign up for expensive baby exercise classes to give her this opportunity. Just head to the park or your own backyard and, most important, baby-proof your home early and often. The fewer no's your daughter hears, the more she will be inspired to follow her own natural curiosity and exuberantly explore every nook and cranny of the world around her.

• **Dads must do their part.** Fathers are more involved with their children today than they've been at any time in the past, but some are more involved than others. There is no question that babies learn a great deal about gender roles, even in the first few months of life. In other words, they notice who is taking care of them, and this awareness influences their own emerging gender identity and expectations about the nurturing tendencies of males and females.

Fathers have their own unique way of interacting with infants, which tends to be more fun, active, and physical than mothers. Fathers are very stimulating! But they also have a tendency to engage their infant sons more than their daughters in this type of play. So once again, fathers need to watch out for their own stereotyping and use their unique nurturing style with their daughters every bit as much as with their sons.

• **Breastfeed for one year.** The American Academy of Pediatrics recommends exclusive breastfeeding for the first six months of life and continued breastfeeding (after the baby is started on solid food) until one year of age. Sadly, only a third of six-month-old babies in the United States receive any breast milk, and only 17 percent of all six-month-olds are exclusively breastfed—that is, receive no formula at all.

Boys, with their greater vulnerability to infections, can especially benefit from the immune components in breast milk. Unlike a formula factory, the factory inside a mother's mammary glands pumps all kinds of antibodies and immune cells into her milk, transmitting to her baby protection against many of the common infections of infancy. Breastfeeding is known to prevent ear infections, which boys are more susceptible to and whose onset in the first six months is known to increase babies' chances for repeated or persistent ear infections during the toddler years. Breastfeeding also protects babies from respiratory and gastrointestinal infections, promotes eye contact, increases physical contact, and even makes babies smarter than formula feeding does.

Of course, girls deserve to be breastfed every bit as much as boys. The same immune, sensory, and bonding advantages apply equally, while the cognitive advantages of breast milk have been found to extend well out to school age and adolescence.

• **Choose glass or safer plastics.** This reduces babies' exposure to phthalates and bisphenol A. While existing research does not clearly prove that these chemicals are disrupting boys' or girls' early hormone surges, some pediatricians urge a precautionary approach: avoid plastics with the 3, 6, or 7 recycling symbols (1, 2, 4, and 5 appear safer); avoid microwaving food or beverages in plastic containers or with plastic cling wrap; don't wash plastic containers in the dishwasher; choose glass or polyethylene (1) bottles or sippy cups, especially for warm or hot liquids; limit canned foods.

- **Watch out for stereotyping.** This applies to children of all ages, but no age is too young to begin revising expectations for our children. Just because your child is a girl doesn't mean she won't be interested in trucks and trains and rolling a ball across the living room floor with you. Just because your child is a boy doesn't mean he won't be scared of going down the slide or doesn't need lots of cuddling and nurturing. Girls should be praised as much for their strength as their beauty (perhaps even more, considering how society is going to treat them later); boys as much for their tenderness as their ball-handling abilities.

And yet, what is perhaps most striking in today's world is that parents continue to stereotype their infants, beginning even before they are born. Overcoming the tendency requires constant, vigilant effort, but it's best to start when your child is an infant. After all, this is the age when babies are their least sexually differentiated. Well before they start demanding their own pink or blue clothes, boys and girls can be seen more for who they truly are: social, active, fussy, compliant, vocal, quiet, alert, intense, relaxed, funny, curious, or squirmy. Parents of infants should savor this gender-free zone, following babies' own cues as to what they need as individuals rather than as tiny boys and girls.

3

Learning Through Play
in the Preschool Years

IT'S A BIG EVENT: your child's first Happy Meal. You're feeling a little guilty about all the grease but genuinely pleased when you see how excited your three-year-old is to finally be getting her own nuggets, fries, and . . . *free toy!*

But hold on. You've already accepted the nutritional compromise, so what's this about selecting a boy toy or a girl toy? You stopped in for some fast food, not a decision about gender stereotypes. And the options this summer are particularly annoying: Extreme Action (skateboarders and stunt bikers) for the boys and Extreme Fashion (the usual Mattel harem) for the girls.

Thinking quick, you mutter, "Boy," hoping to postpone, at least for a while, your daughter's inevitable focus on clothes and beauty. But it doesn't work. She spots Barbie in the display case and loudly declares her own choice. You capitulate and change the request, not wanting to make a scene and deciding to save your energy for more winnable battles.

The food arrives, you gather your napkins and little cups of ketchup, and make your way to a table, where your little daughter is desperately tugging on the plastic packaging to extract her new toy. You rip it open for her and get your first glimpse at the skinny doll wearing platform shoes, a tube top, heavy eye makeup, and a fur coat.

What is this? you think. *Some kind of streetwalker?*

It's tempting to blame McDonald's for its blatant gender stereotyping, but let's face it: the company wouldn't spend millions to manufacture toys in China without mountains of market testing about what kids really want. Children, even those as young as three, prefer playthings with a gender identity. Girls *want* to be girls; boys *want* to be boys; and if you want to sell them some French fries, you better give 'em what they want.

Ah, but McDonald's also markets to parents, who don't want their kids to grow up so fast. A little too late to save you from this particular meal, you learn about the company's "toddler toys": gender-neutral, age-appropriate offerings for kids under four that allow you to postpone, at least for another year or so, the inevitable descent into gender-typed playthings.

The End of Babyhood

The preschool period comes as a bit of a shock to many first-time parents. After a year or two of pure innocence, a child starts developing a will of his or her own. You're barely getting them out of diapers before having to contend with tantrums, contrariness, and their emerging opinions about nearly everything.

Gender awareness is a big part of the change. The brief, blissful period of gender neutrality has ended, and parents are increasingly confronted with children who reject anything to do with the opposite sex: girls who want to wear dresses only, boys who have an aversion to pink.

What happens during early childhood to rip babies out of their fleeting, genderless existence and plant them on one or the other side of the gender chasm? Is it the inevitable result of all the hormone-infused brain development that began before birth? Is it our pink-versus-blue culture, with toy stores that segregate the sexes before a child can even string three words together in a sentence? Or are we parents to blame for using our children to fulfill our frilly-doll and little-jock fantasies, channeling our kids into the childhood roles we reminisce about, and trying to mold them into our own idealized images?

In this chapter, we'll follow children as they enter the process of gender identification and differentiation. Though small differences are present at birth, the gap between boys and girls widens tremendously between two and six years of age, with some differences more stark than they will be at any later time in life. Hormones are not directly to blame, since the gonads of both boys and girls have settled down and remain quiet between infancy and puberty. But as we've seen, early genetic and hormonal influences do set them on somewhat different paths. Long before children hit our gender-coded culture, their brains are primed to respond to different aspects of their environment. Once they do, they rapidly bud into the pink or blue varieties that in many ways fix their future development.

Parents are like gardeners, helping to provide the optimal sunlight, moisture, and nutrients to cultivate these blossoms. Children learn an enormous amount during this early childhood period, before their formal schooling has even begun, and it is up to parents and the other adults in their lives to help channel this growth in the right direction. Adults cannot—nor would most want to—change pink into blue or vice versa, but we can help inoculate each sex against the difficulties that might arise later. Building on each gender's strengths, adults can shape the environment to better round out children's development, giving both boys and girls the tinge of purple that will help them be happier and more successful.

Toddlers and Their Toys

We first noticed it with the trucks back when we lived in Houston: Sam was about a year and a half old when the small digging machine was rolled in to shore up the bayou behind our apartment. He was still so little, but *Bobcat* became his new favorite word. A few months later, he hit the jackpot when an older cousin handed down his sturdy Tonka dump truck and digger. It was the beginning of summer, and Sammy made great use of them at the playground sandbox.

Julia, I'll confess, was never given a real toy truck. (We gave her a stuffed one, but it could hardly haul sand.) By the time the noisy Bobcat turned up outside her bedroom window, she was already a very feminine four-year-old and not much interested. I can't say for sure how she would have responded at age two, but judging by her other passions as a toddler—books and Beanie Babies—it seems the trucks wouldn't have made much of an impact.

A friend recounts a similar memory of her daughter at age three: "I wasn't wearing pink ribbons in *my* hair," she recalls, "so why did Lucy, when she was itty-bitty, beg me to fix up her hair all the time?"

Most parents can tell similar stories about "gender-typical" play in their very young children, and research confirms that the difference is strikingly universal. Whether they are raised in the United States, Europe, Japan, or probably anywhere else, boys between two and five years old overwhelmingly select a toy truck, Hot Wheels car, ball, or other suitably male toy when given a choice between one of those and a doll. Three-year-old girls opt strongly for the baby doll, toy kitchen utensils, or toy beauty set (especially if any of the toys is pink). Preschoolers even make gender-appropriate toy choices when picking presents for children of the opposite sex, showing they are acting not only on their unconscious preferences but also on their conscious understanding and expectation that boys and girls prefer different types of toys.

There's a good reason why parents so often comment on boys' and girls' different toy preferences: this is one of the largest differences between the sexes that psychologists have uncovered. In one Swedish study, the effect size, or d value, was as high as 1.9 for three-year-olds choosing between stereotypically boy and girl toys. This means that 97 percent of boys were likelier to spend their time playing with toy vehicles, balls, and weapons than the average girl, when given a choice between those or stereotypical feminine toys. This difference is much, much larger than any of the cognitive or personality differences between the sexes, which makes it fascinating, but also misleading, should parents mistakenly assume that boys' and girls' toy interests predict comparable differences in their later verbal, social, mechanical, competitive, or other tendencies.

"There must be a wheel gene on the Y chromosome!" That's how many respond to the strong, universal fact of their sons' and daughters' different toy preferences. Of course, neither trucks nor dolls existed a hundred thousand years ago, when the human genome stabilized into its current sequence. But there is some reason to believe that the different properties of "boy" and "girl" toys are innately appealing to each gender, even if a wheel gene or a doll gene is too specific for any particular stretch of DNA.

The counterargument is that it is not innate, that parents ourselves teach such preferences through both conscious purchases and the kind of unconscious assumptions about boys and girls I discussed in the last chapter. This nurture view is not nearly so popular now as it was a few decades ago, and it flies in the face of many parents' *attempts* to interest their sons in dolls or daughters in trucks. Nonetheless, the truth lies somewhere in the middle: boys' and girls' different play preferences are clearly biased by innate tendencies but are further augmented by social factors—especially the child's own growing awareness of actually *being* a boy or a girl.

Surprisingly, a difference between the sexes in toy preference has *not* been reliably shown in infants. In one study of six-month-olds, parents were asked about their babies' favorite objects, and parents of boys were just as likely as parents of girls to report that the preferred object was a doll. Another study of six-month-olds found that boys were just as likely to look at a picture of a blue truck as a pink doll, although the girls in this study showed a clear preference for the doll. By contrast, a Canadian study found that both twelve-month-old boys *and* twelve-month-old girls preferred to look at pictures of dolls over pictures of toy vehicles. This changed substantially by eighteen months, and by twenty-three months of age, boys barely glanced at the dolls while girls looked about equally at both kinds of toys.

The preference for gender-typical toys emerges somewhere around the first birthday. A Dutch study found a small difference in boys' and girls' toy preferences in thirteen-month-olds. Unlike the Canadian study, in the Dutch study the children actually got to play with the toys, which may measure preference more accurately than when they just looked at them. Then there is the Swedish study mentioned above, which tested children the same way at one, three, and five years old. While the differences in preference were most dramatic at three years, they were already significant by twelve months. By the age of five, the boys spent less than one-tenth of their time playing with feminine toys, while girls divided their time about equally between the two categories of toys (a fascinating shift I'll discuss shortly).

On the one hand, the gradual emergence of boys' and girls' different toy preferences negates the idea that the preference is absolutely hard-wired. On the other hand, the strength of toddlers' preference once it does emerge and the universal nature of this difference across cultures tells us that genes and hormones are important factors.

We've already seen that toy preference is one of the psychological differences most clearly influenced by prenatal testosterone. Recall that girls with CAH, who are exposed to high levels of testosterone and other androgens before birth, opt more for "boy" toys than for "girl" toys, in dramatic contrast to hormonally normal girls. Even among normal girls, small variations in prenatal testosterone exposure (measured from stored amniocentesis samples) were found to correlate with more masculine play interests, according to one British study of three-and-a-half-year-olds.

Other striking evidence for an innate basis of toy preference comes from research on monkeys, who obviously grow up without the intense Toys "R" Us indoctrination human children receive. Two recent studies—one in the small, pretty vervet monkeys, the other in the more commonly studied rhesus monkeys—found that males and females exhibited a similar sex difference in toy preference. The first study, carried out at UCLA by Gerianne Alexander and Melissa Hines, tested one-year-old vervets for their interest in a number of conventional human toys. The males spent more time exploring a ball and a toy police car, while the females spent more time with a rag doll and—more mysteriously—a red cooking pot. But the two sexes did not differ in the amount of time they spent exploring two gender-neutral toys (a stuffed dog and a picture book). The results were similar in the study of rhesus monkeys that was carried out at the Yerkes Primate Research Center of Emory University. In both studies, the monkeys had no prior experience with any of the toys and certainly no concept of "boy" or "girl" toys, so these results make a strong case that such preferences are innately biased.

If it's not gender indoctrination, then what is it that attracts males and females to different types of toys? As we saw in the last chapter, baby boys and girls do differ in subtle ways that may influence their early toy preference. Males, with their higher activity level, may be more drawn to moving objects, like balls and vehicles, which they can manipulate and control in a more physical way. Females may find dolls more appealing because of their stronger social orientation or, perhaps, because of a truly instinctual attraction to infants. This is another large sex difference, seen not only in rhesus and vervet monkeys but also in baboons, macaques, and our own little girls. (We'll examine this more closely below.) The draw toward babies may also explain the female vervets' seemingly inexplicable interest in the cooking pot. While human girls may perceive cooking as a female task, vervet monkeys obviously don't. However, the red pot in this study is close to the facial hue of vervet infants, and so it may be the color that lures them.

Then again, the surprising fact about human boys is that until they're about twelve months old, they are just as interested in dolls as girls are. (The monkeys tested were mostly adults, so we don't know how young male monkeys would respond to dolls.) It's possible, therefore, that boys and girls *both* start out liking dolls—just as most infants instinctively prefer the human face over inanimate stimuli—but then boys are drawn toward balls and vehicles during the second year as their higher activity level and physicality kicks in.

There's no denying that boys and girls are naturally impelled toward different playthings. But this doesn't mean that social factors don't also shape their preferences. Nor does it mean their other psychological differences are equally large. When parents see their sons opting for Tonka and their daughters for American Girl at an early age, it's easy to be fooled into thinking they differ just as much in empathy or mechanical ability. Parents must work hard to avoid prejudging boys and girls based on their often dramatic differences in toy choices and instead focus on how to compensate for these play preferences to promote other aspects of each sex's development.

The Nurture Side: "Help, I Was Touched by a Doll!"

Every now and then, I tease Toby about his taste in toys. "Come on," I'll say, "let's play with Julia's Barbies!" This catches him off-guard, and he'll scrutinize me closely, trying to figure out if I could possibly be serious. Then, seeing the wink in my expression, he'll blush and say, shocked and half embarrassed, "Mom! No way!"

"I'm just kidding," I say, to assure him I'm not really challenging his self-image.

Still, I think it is worth an occasional nudge to make children aware of their biases. For although innate factors clearly slant boys and girls toward different types of toys, there is also plenty of evidence that such choices are further amplified by parents, peers, and, especially, their own emerging awareness of actually *being* a boy or a girl.

One trick researchers use to evaluate the relative roles of nature and nurture in shaping behavior is to study twins. By comparing identical twins (who share 100 percent of their genes) with fraternal twins (who share 50 percent, like regular siblings), behavioral geneticists can determine the degree to which a particular trait is controlled by genes. A large study of three- and four-year-old twins led by Robert Plomin at

King's College London determined that gender-typical play is roughly 50 percent heritable. This leaves the other half, of course, determined by nurture, or the various social influences that nudge boys and girls to their extreme differences in play choices.

Another reason to suspect the role of cultural influences is the varying strength of each sex's preferences. As the Swedish study and many others have revealed, boys grow increasingly adamant over the preschool years about avoiding girl toys, whereas girls start doing a reversal in preference about midway through this period. Like Toby, most boys four or older express a strong taboo against anything pink or girlie, like a baby doll or Polly Pocket set. Girls, however, begin opening up around four or five, experimenting more with toy vehicles, sports games, and other blue- or black-colored toys that they avoided a year or two earlier.

The fact that girls grow more flexible and boys less so in their toy preferences reflects the different roles of men and women in Western culture. Compared to men, women have a wider range of options as adults—from stay-at-home mom to businesswoman, soldier, doctor, bus driver, and more. Girls see these roles in real life and, increasingly, in newer picture books that depict female firefighters, construction workers, and pilots. If they're lucky, girls are also taught that they can do anything boys can do. All of this makes boy toys more palatable to girls as they grow beyond simple male-female dichotomies in understanding gender.

Sadly, boys don't get this same message. As much as many people would like to see the same range of opportunities for men, there is still considerable stigma against those men who engage in traditionally female occupations, such as a stay-at-home parent, nurse, or preschool teacher. (If you're not sure about this, go rent the movie *Meet the Parents,* in which Ben Stiller plays a male nurse hounded mercilessly—and quite comically—by his future father-in-law, Robert De Niro.) Boys are rarely told that they can do anything girls can do.

The problem is status. Even today, in the twenty-first century, jobs traditionally held by females are considered lower status (and in many cases are lower paid) than traditional male jobs. And even in families where parents share a more egalitarian relationship, the asymmetry in power is typically evident in some fashion, as in who gets the last word at the dinner table or who is the ultimate disciplinarian. It is also pervasive in the media. Children begin picking up on this status difference early on, and it is reflected in their toy choices: boys increasingly avoid feminine toys, while girls increasingly explore masculine ones.

Parents, alas, play an even more direct role in reinforcing the different toy preferences of boys and girls. While a comprehensive review by Canadian psychologists Hugh Lytton and David Romney found that parents do not differentiate the amount of time, warmth, or restrictiveness showered on sons and daughters, we do react differently to the kinds of toys we see children enjoying. Subtly or not, parents discourage boys from playing with girl toys and, to a lesser extent, girls from playing with boy toys.

In dozens of studies, parents have been found to respond more positively when their young child picks up a sex-appropriate plaything, such as when a boy uses a hammer or a girl pushes a toy shopping cart. And they are likelier to bristle when their child plays with the "wrong" type of toy, like a boy cuddling a baby doll or a girl brandishing a toy sword. Fathers react more strongly than mothers, especially when they see their sons engaging in typical female play. By four years old, boys themselves voice the awareness that their fathers would think it "bad" if they played with toys for girls, even something as seemingly neutral as toy dishes, which the boys in this study found quite attractive.

Daughters get more leeway. Just as tomboy daughters are more acceptable to most parents than effeminate sons are, parents generally aren't upset by seeing their daughter playing with Hot Wheels cars or shooting Nerf basketball hoops. Still, it is surprising how much parents continue to promote stereotypes through their reactions to children's play, especially when you consider how much children can learn from *both* traditional boy and traditional girl games.

We can't do much about children's instinctive biases, but we can help fight toy stereotypes. Encourage your daughter to play ball or build with a toy tool set, your son to cuddle a doll or stuffed animal and help fold laundry. Be especially careful too to avoid gender labels—calling something a boy toy or girl toy is the surest way to dampen children's motivation to explore anything with the wrong label.

Better yet, search for gender-neutral playthings, such as art supplies, board games, and building toys. Such picks are harder to locate in the typical toy superstore, with its segregated aisles of pink or black offerings, but prominent in many smaller, independent or educational toy stores and websites. While they're unlikely to cure children's desire for gender-appropriate toys, such playthings will at least broaden their gender categories and, more important, help them develop a wider range of skills during these crucial brain-building years.

Gender Nonconformity

There's a simple reason why parents hesitate to promote nontraditional play: fear. A mother might worry, "My son is already so attached to me. If I let him play with dolls and my high heels, will this affect his masculinity? Will it make him gay?"

The answer is no. Putting aside whether one believes homosexuality is wrong, most evidence tells us that a child's sexual orientation is not something parents can control even if they want to. It's true, as we saw in chapter 1, that parents can shape the gender identity of the rare intersex children whose external appearances don't match their genetic or hormonal sex. That is, genetic boys *can* grow up to think of themselves as girls if parents and surgeons make an all-out effort to switch them at birth, as pediatricians have often advised for children with rare birth defects involving the genitalia. But when they grow up, such girls often show a male-typical sexual orientation—that is, they are sexually attracted to women—which tells us that something innate is shaping their sexual orientation.*

Another group of children provides even stronger evidence that sexual orientation and gender identity are hard-wired. These are so-called gender nonconformers: boys and girls who persist from a young age in trying to be like the opposite sex, often in spite of strong discouragement from their parents and others. While most children exhibit occasional opposite-gender behaviors at some point, full gender nonconformity is rare and suggests that gender identity is programmed early and potently into the young brain.

Consider Carrie, a classmate of Sam's during his prekindergarten year. One afternoon I was walking over to the building-block corner of their classroom to pick up Sam when one of the teachers called Carrie's name. A small head next to Sam popped up, and I realized for the first time that

---

* Recall that women with CAH, who are exposed to high levels of male hormones before birth, similarly show a higher incidence of bi- and homosexuality, even though most are heterosexual.

this child who was always playing with the boys was not actually one of them. Carrie had short hair and usually wore pants and a T-shirt, so all the normal clues were missing. But, as I noticed that day and for the rest of the school year, Carrie clearly preferred to hang with the guys—at the tire swing, on the soccer field, or playing with the large blocks and LEGOs during free-choice time. Preschool culture offered two largely segregated ways to play, and the boy style was clearly the more comfortable one for her.

Happily, Carrie was easily accepted at that age. Sam matter-of-factly told me, "Carrie's a tomboy," and left it at that. He and their other four- and five-year-old buddies didn't seem bothered by her missing Y chromosome as long as she was keeping up and playing by their rules.

Carrie lives in another town, so we lost touch with her when the kids moved on to elementary school. I don't know if she's still a gender-non-conforming child. Some children do drop such behaviors, especially in public. But if she's still as uninterested in female friends and activities as she was at age four, it saddens me to think of her now in middle school, among early pubescent peers who are almost certainly not so accepting as her preschool clan.

Gender nonconformity isn't common, though most children do exhibit at least a few opposite-gender behaviors on occasion. The best data come from children between six and ten years of age. In one large study, Dutch researchers found that just 1 percent of both girls and boys were judged by their mothers as "[wishing] to be of the opposite sex," while a larger proportion, about 3 percent of boys and 5 percent of girls, "behave like the opposite sex." Similar numbers were reported in a large U.S. study. As with the Dutch sample, American girls exhibited more cross-gender behaviors than boys did (that is, there were more girls playing sports and avoiding frilly clothes than there were boys wearing makeup and playing with dolls). But while few children engage predominantly in cross-gender play, a certain amount of experimentation with opposite-gender behavior is normal: according to this survey, 23 percent of boys and 39 percent of girls exhibited at least ten different cross-gender behaviors on various occasions.

Still, there does seem to be something different about girls like Carrie and boys like my friend's nephew Stephen, who loves polishing his nails, wearing his mom's lingerie, and playing with Bratz dolls. In an odd way, such children prove that gender identity is largely innate. The cost is so high, and yet tomboys and sissies (as researchers actually refer to them) endure against every cultural proscription, including strong family disapproval and merciless peer teasing. Boys, of course, have it worse; girls can (up to a point) gain status by engaging in male activities, but the reverse is not true. That's probably why studies that rely on parental reporting find higher numbers of masculine girls than feminine boys, and why, when a boy *does* exhibit many girl-typical behaviors, he is about six times likelier to be sent to a psychiatrist than a girl with a comparable degree of masculine behaviors. Still, it's a tough way to be for either sex, and the fact that such children persist in their gender-inappropriate play tells us there's something in their neurologic hardware that resonates with either "male" or "female," in spite of the way the rest of their bodies are put together.

Twin studies confirm that children's gender nonconformity has a strong genetic component. The Dutch study was actually based on twins and found that some 70 percent of the variability in children's cross-gender behavior could be accounted for by genes. A smaller U.S. study puts the number at 62 percent heritable. This is a substantial genetic contribution but still tells us that something in children's early environment—during either prenatal or early postnatal development—also contributes to gender nonconformity. No one knows what such environmental influences might be, but given most parents' tendency to gender-type children from the earliest age, it's unlikely these children's gender nonconformity is caused by the way their moms and dads treat them.

For the 1 percent or so of children who do persist in preferring opposite-sex clothes and activities beyond the preschool years, it's a difficult road. Few parents find it easy to accept such differences, even if they acknowledge that the tendency is inborn. They fear the teasing, or they mourn the loss of the boy or girl they expected. But mostly, they worry about homosexuality. According to one study by psychiatrist Richard Green, then at UCLA, about two-thirds of boys who were referred for psychiatric treatment for gender-identity issues were found to lean toward a homosexual or bisexual orientation as teens or young adults, with some analyses putting the figure as high as 80 percent.

So cross-gender behavior does appear to be a strong predictor of later orientation, at least in boys. (For girls, there are no comparable long-term studies, although it's clear that most women who thought of themselves as tomboys in childhood actually end up heterosexual.) Gender nonconformity seems to be determined early in life, and while parents can train away some of the behaviors, there's no evidence such treatment reduces a child's chance of becoming homosexual. According to Dr. Green, boys who received psychological treatment for their "disorder" were just as likely to turn out gay as those who were left alone. In his fascinating book *Queer Science*, neurobiologist Simon LeVay describes the cruel training a boy named Kyle was forced to endure—blue tokens for masculine behavior, entitling him to ice cream and other rewards; red tokens for feminine behaviors, which earned him confinements, spankings, and the loss of his blue tokens. Sure enough, Kyle quickly learned to suppress any outward sign of his femininity. However, by age eighteen, he was a deeply unhappy closeted homosexual on the verge of suicide and unable to form friendships with either gender. We can only wonder how many other gay teenagers have suffered through depression, anxiety, and suicidal thoughts from their families' efforts to change them.

The Power of Gender Identity

Most children do not struggle with their gender identity. In fact, gender identification is a strikingly universal milestone, emerging at similar ages among children from very different cultures. And once acquired, it is a potent, irreversible piece of self-knowledge that crystallizes children's perceptions and choices about much in their world, creating pink or blue barriers that parents find difficult to maneuver around.

Gender identity emerges between two and three years of age. At two, most children are still blissfully unaware of their gender status. They go either way when responding to the question "Are you a girl?" or "Are you a boy?" Nor can two-year-olds sort pictures of men and women into appropriate piles. This despite the fact that infants *unconsciously* understand a great deal about the differences between male and female categories (chapter 2) and, as we just saw, begin opting for gender-appropriate toys as early as twelve months. By two and a half, however, a majority of children succeed at gender-identification tasks, and by the third birthday, nearly every child passes such tests of gender awareness or identity. But while gender awareness is universal, its exact timing can be influenced by a child's environment: researchers at the University of Oregon found that two-year-olds from more traditional homes, whose mothers endorse gender stereotypes more explicitly, crossed this threshold of gender awareness several months earlier than children with less traditional mothers.

As their gender identity firms up, preschoolers grow increasingly adamant about avoiding toys, clothes, and peers of the wrong gender. According to one study, two-year-old boys who could not yet pass a test of gender labeling (pointing to the correct picture when asked to identify either a man or a woman) spent as much time playing with dolls as girls, but boys who had passed the test showed virtually no interest in dolls. In another study of children between two and three years of age, those who could already say for sure that they were girls or boys were significantly more likely to play with a gender-correct toy (a cement mixer, racecar, or tool set versus a baby doll, toy crib, or feeding set) than children of the same age who hadn't yet passed this milestone.

When Julia was four, she went through a definite dress phase, always opting for a skirt or dress over pants or shorts. Boys this age begin shying away from anything pink. By the time he was five, Sam seemed happier when I complimented his clothes as being "cool" rather than "cute." As is the case with toys, boys more fervently reject girl-like clothes than girls reject boylike clothes, both because girls see a wider spectrum of dress in adult women and because the status difference between men and women makes boys realize quite early that they have much to lose by being mistaken for females.

It's another two or three years before children get the concept of gender constancy—that they not only *have* a specific gender but will *stay* that gender for life. It's a bit odd that this understanding should take so much longer, but young children have a very fuzzy notion of time in general and—perhaps because they themselves are changing so rapidly—don't really get the concept of permanence in anything. (Death is another tough concept at this age.)

Toby memorably revealed his confusion about gender permanence when he was four. Though well aware that he was a boy, he was excited about the prospect of having a baby in his tummy when he grew up. It was tough to break it to him that this was impossible. But he has now adapted to the idea. Just the other day at the grocery store, he laughed when he heard a smaller child, probably three or four years old, ask her mom if she would be a mommy or a daddy when she grew up.

Most children achieve gender constancy or permanence by six or seven, which happens to be the age when children form their most stereotyped views of males and females. While younger children can be quite fervent about rejecting pink dolls or black Power Rangers, six- and seven-year-olds are the sternest enforcers. Six is when the teasing kicks in, when boys and girls not only tell each other what they can and can't do but also begin mocking each other for what they see as gender-inappropriate behavior: "Jeff takes ballet lessons. He must be a *giiiiirrrrlllll!*"

So children's awareness of gender stereotypes is quite solid by the time they begin elementary school. The awareness emerges out of the categories babies start discerning in the first few months of life—obvious differences between male and female that help organize their understanding of their world. Toddlers' awareness of gender stereotypes is *not* the main reason they initially gravitate toward different toys, peers, and types of play, but it undoubtedly strengthens their preferences over these formative years.

Children's gender knowledge continues to grow throughout the preschool period. At three, children are still blessedly nonjudgmental—they will ignore another child's selection of the wrong type of toy, even if their own preferences are already strong. By four, however, many children enter an inflexible stage in which they start viewing gender choices as a matter of right and wrong, a phase that peaks around first grade. The remarkable thing is that young children are so much more vigilant about enforcing gender norms—what to wear, what to play, who to play with—than any adult.

Young children's gender intolerance is surprising: How can my son refuse to wear purple shirts if his father wears them? Where did my daughter get the idea that Star Wars LEGOs are only for boys? But in reality, their gender reasoning is not unlike their moral reasoning at this stage. Preschoolers can think in terms of right and wrong, or contrasts and opposites, but they can't yet shades of gray. They need to master the inflexible categories before they can grasp the many exceptions that blur the line between male and female.

Six or seven is not the end of children's growth in understanding of gender stereotypes. School-age children continue to expand such stereotypes, adding to the list of occupations, hobbies, personality traits, and even movies and music that are acceptable for boys only or for girls only. However, they also develop a more flexible understanding of stereotypes, appreciating them as general characteristics but not as defining features of males and females. Preschoolers, who lack this sophistication, take a harder line, internalizing stereotypes as absolute standards for their own and their peers' behavior. According to five-year-olds, you simply *are* male or female based on your choice of clothes, toys, hair length, and favorite color.

Psychologist Sandra Bem cites a perfect example of such gender-defining stereotypes in the experience of her own son, Jeremy. She and her husband had gone to great lengths to raise their two children in a gender-neutral way, so when Jeremy announced one day that he wanted to wear barrettes to nursery school, she simply put them in his hair and let him go. Expecting him to be teased, she was surprised that he said nothing about it when he came home that day. Later, however, she learned from his teacher that Jeremy had indeed been hounded by one boy, who kept asserting that Jeremy must be a girl "because only girls wear barrettes."

"No," Professor Bem's well-taught son had countered, going on to insist that he was indeed a boy because he had "a penis and testicles." To prove the point, Jeremy even pulled down his pants.

But the other boy was not persuaded and replied: "Everyone has a penis; only girls wear barrettes."

Peers

Walk into any preschool classroom and you'll see the immediate consequence of children's growing gender awareness: segregation. The girls will be around the art or writing table, in the dramatic play center, or quietly listening to a teacher reading a story. The boys cluster farther away from the teacher, if they're able, to the large building-block corner, the sensory table, or any activity that doesn't require them to sit still.

It's hard to believe how early kids start doing this, but segregate they do. Under the age of two, children show little preference for male or female playmates—a toddler will happily play beside a child of either gender, neither noticing nor caring if it's a boy or a girl. But this changes rapidly during the third year if children are in a group setting and have a choice of playmates.

Girls initiate the segregation, as they are naturally drawn toward others with similar interests and interaction styles. Two-year-old boys engage in more conflict and negative assertiveness than girls do, such as grabbing toys and resisting peers' attempts to take them back. Girls witness this behavior and pull away, forming their own separate playgroups by about two and a half years old. Boys segregate a bit later—around their third birthday—reflecting their somewhat slower maturation as well as the fact that they have fewer negative interactions with girl toddlers and so less need to separate themselves. From this point on, boys' preference for same-sex playmates grows stronger throughout the preschool period—more so than girls'. By the end of kindergarten, children spend a mere 9 percent of their play time with peers of the opposite sex, forging the two separate cultures that persist throughout childhood.

Gender segregation is ubiquitous among children. Whether in Africa, Mexico, India, Japan, Switzerland, or the United States, boys and girls separate themselves into different groups, play different games, and interact in very different ways. Separation is a fact of human childhood and is equally common among young monkeys and apes. As long as there are enough kids to form a group, the first split will be down gender lines.

To children, the separation seems obvious, but to adults, it's somewhat more mysterious. Sure, I enjoy my female friends, but there are plenty of men who are equally pleasant and with whom I'd rather chat than certain women. But when my daughter was in elementary school, she wouldn't even consider initiating a conversation with the boys on her school bus (especially with her female friends present); neither of my boys has ever rushed home asking for a play date with a female classmate. (Nonetheless, I did manage to get Toby to invite a bunch of girls to his last birthday party.) Why are children so discriminating?

The fact that segregation is so universal, even among our nonhuman relatives, suggests it is somehow evolutionarily programmed into mental development. But how that programming works isn't yet clear. While the different interests of boys and girls obviously drive them apart, at least initially, the separation clearly goes beyond a taste for roughhousing or a penchant for tea parties. There are lots of girls who play sports and plenty of boys who like to draw. Something more must be triggering all the taboos, as when one boy chides another, "How can you use a *pink* ruler? That's for *girls*," or when a girl ducks for cover when her mom drags her into the boys' clothes section to buy a sweatshirt for a school play.

One theory is that children's gender segregation arises as a byproduct of universal incest taboos. Anthropologists find that boys and girls who are reared together (siblings, for example)* rarely become sexually at-

---

* Incest taboo also occurs among biologically unrelated children who are reared together from early life, such as adopted siblings and children reared in Israeli kibbutzim.

tracted to each other. This psychological defense prevents the dangerous inbreeding that would occur if siblings mated (and helps explain the
legal prohibition on sister-brother marriages in most cultures). But unrelated boys and girls don't need to avoid seeing each other as potential
future mates. So segregating in early childhood—and staying apart, until
hormones dictate otherwise—may be a smart strategy for avoiding the
familiarity-breeds-contempt problem that would arise if unrelated boys
and girls grew up like siblings.

Another theory is that evolution drives children to segregate themselves precisely to cultivate sex differences in adults. According to this
view, the main purpose of childhood is not to learn to read and write
but to prepare for mating and child rearing. By separating into different groups, boys and girls create different environments for developing
the attributes that will make them most attractive to the opposite sex.
For males, these attributes are: strong, brave, accomplished, and able to
outcompete rivals for females' affection. For females, the attributes are:
nurturing, compliant, and—don't forget—voluptuous, traits that would
make them appear physically and emotionally fit to be good mothers.

It's a popular theory, and there is almost certainly some truth in it,
except that such stereotypic qualities aren't the only ones that appeal to
the opposite sex. In fact, a survey of more than ten thousand people in
thirty-seven different cultures found that *both* men and women rank the
same two traits, kindness and intelligence, highest on the list of qualities
they value in a potential mate. Nor do the stereotypes seem to be holding
up so well in an era when men are attracted to Lara Croft, and women
appreciate they are better off with a man who'll be a dedicated father
rather than a chest-beating warrior who is rarely around to tuck the kids
in at night. But children, it could be argued, are acting more on their innate impulses than are modern adults choosing mates, and so it likely is
true that the universal segregation and different play styles of boys and
girls reflect such evolutionary heritage.

Whatever its ultimate cause, children's self-imposed gender segregation is a powerful force in their development. It limits their peer role
models, reinforces their emerging sense of male and female, and inevitably leads to the us-versus-them mentality that makes children the greatest gender enforcers of all. For all their unconscious gender shaping,
parents rarely utter the kind of things kids say all the time: "You can't
play dress-up, you're a *boy*" and "Girls aren't allowed to play football!"

A classic study performed at SUNY Binghamton reveals just how strong the influence of peers is, even at this young age. Psychologists brought three- and four-year-olds into a special playroom at the university's child-care center. Arranged in a row were six appealing toys, three of them boy toys (plastic soldiers, miniature fire trucks, toy airplanes) and three of them girl toys (small dolls and furniture, a plastic tea set, a child's ironing board and toy iron). Sometimes the child was alone, but at other times the child was in the presence of a male or female peer who was instructed to sit at a desk in the corner and color with crayons while awaiting a turn to play.

Not surprisingly, boys migrated to the boy toys and girls to the girl toys for the majority of the three minutes they were in the room. And once again, boys' preferences were stronger than girls': the boys spent only 21 percent of their time exploring the dolls, tea set, or ironing board, while the girls spent a bit longer (34 percent of their time) playing with the soldiers, fire trucks, or airplanes. More remarkable, however, was that the strength of both boys' and girls' preferences appeared to increase when another child was present, even though they weren't talking or playing together. Boys spent only 10 percent of their time playing with the girl toys when a girl was in the room; girls spent only 24 percent of their time playing with boy toys when a boy was at the coloring table. Interestingly, it wasn't the negative glances or remarks from the child coloring at the table so much as the self-conscious peeks from the child doing the playing that seemed to redirect him or her to an appropriate toy.

So peer pressure begins early! Even if children are not overtly bossing one another around, preschoolers are already aware of what's acceptable to their peers and what's not, and they adapt their behavior to fit those expectations.

Siblings

You can choose your friends, but not your family. The fact that children migrate toward same-sex peers means that, given a choice, they will have friends like themselves who will enforce and reinforce their own tendencies in play, appearance, communication, and thinking style. In a family, however, things are different. A child may have brothers, sisters, both, or neither. What's the impact of different sibling arrangements on children's early development?

It's significant, especially for the younger siblings in a family. In one important study, British researchers measured the interests, characteristics, and toy preferences of more than five thousand three-year-olds who each had either no siblings, one older brother, or one older sister. (The interests and characteristics of the older siblings were not evaluated.) As expected, boys were likelier to score on the high end for items such as "interest in snakes, spiders, or insects," while girls earned higher scores on attributes like "wears jewelry." More interesting was the influence of the older siblings. Girls with older brothers measured substantially more masculine than girls without siblings, who in turn were more masculine than girls with older sisters. Similarly, boys with older sisters scored more feminine than singleton boys, who in turn were more feminine than boys with older brothers.

Unlike most relationships, sibling influence is largely one-way: younger siblings have very little impact on the interests and activities of their older siblings, and even twins have less influence on each other than their older siblings have on them. Little ones look up to big ones, wanting to dress like they dress, play with their toys, tussle or chase or chatter just like their powerful older brothers or sisters. Older children often adore their younger siblings, but they rarely copy them or treat them as role models.

I was thinking about this data when I opened my local AAA (American Automobile Association) magazine and saw a picture of sixteen-year-old Monica Kunkel, winner of its 2008 student automotive-skills competition. Struck that a female would win in such a heavily male-dominated field, I read further and learned that Monica was the youngest child and only girl in her family, with five older brothers and a father who himself was an avid automobile enthusiast.

I too was the only girl in a family with three older brothers, and while I had some great female friends growing up, most of my time at home was spent horning in on my brothers' ball games, board games, or building projects. The tagging-along earned me a horrible nickname (the Brat) for more years than I care to admit, but it probably also helped shape several of my more masculine abilities and interests. My brothers' influence seems especially clear now that I have a daughter of my own who doesn't have older siblings and who spends her time in decidedly less physical and competitive ways than I did at her age.

Of course, every family is different, and it's important to point out that researchers have had a more difficult time generalizing about sibling influence in large families. Compared to two-children families, where there is just one sibling relationship at work, larger families provide several potential role models for younger siblings and also raise the issue of differentiation—the fact that children with several siblings of the same gender often tend to be *less* sex typed than those from more mixed families. We all know families where this happens, such as the tomboy sandwiched between very feminine sisters or the four brothers who span a wide range of jockiness and sensitivity despite their male-centered upbringing. Children seek out unique identities as they grow, and parents, reveling in the strengths of each child, probably encourage such differentiation in larger families.

Still, it's clear that siblings and birth order are important in gender development. In an ideal world, each of us would grow up with an older brother and an older sister—preferably twins, so one is not closer in age than the other one is—to bring out both male and female strengths.

Then again, someone has to be the firstborn!

Rough-and-Tumble Play

One very generous mom called them "a couple of bear cubs." She was watching Sam and Toby, then eight and five years old, rolling around on the sidelines at Julia's soccer game. I'll admit it's cute at times, but the incessant wrestling, kicking, tackling, and sparring among brothers is a major annoyance to parents (and, worse, to grandparents).

Boys are very, very physical in their play. They seemingly can't help but test their strength, endurance, and dominance through play fighting. If a boy has a brother or close male friend, they'll go after each other, but even without such partners, boys are likelier to play in exuberantly physical ways—running, jumping, and generally shaking the rafters more than girls do. Boys' higher activity level begins in infancy and grows throughout the preschool years and beyond.

Throughout the world, boys express this physicality in the form of roughhousing, or rough-and-tumble play. It shows up most often in groups of three or more boys, and especially when adults are not present. Rough-and-tumble play begins around three years of age, right when boys and girls are first segregating into different play groups. And it continues for many years thereafter, especially if you include the various competitive sports that boys still play to a greater extent than girls do.

Rough-and-tumble play is not unique to our species. Kittens, puppies, lab rats, and all those adorable cubs on PBS nature shows are also fond of mauling one another, claws held in just enough to avoid causing lethal injury. And while females of all species (including our own) engage in some wrestling and roughhousing, this behavior is much more common in male mammals.

This similarity to other mammals makes a strong argument that boys' propensity for rough-and-tumble play is innately programmed. In earlier chapters, we saw how exposure to testosterone around the time of birth promotes rough-and-tumble play in rats. In monkeys, whose developmental timing is more similar to humans', it is the testosterone *before* birth that promotes rough-and-tumble play; alterations in testosterone levels *after* birth have no effect. So as much as mothers like to blame the testosterone when their boys are roughhousing, this behavior is actually a reflection of the hormone's effect from a much earlier time, before they were even born, and not a matter of chemicals rushing through their blood as they race and tackle one another. (As we've seen, testosterone levels differ very little between boys and girls between about six months of age and puberty.)

But while there is good evidence that prenatal testosterone directs boys toward a rougher, more active play style, researchers have had a harder time linking prenatal testosterone to variations in such play *within* either sex. Though one study did find a modest correlation between the level of testosterone in mothers' blood measured at midpregnancy and their daughters' later propensity for male-type play, there was no such correlation for boys. More recently, Simon Baron-Cohen's team compared testosterone levels in amniotic fluid samples to children's play interests and activity levels at five years of age. Though boys as a group were exposed to much higher levels of testosterone than girls were (about two and a half times more), variations in prenatal testosterone did not predict how masculine or how feminine individual boys or girls would be compared to others of the same sex. So the jury is still out on whether prenatal testosterone shapes individual differences in rough-and-tumble play.

As annoying as it can sometimes be to parents, rough-and-tumble play is not without value. Boys bond through this kind of intense physical contact. They provoke one another, tussle it out, and then laugh together when they're done. Rough-and-tumble play promotes physical strength, helps vent boys' high energy, and can teach important social skills like negotiation, turn-taking, coalition-building, and how to compete fairly and graciously. But it can also be harmful and scary, and it can desensitize boys to other forms of aggression and the distress of other people.

So what should parents do about roughhousing? Stand back and let them go at it, because boys will be boys and there's not much we *can* do about it? To some degree, yes. After all, it is play—fun, interactive, and only rarely seriously harmful. The drive to engage in it is obviously innate and very strong in boys. Parents and, especially, teachers of young children must make greater efforts to accommodate boys' physical needs as opposed to constantly suppressing them. Otherwise, boys grow up hearing *no* and *bad* all the time, messages girls hear more rarely.

But we also need to appreciate that there is a lot of flexibility in this behavior. Obviously, physical fighting—even of the playful sort—is much less common among grown men, the vast majority of whom eventually learn other ways to compete, negotiate, and bond with one another. Most boys do grow into "gentlemen," the eternal goal in raising sons. Surely, it's no coincidence that we address them this way when trying to encouraging their self-control and mature behavior.

Then there is the fact that not all boys are equally disposed to play fighting. I know my own boys certainly differ in this regard. Toby, who happens to have a stockier build, is usually more game to wrestle and kick than Sam, who despite being older and taller is rarely the aggressor in their frequent matches. Sam tolerates Toby's provocations but is increasingly fed up by the constant assault, leaving me to de-escalate the warfare.

Research with monkeys is again revealing here. In addition to prenatal testosterone, scientists have discovered important social variables that influence the frequency of rough play in juvenile males. In a natural setting, most males grow up in close contact with their mothers but engage increasingly with other juveniles, usually males, with whom they fight and wrestle to their hearts' content. However, monkeys reared in a kind of male orphanage—without their mothers but with several male peers—engage in more rough-and-tumble play than males reared by their mothers, while males reared in isolation—without their mother or any peers—engage in much less play fighting once they are integrated back into a social setting and have a chance to try it out.

These findings will resonate with many parents, particularly those with two or more sons in the house. The more boys, the more physical the fighting, particularly if the boys are close in age, without sisters to break up the pecking order. Long before my husband and I had kids, we witnessed this close up: friends who had had three sons within the space of four years. It was "all wrestling, all the time" at that house, with the parents clearly worse for the wear. Since then, I've known many other parents with three or more boys, and most tell similar war stories. As with monkeys, the presence of other males makes a difference in the amount of rough-and-tumble play. I often wonder whether Sam's more gentle nature has to do with his older sister's influence, as compared to Toby, who spends most of his waking hours alternately emulating and provoking Sam.

What all this tells us is that boys' roughhousing isn't solely a matter of genes or prenatal hormones. Environment is also important. Boys are inclined to play rougher than girls, but without partners or parents who will tolerate it, this tendency won't be as strongly expressed. Cross-cultural comparisons tell the same story. While rough-and-tumble play is universally more common in boys than in girls, it is also more common, and more violent, in some societies than in others. Generally speaking, the more aggressive the adult males in a culture, the rougher and more frequent the play fighting of their boys. Boys of the Yanomami tribe—a dwindling population of preindustrial Amazonians—are said to "play" fight with clubs and real bows and arrows, while the Sioux tribe of North America was notorious for the bloodiness of its boys' games.

So take heart, parents, it could be worse. Boys will be boys, but only as far as they are allowed or encouraged to be. Boys do need the opportunity to express this challenging, exuberant side of themselves—to test their physicality, cement their friendships, and bond with their brothers—but within limits. We should permit boys (and girls!) to wrestle and roughhouse, but to keep it in view, with more, not less, parental supervision. Boys also need rules (no teeth or weapons) and, especially, reminders and conversations about their own and their opponents' feelings about these encounters.

Aggression nurturance is the term that boys' advocate Michael Gurian uses to describe this philosophy. It may sound like an oxymoron, but it is probably the best way to let boys fulfill this obviously strong need. Rather than constantly suppressing or disapproving of boys' physicality, we need to adapt to it and structure their environment so young boys can express this drive in safe, respectful ways.

War Toys

Kittens and bear cubs have sharp claws, though they rarely inflict them on one another. Our species is less well armed by nature, though technology has more than made up for this evolutionary shortcoming: we have weapons—guns, swords, knives, bows and arrows, slingshots, light sabers—and what could be more attractive to a young child seeking power and control over his brother or preschool playmate?

Boys, as most parents discover, love their weapons. They are the male plaything of choice, and for obvious reasons: they can be aimed, triggered, and, best of all, used to dominate other people. Guns are power, and boys figure this out early. Guns create the ultimate victors in the rough-and-tumble world that boys gravitate toward.

But guns are also a scary commodity in modern society. Every time another Columbine or Paducah comes along, accusations fly that toy guns and violent video games are to blame. Boys spend their early youth pretending to shoot things, their middle childhood honing their video targeting skills, and then, as some argue, they start pulling real triggers in adolescence because it is the only way they know to solve their problems.

The flaw in this logic is that most boys don't end up as violent offenders though most do play with pretend guns at some point, even if it is just a stick or their own thumbs and forefingers. Parents argue vehemently about the wisdom of letting their sons play with cap guns, squirt guns, Nerf darts, laser tag, bug launchers, and (my personal favorite) potato guns. Many ban them. I have mixed feelings, but my husband was a champion target shooter in high school. How can we disallow the sport he scored a varsity letter in? So yes, we've collected a diverse arsenal of toy weapons over the years, mostly purchased by the boys themselves with birthday or allowance money. (One of the joys of our plastic society is that such toys never last long and can be discreetly discarded after they have been dropped and cracked a few dozen times.)

The issue gets stickier when it comes to preschool. Most classrooms disallow toy weapons. Teachers state, reasonably enough, that children have plenty of opportunity to play with these at home and that bringing them to school—like any other highly stimulating toy—creates too much distraction. But what about the homemade weapons boys fashion out of LEGOs, clay, and other art-of-the-day projects? What about the light saber or the sword that a boy proudly dons to complete a favorite Halloween costume? What about superhero games? Why can little girls play and wear whatever they want but not little boys?

Here is where some early childhood educators are beginning to re-think the war-play ban of recent years. Clearly, many boys have a strong developmental need to act out the roles of warriors, heroes, allies, and leaders. There really are good guys and bad guys in the world, and how are children going to learn the difference if they don't have a chance to role-play and pretend? War play has many of the same elements as rough-and-tumble play and is probably as universal among boys. This is where boys bond, beat the bad guys, and start dreaming about doing great deeds.

The counterview is that such play is too aggressive and may fuel the development of violence. However, according to one study, teachers are more likely than other adults to see superhero play as aggressive; non-teachers are likelier to call it playful. Maybe we should permit such play in preschool classrooms, but with limits and with adult supervision to help children interpret what they're experiencing. In other words, war play can be used as an opportunity to talk about disputes between peo-ple, how they can be resolved, and the need to respect and protect indi-viduals. By letting children experience firsthand what it feels like to beat another or be defeated, it may be possible to build empathy, negotiation skills, and an appreciation of nonviolent resolution.

This is the view taken by Nancy Carlsson-Paige and Diane Levin in their book *The War Play Dilemma: Balancing Needs and Values in the Early Childhood Classroom.* These authors lay out several useful guidelines for managing and learning from war play, including encouraging children to trade off being the good guys and the bad guys; addressing the same concepts of conflict and danger in other parts of the curriculum, as at story time; offering alternative props for dramatic play that allow boys to be equally powerful but engaged in more humanistic roles (such as firefighters, construction workers, mountain rescue teams); and, inter-estingly, encouraging *girls too* to engage in war play—just as women now participate in all branches of the military.

However adults decide to manage boys' desire for war play, Carlsson-Paige and Levin make one point quite firmly: they need to limit children's exposure to heavily commercial TV and movie-tie-in toys. Squirt guns are one thing, but there are all kinds of gruesome, scary, and lethal-looking toys marketed to children based on movies they are presumably too young to see. Consider these: a *Lord of the Rings* twenty-piece figure set (all twenty with swords, chain mail, or both) marketed to children ages three and up; a *Star Wars Episode III:* Darth Vader Voice-Changer Helmet (for ages four and up); and a fully armed *Batman Begins* Deluxe Batmobile (ages five and up). Each of these serves as direct advertising for a PG-13 film, making boys desperate to see movies that even the studios agree are much too old for them. What's more, unlike the more creative, open-ended play that children learn the most from, these detailed commercial toy weapons channel their play into violent, predictable scripts (especially if the child was also permitted to see the movie). Children learn from such play, but unfortunately, what they learn is that violence is fun, cool, glamorous, and the ultimate way to solve conflicts.

In Sweden and Finland, the sale of such toys is actually prohibited under a voluntary agreement between the government and toy trade organizations. This kind of social engineering does not go over well in the United States, but parents can vote with their credit cards against such age-inappropriate toys.

Gals and Dolls

If boys like their guns, girls like their dolls. Play-parenting is universal among girls. Just as some boys who are denied toy guns or swords by well-meaning parents will turn anything into a weapon, girls will turn anything into a baby: a doll, a stuffed turtle, even a toy truck (if it's the smallest among a group of trucks). Girls' play often emulates family life. They like to play house, taking turns in the coveted role of Mommy, who gets to hold, feed, and rock the baby doll as well as order around everyone else (an imitation that comes as quite a shock to the real mom in the house!). This focus on mothering makes sense in cultures where even young girls are expected to help care for younger siblings, but it's also common in societies like ours, where five-year-olds are practically forbidden to be alone with a new baby.

There does seem to be something instinctive about girls' fascination with dolls and babies. As we've seen, female monkeys show a greater interest in human dolls than their male counterparts do. There's evidence that prenatal hormones may contribute to the difference, since girls with CAH (exposed to high levels of prenatal androgens) show significantly less interest in babies than their unaffected sisters do. Thus far, however, researchers have been unable to alter the sex differences in rhesus monkeys' interest in babies by manipulating prenatal testosterone levels in either males or females.

Another reason to doubt that girls' greater interest in babies is strictly innate comes from the fact that this sex difference is not found in the youngest children. As we saw earlier, boys and girls up to one year of age are about equally attracted to dolls. And it is not until age two or three that girls appear more strongly attracted than boys to real babies. From then on, girls of all ages respond more enthusiastically than boys to pictures of babies, to playing with babies, and to taking care of their infant siblings. Many parents also state that their daughters are more nurturing toward infants than their sons are.

Are parents responsible for the sex difference in preschoolers' interest in babies? This issue has been studied by Judith Blakemore, a psychologist at Indiana University–Purdue University Fort Wayne. Although her research revealed no difference in the amount of praise mothers bestow on sons and daughters for lovingly attending to baby siblings, it did find that children from more traditional homes—those whose parents subscribe to more stereotyped gender roles—show a greater sex difference in nurturing behavior than children from more egalitarian homes. In fact, four-year-old boys with more egalitarian parents were virtually indistinguishable from girls of the same age in the amount of interest they demonstrated toward babies. So this research suggests that parents' actions (role modeling) speak louder than our words (praise or disapproval) in shaping children's nurturing behavior.

Parents of boys should take note, because there's a lot to be gained from playing with dolls or attending to babies. Doll play and pretend parenting reinforce social-emotional skills: caring for other people, considering and accommodating their needs, and appreciating what they may be feeling. In short, this kind of play fosters the development of empathy. Playing house or family also involves lots of verbal and communication practice, even if the child is playing by him- or herself. (Most children—boys and girls—do talk to themselves, especially between four and six years of age, an audible "private speech" that helps guide their behavior during new or challenging tasks.)

As we saw in the last chapter, infant girls are already ahead of boys in both the social and verbal realms, a bias that presumably attracts them to such play in the first place. Then, once their play parenting begins, all the extra verbal and emotional practice further reinforces girls' communicative and empathetic tendencies. It's known that preschool-age girls are better able to identify another person's emotional expression. In one of the more dramatic studies, young girls performed much better than boys on a test of emotion recognition: picking from among several photographs of different facial expressions the one best matching a particular vignette, such as "Tommy's puppy ran away. Tommy is sad." To give the boys a fighting chance in this experiment, the researchers even limited the photos to pictures of boys. But girls still outperformed boys, with girls just three and a half years old performing as well as boys of five.

The gap in verbal skills between boys and girls also grows more distinct over the toddler and preschool periods. Girls start talking a month or two earlier than boys, and their expressive vocabulary remains larger through much of this period. They also speak more clearly than boys; in longer, more complex sentences; and with fewer grammatical errors, stuttering, and articulation problems. Girls with CAH are no less verbal than normally developing girls, so it doesn't appear that prenatal androgens are responsible. Still, it's likely that the small, innate differences in language processing I described in the last chapter kick off girls' verbal advantage, which in turn strengthens their inclination to vocalize and converse while playing with their dolls, friends, siblings, and parents.

Contrast this to the kind of practice boys get with their preferred toys. Balls, bats, trucks, Hot Wheels—not to mention all their shoot-'em-up play—don't exactly bring out boys' vocal, empathetic side. Boy play is great for developing a child's visuospatial sense, gross motor skills, and movement perception abilities, but there just isn't as much to discuss when building Duplos, kicking a football, or blowing away the bad guys than when feeding a baby doll, playing house, or chatting up your Webkinz pet.

Fortunately, there are plenty of ways to bring out boys' verbal side, and there's evidence that such stimulation can actually improve their verbal and literacy skills. Similarly, boys' nurturing impulses can be encouraged and strengthened through play. While most lack girls' interest in dolls, young boys do relate very well to babies. One prominent study of social development in twelve very different cultures (from Africa, East Asia, India, North America, and Mexico) found boys of all ages to be quite responsive to infants and toddlers. Though rarely assigned child-care duties, as their sisters are, boys in every culture were observed to treat babies and toddlers in an overwhelmingly positive, friendly, and nurturing way. Young boys in Western cultures also do plenty of cuddling, cradling, and sweet-talking to their stuffed animals, even if they avidly avoid their sisters' dolls. I know I will happily wait at bedtime while Toby tucks in his large menagerie of stuffed animals all around him. Even in third grade, Sam would warm my heart by gently lining up his various Beanie Babies at the foot of his bed.

And finally, after they've grown up and become fathers, most men today show a tender, sensitive side that they (and their wives) may not even have known they possessed. Nurturing is not an either/or trait but a continuum that can be strengthened or weakened, depending on circumstances.

Learning Through Play

Emma is overjoyed! She just got her first American Girl doll, Samantha, complete with two different outfits, a storybook, and all sorts of adorable accessories. Her mom has been itching to take her to the big store for a couple of years but wanted to wait until Emma could truly appreciate such a valuable possession. Now that Emma's five, Mom decided she's old enough. They shopped for Samantha and her clothes, brought her to tea at the café, got her hair done in the Doll Salon, and now get to bring her home and show her to Emma's friends. Okay, so Mom's out almost four hundred bucks, but what could be more fun for her special little girl?

Well, many things, especially when you factor in the price. Of course, dolls are wonderful playthings—for promoting all the verbal, nurturing, and empathetic skills we just discussed. I am *not* arguing that girls should be deprived of dolls, any more than I believe boys should be denied their Tonka trucks, squirt guns, and superhero figures. (Well, a prohibition on PG-13 toy weapons isn't a bad idea.) Still, there are other experiences that would probably be more valuable to Emma at this stage. How about a visit to the planetarium or a week of soccer camp? What about a comparably priced Rokenbok or Harry Potter LEGO set (both of which, in my highly scientific sample of one daughter, are pretty appealing to girls)?

Many parents think play doesn't matter. Sure, kids should go to preschool and learn their numbers and letters, but play time is really about having fun and making friends. Who cares if girls are playing with Barbies and dressing up in princess costumes while boys are wrestling and pushing their trucks through the sandbox? It's all good wholesome fun, and after all: they're just *little* kids. They have precious few years to play before all the homework, lessons, sports, and expectations start kicking in.

But play is very, very important. It is "the work of the child," as developmentalists since Friedrich Froebel, the inventor of kindergarten, have declared. Play is a natural route of exploration; when children are absorbed in play, they achieve a sense of mastery and well-being. And the youngest children, as Jean Piaget and Maria Montessori later taught, learn through sensory-motor play, by exploring the world using all their senses and budding motor skills. Children need to get their hands on things—toys, yes, but also tools, books, plants, pets, sand, water, soil, paint, clay, soap (anything safe)—and certainly more than a computer mouse and a TV remote control with their addictive strangleholds on their attention.

Children also need to interact: with their parents, siblings, peers, teachers, and anyone else who can guide their emerging skills, model more mature behavior, motivate them to communicate, help them express their feelings, and teach them to understand and care for others. Indeed, this intense social drive is what fuels much of children's attraction to TV, videos, and computer games. But while such entertainment provides the human voices and faces children crave, it can't provide the crucial feedback and interaction they need.

Hands-on play is much better. Even the most traditional boy and girl toys let children explore, manipulate, move, feel, talk, and negotiate to some extent. The problem is in degree. Boy play and girl play each tend to exercise some mental skills more than others; if each sex sticks to its own gender-appropriate play, children wind up strengthening those same brain areas that were already biased to work better from birth.

Dolls, dress-up, and drawing (why is it all the girl stuff begins with *d*?) promote fine motor, verbal, preliteracy, nurturing, and empathetic skills. Trucks, balls, and wrestling promote gross motor, spatial, visual, competitive, and risk-taking skills. Each of these abilities begins diverging by the end of the preschool period. We've just seen how girls pull ahead of boys in language and social perception by the time they're three or four years old. Boys, for their part, begin outdistancing girls in physical strength, speed, and spatial abilities by the end of this period. In one study conducted at the University of Chicago, boys were found to outperform girls at four and a half years old on spatial tasks like the one shown in Figure 3.1.

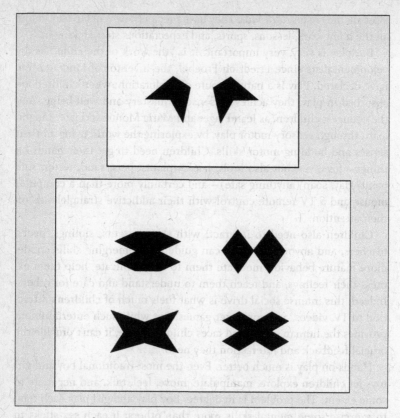

Figure 3.1. Test of mental rotation ability in four- to six-year-olds. Boys and girls were shown these two cards and asked to select the one of the four items in the lower card that could be constructed by putting together the pieces in the top card.

Answering correctly requires mentally picturing the halves fused and then rotated to match one of the potential targets. When four-year-olds were given thirty-two problems like this, both boys and girls found the task equally difficult, answering just ten correctly. However, in the next older group of children, four-and-a-half-year-olds, boys solved more problems than the girls by a small but significant margin. On average, boys between four and a half and seven years old answered about two more problems correctly than the girls did.

The difference is small (d = 0.25), but it illustrates how cognitive skills begin diverging toward the end of the preschool years as the different play and social experiences of boys and girls start adding up. Think about it: kids who spend more of their time launching cars along looping tracks, building and knocking down block towers, fencing with mock light sabers, racing around on scooters, and catching spiraling footballs are simply going to have a better sense of how objects move through space than kids who spend their time coloring, reading, brushing a doll's hair, arranging doll furniture, and getting their nails painted.

The same is true for other skills that start setting boys and girls apart. Girls are slightly ahead of boys in fine motor skills as infants but they're considerably more ahead by age five, after all those years of drawing, stringing beads, and fastening doll clothes. Boys are not especially prone to take risks in infancy (remember the ramp-crawling experiment in chapter 2), but they are much more so by the end of preschool, when they've learned that most of the time climbing on chairs or standing on a teeter-totter is not going to cause fatal injury. (Parents also contribute to the sex difference in risk taking, according to several studies by Canadian psychologist Barbara Morrongiello, who found that mothers caution daughters more often than sons about the potential for harm, and they encourage sons more often than daughters to push the limits in testing their strength and courage.)

By the end of preschool, these small differences become significant. But they don't have to be so dramatic. No matter what the skill, it can be enhanced by practice, especially in early life. Very young children can learn to speak multiple languages, play violin, swim, and even read if they are immersed (literally or figuratively) and taught in a developmentally appropriate way. I don't actually advocate such a hothouse curriculum for young children, which leaves little time for creative play-based learning and exploration. Still, it's clear that parents and preschool teachers can do more to help both boys and girls nurture the specific cognitive and emotional skills, or intelligences, that will land them on a firmer footing by the end of this wonderful age.

Putting Nurture to Work

The differences between boys and girls seem dramatic by the end of preschool. But while children are in many ways *more* rigid about boy-girl differences during this phase, and while they are beginning to be influenced by peers and media images, they are still more easily influenced by their parents and other teachers than at any later time in childhood.

Gender differences begin as little seeds, planted by genes and hormones but nurtured through social learning, gender identification, and children's strong urge to conform. Consider, for instance, the CAH girls who are bathed in high doses of androgens before birth, easily as high as most boys are exposed to. Yes, they gravitate toward traditional boy toys and a more active play style than the average girl, but their gender identity remains solidly female. Raised as girls, they almost never convert to a male identity, in spite of their more masculine style of thought and play.

Nurture is important. Maybe not as important as an earlier generation had hoped, but certainly more important than current "nature" enthusiasts are pushing. As we've seen in this chapter, children from more egalitarian families exhibit less stereotypical behavior and views than children from more traditional families. Boys in more violent cultures play rougher than boys in more peaceful cultures. Children who grow up with older siblings of the opposite sex will play in less gender-traditional ways than children with older siblings of the same sex. In another revealing study, toddlers' activity levels were found to vary by the type of toys they played with. While boys naturally opted more for boy toys (a train, a set of tools, or a truck) and girls for girl toys (a doll, a dollhouse, or a tea set), the amount of physical activity each child engaged in (climbing, running, banging, chasing, and so on) was more influenced by the type of toy being played with than by whether it was a boy or girl doing the playing.

In other words, toys and play do make a difference, along with playmates and the encouragement parents express. Obviously, we can't just switch the trucks and dolls. Many parents have tried this, to little effect. Girls turned the trucks into families, boys played catch with the dolls, and both sexes knew there was something fishy going on.

Rather, the trick is to find engaging toys and activities that will coax members of each sex to practice the skills they are not otherwise inclined to. For girls, it may be as simple as finding pink or other colorful toys that engage their movement and spatial skills (such as balls, puzzles, LEGOs, large cardboard boxes, sidewalk chalk, and doll strollers). We can tap boys' fascination with large powerful entities such as dinosaurs, astronomy, heavy machinery, and even army guys to get them reading, coloring, and communicating with others. What matters most is not the topic of their play but the skills they're using while engaged in it.

Finally, if we really want to help young children, we have to deal with the issue of *choice*. Homes, child-care centers, and preschools are now chock-full of toys—lots and lots of them—an unfortunate legacy of our Western affluence and the ascendance of cheap Chinese manufacturing. While choice is usually thought of as a good thing because it gives children a sense of control over their environment, too much choice can be overwhelming and makes it likelier they will default to their biases. The more freedom boys and girls have to choose, the greater the chance they will pick the most stereotyped items and reinforce the skills and interests they have already solidified. So this explosion of toys, DVDs, and extracurricular learning experiences appears to be backfiring on our kids, driving them farther apart than they might otherwise have been. Given a choice in preschool between sitting and drawing versus building with large blocks, girls and boys will naturally segregate themselves and miss out on the kind of cross-training they need most at this age.

I suggest that all the choices be within a single cognitive domain, such as literacy, spatial skills, quantitative skills, music, art, movement, and so on. Clear out the room except for several varieties of building toys, and what's a girl going to do? Get practice at three-dimensional imagery and spatial rotation. Dedicate preschool center time to preliteracy skills, such as a journal table, an ABC easel (for painting big letters while standing up), a listening corner (with headphones for children to follow books on tape), and a dramatic center for acting out the scenes in a favorite book, and boys will naturally hone their skills with words, letters, and communication.

There can be no denying the different interests and biases of young boys and girls. But these differences are not as fixed and permanent as all the flashy brain scans suggest. We can help bring out boys' verbal, empathetic, and fine motor skills; we can find better ways to encourage girls' courage, strength, and spatial skills. The two sexes will never be identical, but both will have a fighting chance when they enter school.

Other Tips for Young Boys and Girls

Children's brains will never again be more malleable than they are during the preschool years. This presents a great opportunity to inoculate boys and girls against the potential trouble spots in their development. As a society, the best thing we can do for children this age is to implement universal, high-quality preschool, which abundant research has shown improves both sexes' future school success. *High-quality* refers to various measures of teacher training and classroom size, but it emphasizes the sensitivity to young children's physical and emotional needs and a play-based, as opposed to drill-and-kill, immersion in numbers, shapes, measurement, scientific thinking, letter sounds, rhyming skills, writing, and drawing. Cost-benefit analyses have shown that for each public dollar spent on education at the youngest ages, society earns back some eight dollars when those children complete their education and become more productive taxpayers. The issue is most urgent for lower- and middle-income children, many of whom currently lack access to the high-quality preschools that could get them off on the best footing.

Meanwhile, here are some additional ideas, specific for each sex:

For boys

- **Language and literacy enrichment.** As in infancy, toddler and preschool-age boys continue to need lots of verbal interaction to boost their vocabulary and other language skills, which in turn predict later reading ability. Reading to them is probably the best way to do this. Boys often have an especially strong interest in nonfiction—books about vehicles, sports, animals, outer space, and the like—which may not always be a parent's first inclination when browsing at the library. Another boy-friendly way of increasing literary exposure is having them listen to books on tape or on CD. It's amazing how adding Play and Pause buttons, along with a set of headphones, can entice some boys to sit still and follow a story.

- **ABCs and letter sounds.** As we'll see in chapter 5, children who know their letters and can recognize the sounds at the beginning and end of words have a much easier time transitioning to independent reading than children without these skills. Parents can emphasize such skills by reading ABC books, emphasizing letter sounds, playing games involving rhyming and alliteration, and encouraging their sons to practice printing their names and other words.

- **Preliteracy computer games.** Many early childhood experts dislike computers, arguing that children should be acquainted with real reality before virtual reality. (After all, what is the point of playing a LEGO computer game when you can play with the real things in multisensory 3-D?) I agree that computer time should be strictly limited. But computers are also powerful learning tools, and boys, in particular, are strongly attracted to them. With their strong visuospatial input and fun, game-based learning activities, preliteracy computer programs can give boys extra practice learning their letters, letter sounds (phonics), rhyming, and other reading-readiness skills.

• **Fine motor skills.** These don't come as easily to boys as girls, but they are essential for the many pencil-and-paper tasks of grade school. If they don't gravitate toward writing and drawing, preschool-age boys can nonetheless work on their hand skill and co-ordination through tasks like cutting, stamping, and building with small construction toys. Other ways to encourage writing skills in boys include painting or drawing at easels (which some boys may find easier than sitting still at a desk or table), typing (either on the computer or, more fascinating still, a real typewriter), and clipboarding (walking around tallying or charting objects in the environment).

• **More movement.** Boys are more physically active than girls and need ample opportunity to move, both at home and in a preschool or child-care setting. At home, this means shutting off the TV or game system (or, better yet, not purchasing one in the first place) and getting them outside on swings, scooters, tricycles, or playing other sports. In daycare or preschool, it means frequent recess and other physical breaks, both inside and outside the classroom.

• **Rough-and-tumble time.** Boys need this kind of play, and most parents are comfortable letting brothers wrestle at home, but what about preschool? While boys can certainly learn to leave it at home, some may need the physical release more than others, especially if they're in a full-day preschool class. One early childhood expert, Lisa Donovan from Cuyahoga Community College in Cleveland, offers an innovative solution: a wrestling center in the classroom, where pairs of children can grapple for brief periods of time, using firm rules (no kicking, biting, pinching, or holding). She found that the wrestling mat soon became a favorite activity for the boys (and even the occasional girl) and had the added benefit of settling them down for other forms of learning.

• **Focus on feelings.** Males of all ages tend to be less accurate at identifying emotions, both in themselves and in others. Parents and other caregivers can help boys give voice to their feelings from a young age, distinguishing happiness, sadness, anger, fear, disappointment, jealousy, embarrassment, and shame. By nurturing the habit and vocabulary of emotional expression, parents can give boys a verbal outlet for their feelings and promote the empathy skills that tend to come a little less easily to them. Look for books and other activities to make this learning fun, such as the Thomas the Tank Engine stories—animated train engines with vivid personalities and feelings about their jobs and one another.

• **Pet care.** This is a great way to teach young boys nurturing skills. If a dog or cat is too much for you to handle, smaller pets, like goldfish, lizards, or gerbils, are a great place to start. Most kids crave the chance to own their own animals. Taking responsibility for the family pet is a wonderful way to cultivate a boy's sensitive, caring side.

For girls

• **Vestibular stimulation.** Movement stimulates the brain's vestibular system, the inner ear sense that detects a body's motion and position with respect to gravity. There is some evidence that vestibular stimulation—spinning, swinging, jumping, cartwheeling—enhances reflexes and gross motor development. Although girls do not lag behind boys in gross motor skills during the first year, they are slower and weaker from the preschool years onward. The benefits of vestibular stimulation are well enough proven to justify extra movement experience for young girls.

• **Ball games.** Girls begin falling behind boys in certain spatial skills by the end of the preschool period. One theory is that boys' greater movement and experience with projectiles—balls, darts, paper airplanes—begin training their brains by this age to better perceive three-dimensional moving objects, a sex difference that continues to grow throughout childhood. Girls' spatial skills may therefore benefit from more movement and practice at ball games, targeting, and other hand-eye challenges.

• **Sports.** Preschool is a bit young for organized athletics, but some girls may benefit by age four or five from joining a peewee gymnastics, soccer, or T-ball league, especially if they have little opportunity to run and play at home. In addition to spatial skills and hand-eye coordination, sports help children develop gross motor skills and are good for mind and body alike. Ideally, girls and boys would get all the running, kicking, and batting practice they need among small groups of neighborhood friends (which tend to mix the sexes better than at school or in organized activities), but the reality today is that most kids need to be signed up to get started on a sport, and younger girls are likelier to try out a new sport than those who are approaching the teen years.

• **Puzzles, mazes, and other visuospatial games.** Jigsaw puzzles epitomize the kind of spatial and mental rotation tasks at which males typically outperform females. And yet, there are many spatial toys available for preschool-age girls to hone these skills. I'm fond of a refrigerator magnet toy, Gear-a-tion, that allows kids to experiment with gear movement while keeping Mom or Dad company in the kitchen.

• **Building toys.** Many girls love these, but they are often not marketed to appeal to them. When they were both young, my son and daughter played about equally with LEGOs, but Julia rarely asked me to buy them for her. I think this is because most of the themes (for example, Star Wars, Jack Stone, Life on Mars) and colors (black, yellow, gray, beige, and army fatigue green) are created with only boys in mind. Lincoln Logs, Marble Works, K'NEX, and a big collection of old-fashioned wooden blocks are less-gendered options that girls seem to enjoy. Whatever the choice, playing with building toys and, specifically, translating a series of instructional diagrams into a three-dimensional structure provide excellent practice at the kind of visuospatial skill that is linked to higher mathematic achievement.

• **Hand her a tool.** Young boys often race to dad's side the minute he pulls out a screwdriver, but few girls exhibit the same hardware magnetism. Fathers and mothers should both encourage their daughters to participate in simple home-repair projects, such as changing furnace filters or assembling a piece of new furniture. Many girls who show no interest in toy tools become much more enthusiastic when they can help fix or create something real for their household. And dads should not be the only ones setting an example of mechanical aptitude; mothers need to tackle more of these jobs if we are going to help girls grow up comfortable and confident in their mechanical skills.

• **Visuospatial computer games.** Guess what: video games are actually good for something. Several studies have now found that computer games involving spatial manipulation improve children's ability to mentally visualize and rotate objects (see chapter 6). Most of the studies have been done with older children and adolescents, and the gains are similar for both boys and girls. Considering, however, that this kind of spatial ability shows the largest sex difference of any cognitive skill, girls may especially benefit from being *encouraged* to play spatially oriented computer games from a young age, particularly those depicting three-dimensional objects or virtual navigation.

• **Musical keyboard training.** No, girls (or boys) won't get smarter from passively listening to Mozart, but there is intriguing evidence that actively *making* music—specifically, learning to play the piano or electric keyboard—does increase spatiotemporal reasoning skills in preschool-age children. Because the musical scale is experienced as a visual pattern on a piano or a xylophone, learning to play such instruments may train the brain to recognize patterns in both space and time. This kind of learning is thought to be especially helpful for mastering mathematical concepts such as fractions, proportionality, and geometry.

4

Starting School

MATTHEW ISN'T READY FOR SCHOOL. Or so his mom believes. He's a bit small for his age, has a late-summer birthday and so will be younger than many of his classmates, and just can't seem to sit still for anything except TV or computer games. Maybe it's because he's the baby in the family, with two older sisters who love school and were eager and ready to start kindergarten when they were five. But Matthew seems different—more active, less focused, and less likely to mind his mother when she asks him to put away his toys or help set the table. She's worried he will get off on the wrong foot and more than a little sad at the prospect of sending her last baby away for a big chunk of the day. So she's having Matthew tested, meeting with the principal, and trying to figure out if she can wait another year before enrolling him.

Like Matthew, some boys do mature noticeably slower than girls during the first few years of life. The differences are in many ways most pronounced at this crucial stage, the transition to school, when the stakes start rising and parents begin worrying about how their children are going to fare in the demanding environment of many early elementary classrooms.

But it doesn't have to be this way. As we'll see, boys develop more gradually in some ways but not all. They can have a tougher time than girls sitting still, writing and drawing, learning their letter sounds, and negotiating with their five- and six-year-old peers. But they have other skills that are equal or better than girls' and which add a lot to a kindergarten classroom. The problem posed by boys' maturity is not going to be solved by keeping male five-year-olds out of school but by making schools adapt to the needs of all five-year-olds, male and female.

The Dirt on Redshirting

The practice of holding children—usually boys—back a year in school is called redshirting and gets its name from a long-standing trick colleges have used to allow athletes to compete an extra year on sports teams. Some 9 percent of kindergartners are redshirted each year, more than twice as many boys as girls. And the trend seems to be growing. Between redshirting and the shift in school cutoff deadlines from December to September, the proportion of six-year-olds entering kindergarten instead of first grade at the beginning of the academic year has increased from 4 percent to 16 percent over the last forty years.

While parents typically state developmentally correct reasons for holding boys back from kindergarten, their true reasons are probably not much different than the athletes': competition. Redshirting is much more common among white children than minorities and in private schools than in public schools, despite the fact that these children already have higher odds of school success. Parents are hoping that the extra year will spurt Matthew from shrimp to shark and also give him the greater attention span, verbal skills, fine motor control, and social adeptness needed to be a class leader instead of one of the herd. Certainly, there is truth in this maturation claim: boys do develop more slowly in these specific areas. But boys are actually ahead of girls in some skills (visuospatial, mechanical, gross motor) and at par in others (mathematics), so it may be that parents hold boys back more often simply because they care more about their sons' competitive advantage.

Whatever the motives, most research finds the practice of redshirting misguided. Although the older children in a class may have a modest advantage in kindergarten and the first few grades, their academic boost typically fades by later elementary school. There is also some evidence that children who were held back are more vulnerable to risk taking and other emotional and behavioral problems when they reach adolescence ahead of their classmates. Some experts argue that delaying kindergarten may mask any true developmental delays or learning disabilities—the same problems that persuaded parents to hold their children back in the first place. Such issues would be better treated in early elementary school than in the older grades, when the academic workload intensifies.

The fact is, school makes kids smarter. Children who attend more preschool, or longer kindergarten days, or more years of formal school for their age score higher on IQ tests than those with less schooling. According to one large study, an extra year of school raises a child's IQ about twice as much as an extra year of age alone; in this report, young fifth-graders scored within a point or two of their older classmates but some five points higher than children of the same age who were still in fourth grade. In another study, Canadian researchers compared the youngest first-graders from twenty-six public elementary schools to the oldest kindergartners at the same schools—groups who differed by just two months in age, on average—and found that the young first-graders made far greater progress in reading and math over the academic year than the kindergartners.

Communities that permit redshirting create a vicious cycle: if September is the cutoff date and most of the boys with summer birthdays are held back a year, then parents of sons with spring birthdays start getting nervous. Will their sons be at a disadvantage? And as kindergarten teachers boost their curricula to accommodate more six-year-olds, fewer five-year-olds will be able to keep up, ratcheting up the pressure to hold even more children back.

Another issue is the gender discrepancy redshirting creates down the line when nineteen-year-old males are still in high school with mostly younger females. How does the age gap affect girls' confidence and the already difficult issue of sexual harassment? Can these older boys—or men, really—stay emotionally invested in school among much younger peers? In fact, while parents believe they are giving sons the gift of time by holding them back, recent research suggests that the later average starting age of boys is actually contributing to the gender gap in high-school and college completion rates.

Of course, parents are not the only force driving redshirting. The real issue is in the schools themselves, which are often not as boy friendly as they could be. As the academic requirements of first grade increasingly descend into the kindergarten curriculum, many children are facing unreal demands on their attention spans, memory, fine motor skills, and verbal skills, which can set them up for behavioral problems and school failure. Every five-year-old is actually ready for school, provided the schools are suited to the real needs of five-year-olds.

Sex Differences at Age Five

Just how different are five-year-old boys and girls from each other? Not as much as you might think. While they may insist on wearing pink dresses or blue jeans to get to school, once inside the classroom, boys and girls have a lot in common. Yes, girls are more verbally precocious and tend to speak in longer, clearer sentences during the preschool years. However, verbal ability is actually one of the smaller sex differences, equivalent to a mere two IQ points at this age, and it appears to shrink even further over the elementary school years (no doubt because of the intense language immersion children get once they are in school). In other words, plenty of five-year-old boys have the gift of gab, and while more boys than girls have difficulty learning to read (see the next chapter), we shouldn't assume that every boy will struggle or, worse, that a girl does not need some extra help in language or literacy simply by virtue of her sex. Nor are girls ahead in all measures of language ability. Vocabulary, in particular, is one verbal skill that does not differ between the sexes, at least by age six and in all subsequent years.

Other cognitive differences actually favor boys at this age: boys perform slightly better on tests of visuospatial ability, a difference that grows steadily throughout childhood and adolescence. Boys are also on par with or better than girls in math. While the average five-year-old boy and girl score about equally on math tests, young boys are likelier than girls to be identified as mathematically gifted. This might look like a selection bias—that parents and teachers are expecting boys to do better and are therefore more likely to nominate them when the call for math talent goes out. But one study of mathematically gifted four- and five-year-olds managed to come up with a balanced number of males and females; within this group, boys scored higher on eight of eleven subtests, including number knowledge, problem solving, calculation, word problems, and counting span. The girls in this group, while talented, scored higher than the boys on only the three verbal tests that were included, and this difference was not even statistically significant.

In sum, boys and girls are pretty much equally ready for school when it comes to academic skills. Though the brightest boys and girls may fit the math-versus-language stereotypes, these labels don't apply to the vast majority of five-year-olds as they head off to real school for the first time.

Fine motor skills

Boys and girls do differ, however, in a few key areas that arguably have a great impact on their adjustment to school. One is fine motor skill, or the ability to manipulate small objects with the hand and fingers. Females of all ages are more manually adept than males, though the differences are small. A classic test involves moving pegs around a pegboard, like a cribbage game. When asked to place as many pegs as possible in a fixed amount of time, women prove about 6 percent faster than men, though smaller fingers are one reason for their better performance. Still, a sex difference in this test appears in children as young as five years old (when finger size is more comparable between the sexes). Girls as young as three are also reported to be more accurate than boys at another fine motor task, imitating a series of hand gestures modeled by an adult.

Fine motor skills are a big deal in kindergarten, when children are expected to start producing serious work with their hands: drawing pictures, practicing penmanship, and even writing little stories with their scratchy handwriting and invented spelling. All this practice is crucial for boys, but they are more likely to find it frustrating than do girls, who generally take greater pride in their penmanship and whose handwriting is notoriously neater throughout elementary school.

Fine motor skill is not limited to the hands but extends to speech, which involves the intricate coordination of many small muscles of the throat, lips, and tongue. Verbal articulation differs reliably between the sexes, both in quantity and quality. In the classic test known as verbal fluency, subjects are given thirty seconds to say as many words as possible that begin with a certain letter; for example, *p*. Women spit out more words—like *pig, play, prince, pink, pine, pool, part*—than men do, with the average woman scoring higher than about two-thirds of men. The difference is not large, but it is present as early as six years of age. Girls also enunciate more clearly, whereas boys are three to four times more likely to suffer from stuttering and other speech impediments.

Taken together, these manual and linguistic skills can be an impediment to boys as they enter kindergarten. Fine motor skills underlie most of what kids produce in school—every drawing, bit of writing, and verbal response they give when asked questions. Parents and teachers need to appreciate that many boys are not as neat or fluid as girls in their writing and speaking and should calibrate their expectations accordingly.

Unfortunately, this is not always easy to do, and boys are often unfairly judged on their penmanship, like when my daughter told me: "It just seems like boys don't really try or care about making their handwriting look impressive."

Girls obviously do, even in preschool, when they log loads of practice at writing and drawing while boys are off building block towers or pouring sand at the sensory table. As we've seen, fine motor skills differ little, if at all, in infancy, but they start diverging around three years of age, when girls start honing all the preschool skills that serve them so well. Like all skills, fine motor control is learned, and practice makes perfect. Such skills are not reflective of children's intelligence, but the lack of them can mask it. So parents and teachers need to find more engaging and effective ways to help boys exercise these skills. (See below.)

Inhibitory control

Fine motor skills are important, but they are not the most crucial sex difference at this age. Walk into any kindergarten classroom and what do you see? Quiet, attentive girls, sitting still and doing pretty much exactly what the teacher asks. Boys are more often the fidgety ones, less able to focus on what the teacher is saying, distracting one another with pokes and jabs, and calling out answers when they are supposed to raise their hands. There are, of course, some girls who exhibit these disruptive behaviors and plenty of boys who behave well and thrive in school. Overall, though, boys are slower than girls when it comes to developing the self-control that is essential for success in school and other social situations.

Neuropsychologists have a simple, unifying explanation for these differences: it is known as inhibition. Though we tend to think of development as the gradual *addition* of new skills, in fact, much of growing up involves the ability to *suppress* behavior—to stop moving, or talking, or elbowing the kid next to you when you should be listening to your teacher, reading a book, or completing a math worksheet. Inhibition is well known to depend on the brain's frontal lobes—the most distinctly human but also the slowest region of the cerebral cortex to mature. It is crucial to a child's ability to sit still, concentrate on the task at hand, and get along with other people. Children with attention deficit hyperactivity disorder (ADHD) have underactive frontal lobes (compared to other children of the same age), a difference that can explain all the core symptoms of this syndrome: *hyperactivity* (because they cannot suppress inappropriate movements), *impulsivity* (because they can't suppress other inappropriate actions, like blurting out answers), and *distractibility* or inattention (because they can't block out competing stimuli, such as other noisy children, that interfere with their ability to focus on the task at hand).

Of course, most five-year-old boys do not have ADHD, but this diagnosis is at least twice, and perhaps as much as nine times, more common in boys than girls.* Boys are, by most measures, slower than girls to develop inhibitory control: the difference shows up as early as eight months of age, when baby boys and girls are compared on a classic task that requires inhibiting their urge to uncover a hidden toy. By the toddler years, boys' comparative lack of inhibition is evident in their higher activity level and greater mischievousness: they just can't slow down or heed parents' warnings, which is why boys are more often the ones caught crawling into forbidden cupboards, pulling down curtains, putting objects into electrical outlets, and climbing out of shopping carts.

In one study of two- and three-year-olds, boys came out worse than girls on virtually every measure of self-control, impulsivity, and rule following: they were less able to stop themselves from eating an M&M placed under a glass cup (though asked to wait until a bell rang); more likely to peek over their shoulders at a gift being wrapped; more likely to release a pinball plunger before being given the go signal; and more likely to commit an illegal act, like scribbling in a book, when encouraged by the experimenter.

Differences in inhibitory control persist through childhood and, by certain measures, well into the teens. According to one large summary study, girls' advantage in inhibitory control is about the largest sex difference of any temperamental trait among children between three and thirteen years old. So it is this difference—in the ability to sit still, tune out conflicting impulses, and focus on completing their work—much more than cognitive sex differences that makes boys' adjustment to school more challenging than girls'.

Because inhibition depends on the frontal lobe, neuroscientists pre-

---

* The fact that more boys than girls are diagnosed with ADHD is an inevitable outcome of the fact that ADHD is defined by reference to a "normal" range of inhibitory control for all children of a given age. But because boys develop this control more gradually than girls, a much larger number of them will fall below the normal cutoff. The obvious solution is to have separate norms for boys and girls, but current practice leaves this judgment up to the individual physician.

sume that boys' slower journey to self-control reflects the slower maturation of this part of the brain. But there is surprisingly little evidence to support this idea. One MRI study out of UCLA did find that girls reach their peak level of frontal gray matter about a year before boys (ten versus eleven years old). Other evidence, however, suggests boys' frontal lobes actually mature *earlier* than girls', at least in the lower, or orbitofrontal zone, which is involved in certain forms of decision making. Brain-wave measures (EEGs) paradoxically also show earlier maturation in boys than girls, according to Australian researcher Robert Barry and colleagues. There is, however, one recent study that found a larger frontal-lobe response in girls than boys during a test of inhibitory control. This EEG difference showed up only when the children *failed* to stop themselves from responding inappropriately, suggesting that girls may process such mistakes more efficiently. Perhaps this enhanced frontal-lobe activity allows girls to better learn from their mistakes and catch themselves the next time they are tempted to do something impulsive, such as talking out of turn or poking a neighbor.

Verbal skills may be another reason for the sex difference in self-control at this age. As early as the second year of life, children with stronger verbal skills exhibit better inhibitory control, and girls, as we've seen, are more verbally advanced throughout this period. As their language skills improve, preschoolers increasingly use words to gain awareness and control over what they're doing.

You can hear this when young children talk to themselves. Though worrisome to some parents, such verbalization is perfectly normal and often quite useful. Sometimes children talk to themselves to practice dialogue, as when playing with dolls or action figures. But at other times, children talk to themselves precisely to help regulate their behavior, like stopping themselves from eating that cookie when Grandma told them to wait until after dinner. ("Oh, it looks so good! I reeeeeeeeeally want it.") Talking aloud gradually shifts to *private speech* by the end of the preschool years, when children begin silently reminding themselves how to behave or how to carry out a challenging task. If self-control is aided by talking to oneself, and if girls are better at putting their thoughts into words, it stands to reason that girls have an easier time controlling their behavior than boys do.

Inhibitory control is very important to success in school, and boys, on average, are slower to develop it. But this doesn't mean large numbers of boys are doomed to failure or can only succeed in school by taking medications for ADHD. Like all skills, self-control can be improved through practice and training—which is really what discipline is all about. One innovative method, called Tools of the Mind, successfully enhanced inhibitory control and other frontal-lobe functions in a recent study of five-year-old children. Other research suggests that children can be explicitly trained to improve their attention skills by using a deliberate self-verbalization technique to monitor focus and performance. Most kids figure this out for themselves as they gradually adapt to the growing demands of school, but for some, especially boys, explicit training in verbal self-monitoring may be beneficial.

Boys and Girls on the Playground

It's recess time. Thank goodness! Twenty squirmy children line up, trying hard to keep their hands to themselves. Finally, the fidgety line makes it out the playground door, where forty pent-up legs race for the swings, the slides, the kickball field, and a much-needed break from the classroom.

Recess is crucial in early education. Children, with their immature frontal lobes, can't sit still for long periods and desperately need physical breaks to recharge their mental batteries. While schools around the United States are cutting into recess time in an effort to cram in extra computer, language, and math lessons, perhaps we should take a lesson from Japanese schools, where children break at least ten minutes every hour to help maintain focus on their intense academic work.

Physical activity is critical for mind and body alike. Health experts agree that children today are not getting enough exercise to ward off obesity, diabetes, and heart disease. But the brain benefits as much as the heart and other muscles from vigorous physical activity, according to a growing body of research in rats and mice. Animals that voluntarily hop onto the running wheel every day learn better than their less active littermates. Their advantage is due to several exercise-induced brain changes, including an increased production of nerve cell growth factors, enhanced growth of brain capillaries and cerebral blood flow, and even the birth of new neurons in areas known to participate in memory storage.

Fortunately, children are predisposed to be active: they actually crave all the running, jumping, throwing, and physical exertion that most adults manage to accomplish only through great guilt. Ask any child, boy or girl, about his or her favorite subject at school, and chances are the answer will be "Recess." While boys tend to be more physically active than girls, both sexes are equally glad to hit the playground.

But while both boys and girls look forward to recess, their experiences on the playground are quite different. And these differences, like the gender segregation that makes them possible, have important consequences for their development.

When Toby was in kindergarten, I walked him to school every day, where I got to see some of these differences in action. The kindergartners at his school had their own playground where the kids would gather while waiting for the eight o'clock bell. I would kiss Toby goodbye and watch him place his backpack in line and then slip quietly onto the playground. Toby is a bit shy, not one to race out and immediately join the throng. Still, when he did make up his mind to find company, he never approached the closer crowd of girls in the sandbox. No, he'd head straight for the boys' corner, the zone containing the biggest piece of equipment on the playground, where the rest of the crowd—nearly all boys—would be racing one another up the ladder and down the slide, pushing, teasing, and energetically exerting themselves in the few precious minutes before school started.

We saw in the last chapter that boys and girls begin forming separate play groups as early as preschool. However, the segregation is less absolute among three- and four-year-olds, who can still be found sharing the swings or riding Big Wheels together, compared with kindergartners, in whom the mixing effectively ceases for the duration of elementary school. Sure, the occasional girl may insinuate herself into a kickball game, and the occasional boy may find himself chased by a posse of girls, but for the most part, the line between sexes is etched deep in the bark chips.

Once apart, groups of boys and girls play very differently, reinforcing the same attributes that helped segregate them in the first place. The most obvious difference is that boys play more roughly, which we've seen is clearly influenced by prenatal testosterone. Boys are more active than girls, and their interactions involve considerable physical contact, taunting, and play fighting. This physicality soon sets up a pecking order among boys, which contrasts dramatically to girls' more cooperative, communicative play. While girls strive to promote group harmony, taking turns being Mom in the sandbox kitchen, boys are busy figuring out who is the strongest, or fastest, or has the most accurate aim. Given the current trend to de-emphasize competition among children, the female style sounds superior. And sure enough, girls do receive less negative feedback about their play while boys receive more limits, reprimands, and disapproval when their physical exuberance crosses into the overly aggressive side.

The second difference in play is that girls tend to stay in closer proximity to adults, selecting activities that are structured by adults, while boys tend to play in more removed places, farther from adult supervision. This was obvious on Toby's kindergarten playground: the vast majority of girls clustered in the corner closest to the teachers or by the sidewalk where other parents were dropping off their children. The boys, in contrast, raced to the far end, where, they had undoubtedly learned, their rough-and-tumble fun would come under less scrutiny. Given a choice, boys also play outdoors more than girls, again encouraging greater independence, more physical exercise, and freedom to generate their own rules and standards for behavior.

A third difference between boys' and girls' play is group size. Girls tend to play in pairs or trios, small groupings that allow lots of one-on-one conversation and intimate sharing. Boys hang out in larger groups in which all children do not participate equally. Instead, boys establish dominance hierarchies; certain boys become more influential, while others become their followers. Such hierarchies emerge faster and are more stable the older boys become. In one study from Durham, North Carolina, groups of first-grade boys and groups of third-grade boys were brought together for successive play sessions. Though none of the six boys in each group knew one another at the start, dominance relationships emerged within the first forty-five-minute session; lower-status boys quickly began deferring to higher-status boys, letting them choose what game would be played and cast each person's role in it. The resulting hierarchies were remarkably stable over the remaining play sessions, especially among the third-graders.

Put these differences together and you get two dramatically different peer cultures to grow up in: one emphasizing verbal negotiation, intimacy, and teacher approval; the other fostering physicality, competition, and a disregard for adult authority. Each group reinforces gender differences that were already poking through in early childhood but that now start to diverge with a vengeance. It's a positive-feedback loop, like the spread of an epidemic or like global warming once the polar ice-caps have started to melt. Boys spend their time with other boys, sealing the boys-will-be-boys prophecy; girls hang out with other girls, honing one another's chatty, cautious, and decidedly pink preferences in clothes and accessories. Researchers call it gender intensification, and intensify it does: before long, the small initial differences between the sexes get magnified by peer groups in accelerating spirals that leave boys and girls with few similarities by the time they finish kindergarten.

These are the days when boys and girls are most famously uninterested in each other. They taunt each other with "cootie" inoculations and by proclaiming "Boys rule; girls drool" or (my daughter's favorite) "Girls go to college to get more knowledge; boys go to Jupiter to get more stupider." (Both genders are, of course, routinely interchanged in either taunt.)

Parents know that their children will go to great lengths to avoid being identified with the opposite sex. Researchers have focused on how this aversion affects other aspects of their development. In one study, children between five and ten years old were shown a video of other boys and girls playing with one of two gender-neutral toys: a slick Fisher-Price movie viewer showing a Disney cartoon or a more boring kaleidoscope. Initial testing had shown that both boys and girls strongly preferred the movie viewer to the kaleidoscope, but guess what: boys who had seen a boy playing with the kaleidoscope and a girl playing with the movie viewer spent more time with the kaleidoscope than boys who had seen boys playing with the better toy. Same for girls. In other words, children override their own likes and dislikes to act in ways endorsed by their same-sex peers, even when no other child is watching!

Parents are not, by and large, responsible for this rigid gender segregation. Kids do it to themselves. Still, there is room for both parents and teachers to influence these groupings and give both boys and girls the advantage of interacting with children of the opposite sex. Gender segregation is less dramatic at more progressive schools, where stereotyping is more purposefully downplayed. And such integration can be beneficial. In one study of four- and five-year-olds at school, the higher-energy, more easily aroused boys were better behaved and better liked the more time they spent playing with girls. Unfortunately, the opposite was not true: higher-energy kindergarten and pre-K girls actually behaved worse the more time they spent playing with boys. Nonetheless, girls can benefit in other ways from playing and working with boys. We've already seen how girls with older brothers are likelier to play with building toys and participate in sports, enhancing their spatial and gross motor abilities. Other research similarly indicates a small but significant enhancement of spatial skills in girls who engage in more active, spatially oriented activities, such as building forts or playing with train sets.

Along these lines, I was glad to see our son Sam make a new "girl-friend" back in fourth grade. It took a few months after we moved to the neighborhood, but Sarah from across the street started stopping by, asking him to join the throng raking and jumping into a humongous pile of leaves in her yard. This was at an age when cross-gender play was most uncommon, but happily, standards are a lot more lax around the neighborhood than in the schoolyard.

It's No Picnic for Girls

Ginny is having a tough day. She doesn't get along well with the other girls in her desk group, and the boys are all icky. Reading is still hard work for her, and today's math topic, the difference between odd and even numbers, is kind of confusing. She tried raising her hand for help, but the teacher is too busy with Scott to even notice her.

Sex differences can be useful for understanding children, but we can't lose sight of the fact that the generalizations are far from universal. Yes, the well-documented boy-girl differences can give us an overall snapshot of an age or classroom, but they can also be misleading if applied to every child. Sure, many girls thrive in kindergarten, and some boys have a tough adjustment. But this doesn't mean there aren't plenty of boys who adapt wonderfully and occasional girls, like Ginny, who fail to blossom in kindergarten or first grade.

Here's a child who, contrary to expectations, does not love going to school, or sitting still for six hours, or filling in endless worksheets. Ginny likes music and recess and has made a couple of lukewarm friends, but the more confident girls seem to avoid her, and even her teacher seems to forget she's in the class at times.

Girls do, on average, have an easier time at the beginning of the school years, but this doesn't mean every girl has it made. Some, like Ginny, will need some extra attention and encouragement to keep them on track. Certain of the boy-friendly classroom tips, such as more opportunity to move around the classroom and more hands-on activities, may be helpful for girls like her, but Ginny's case reminds us that what children need most is not any kind of stereotyping but attention to their needs as individual learners.

Even for the typical girl, the early school years can be a time of lost opportunity. While most girls cause little fuss—sitting still and dutifully filling in their penmanship books with neat *r*'s and *m*'s, they may not be getting the best preparation for later, more challenging subjects. Yes, girls generally learn to read and master arithmetic without much difficulty. But while they are carefully completing their math and spelling worksheets, they could also be expanding their spatial, mechanical, and public-speaking abilities. Such skills are rarely taught in early elementary school, but they start figuring importantly in middle school, the same time girls begin trailing behind boys in certain areas of math and science as well as in classroom assertiveness.

Here, then, is a chance to inoculate girls against such disadvantages by stretching them beyond their verbal and fine motor strengths. Decades of research in neuroscience have proven that young brains learn more easily than older ones. Even if a skill, such as map reading, isn't fully flexed until later in life, its roots are laid in the first few years. And yet girls are falling behind in spatial skills as early as age four. Adding more three-dimensional, hands-on activities involving puzzles, gears, globes, and math manipulatives will help girls in the early elementary classroom acquire a deeper understanding of shape, directionality, and math concepts such as fractions and decimals. Innovative software can also help expand children's early understanding of sequencing, proportion, symmetry, and rotation. The U.S. National Science Foundation has supported development of an exemplary program called Building Blocks that combines hands-on materials and computer-based training to enhance young children's higher mathematical thinking. Boys may gravitate more naturally to such manipulatives and computer programs, but girls need to get their hands and eyes engaged in such activities too, so they can better develop their spatial, geometric, and quantitative abilities at this formative stage.

Adults' Expectations and Beliefs

Sometimes I worry that my own kids will start reading one of the dozens of books on male-female differences on my bookshelf. God forbid Julia gets the idea that girls aren't supposed to be good at math, or that Sam starts thinking he'll never read as well as his older sister. Yes, parents should be aware of the different assets and liabilities of each sex, but what if the kids themselves start believing this stuff?

Research on sex differences can be dangerous. As much as researchers emphasize that group differences can never predict an individual child's achievement, such data do build up expectations in parents, teachers, and children themselves. And expectations can be every bit as influential as toys, curricula, and peers in shaping our boys and girls.

One neuroscientist, Melissa Hines, likens gender expectations to the placebo effect in medicine: the well-proven fact that a patient's mere belief is enough to make a sugar pill stop a headache or lower blood pressure. "Expectations and beliefs, as well as hormones, can engender the brain," she writes, reminding us of the potency of self-fulfilling prophecies.

Expectations are already lurking when children begin school. According to one landmark study, mothers of kindergartners in three different countries—the United States, Japan, and China—were likelier to state that boys are better than girls at math and girls are better than boys at reading. In fact, there were no differences in this study between boys' and girls' math achievements in kindergarten, and while girls in all three countries did outperform boys in reading in kindergarten, there were fewer sex differences by third grade and none by fifth grade. As is often the case, stereotypes about sex differences are greater than the actual differences themselves.

Gender expectations are even embedded in the reasons parents give for a child's success. Asked whether their child's achievement in math or reading is due more to effort or to intellect, parents are likelier to credit a son's success to innate talent and a daughter's to hard work. Girls, in other words, are less often assumed to be naturally smart, an attitude that girls themselves soon adopt and that comes to haunt them at later, more challenging points in their academic careers.

Parents, then, are one source of bias regarding the skills boys and girls are assumed to possess. But teachers are imbued with the same stereotypes, so it shouldn't surprise anyone that their expectations similarly influence boys and girls in the classroom.

There's a classic study of teachers' expectations that demonstrated what came be known as the "Pygmalion effect," after George Bernard Shaw's well-known play. In this experiment, Harvard professor Robert

Rosenthal teamed up with elementary-school principal Leonore Jacobson to analyze the effect of teachers' expectations on children's achievement. At the beginning of the school year, Rosenthal and Jacobson randomly selected 20 percent of the students and told their teachers that these children showed "unusual potential for intellectual growth." By the end of the year, these children did make more significant academic progress than their peers. What's more, their teachers rated them as better socially adjusted, more intellectually curious, and even happier than children not singled out for high achievement.

Rosenthal and Jacobson focused on the positive effect teacher expectations can have, but further research has analyzed both the ups and the downs—namely, the ways in which a teacher's knowledge of a child's social class, gender, or even prior test scores can create self-fulfilling prophecies for that child's achievement or failure. The issue is certainly most dire for underprivileged children, but in the case of gender, teacher expectations have undergone a dramatic flip-flop over the past generation.

Back in the 1970s and '80s, it appeared that teachers favored boys in the classrooms. Studies found they would ask questions of boys more frequently than of girls and also responded differently to each sex when boys and girls called out an answer out of turn; girls were likelier to be criticized for not raising their hands, while boys were likelier to be praised for getting the answers right. In other research, boys were given the benefit of a longer wait time than girls—this is the brief pause a teacher permits a student to answer a question before moving on to someone else; though small, this extra half a second can make a difference in whether a student comes up with the correct answer, which has an impact on his or her feeling of competence. Researchers also found that teachers were generally unaware of the greater deference they showed to boys. In one study, teachers were shown a film of a classroom discussion that they overwhelmingly judged to be dominated by girls, but when they subsequently counted utterances, it turned out that the boys were outtalking girls by a ratio of three to one.

Of course, boys also get more attention because they misbehave more, and some of these interactions involve discipline and reprimands. Nonetheless, it seemed the general tendency of teachers, at least during the era when these studies were conducted, matched the behavior of the society at large: to take males more seriously than females.

Studies like this have influenced teacher training. Gender equity can be taught, and studies show that teachers who are made aware of their biases and taught specific techniques for overcoming stereotypes—such as deliberately alternating between calling on boys and girls—do end up treating the sexes more equally. Girls are now doing better than ever in school, a result that probably reflects more equal treatment and, particularly, the greater expectation of parents and teachers that they will actually use their education to pursue valuable job opportunities. There is still room for improvement, particularly in math and science classes, where girls remain vulnerable to lower teacher and parent expectations (see chapter 6). Nonetheless, the expectation gap appears to have closed, or even reversed, between boys and girls during the early grade-school years.

Indeed, girls are doing so well in early schooling that concerns are running in the opposite direction: now boys are the ones for whom teachers see less potential, at least at the outset of formal education. Such a shift has been documented in Britain, where boys' growing disengagement from school has been given a catchy name, laddism, and has come under intense social and political scrutiny. One large British study found that teachers of six- and seven-year-olds do rate boys slightly lower than girls in English, math, and science when compared to their actual performance on standardized school achievement tests. The problem is being vigorously targeted by educational leaders, who have launched a national campaign to improve boys' early school performance through better teacher training and classroom initiatives aimed at making boys feel more comfortable and motivated in school.

A similar boys' movement is under way on this side of the Atlantic, where the debate about which sex is being shortchanged has become quite vitriolic. Many boy advocates believe that the pendulum has swung too far toward the girls' side—that in their effort to support girls, teachers have adopted a boys-are-toxic attitude. They argue that teachers today have little patience for boys' louder, less-controlled behavior and often cope by diagnosing boys—with learning disabilities, conduct disorders, or the ubiquitous ADHD—rather than by actually teaching them.

According to Judith Kleinfeld, a University of Alaska psychologist who has sharply criticized the early studies claiming teacher bias against girls, "The research on classroom interaction does not show consistent favoritism toward boys or girls." Still, there's little arguing with the data on school performance and the fact that girls are getting better grades than boys in most subjects, especially in early elementary school. Performance itself breeds expectations, and we do seem to have entered an era where boys are losing the expectations war—parents and teachers alike have begun exaggerating boys' immaturity and overlooking their considerable strengths at the onset of schooling.

Expectations come and go. The only way to fight such biases is to be ever vigilant and aware of them. Each of us—parent and teacher—must constantly challenge our assumptions when evaluating or encouraging any child. Gender should never be an excuse for a child's low performance in a given area, whether it is reading or math, writing or science. Expectations are important, so we must hold them consistently high for all children.

Creating Boy-Friendly Classrooms

Boy advocates and girl advocates will continue to argue about which gender is being shortchanged in the classroom. But there's no escaping the fact that boys today are not doing as well in school as girls, beginning at this early entrance stage. Thoughtful teachers appreciate that things need to change in order to help boys feel more at home in school and to tap into unique features of their learning style and motivation.

Many factors have converged to make school tougher for boys. One is a change in the role of girls. No doubt boys have suffered some fallout from the women's movement. While six-year-old girls have probably always had the language and attention skills to thrive in elementary school, they didn't always have a reason to work hard and be successful. Now that girls are expected to have careers when they grow up, parents and teachers take them more seriously, pushing girls to excel at ages when some boys can't keep up. Contrast this to the old days, when boys were the only ones with intellectual futures to worry about. Sure, girls needed to learn how to read and do arithmetic, but it was the boys' education that ultimately mattered to society.

The problem is all the greater because of the downward migration of the academic curriculum: the fact that kindergartners now learn what used to be taught in first grade, first-graders are expected to master much of what was once the second-grade curriculum, and so on. As schools continue to raise the academic bar and focus on preparing students for standardized reading and math tests (the benchmark of the No Child Left Behind Act), boys are increasingly the ones left behind in early grades because they don't, on average, have the self-control, fine motor abilities, and verbal skills needed to keep up in classrooms as they are currently structured.

Redshirting is an obvious, though not particularly appealing, solution to this problem. If you hold back the least mature boys at the outset, then you can create a more orderly classroom where the children who are truly ready to learn can thrive. But the problem resurfaces later. As we've seen, the data do not show any long-term academic advantages for children who have been redshirted; compared to age mates, they lose the benefit of one year of schooling and encounter many of the same behavioral problems of children who are held back in later grades.

Another seemingly simple way to make classrooms more boy friendly is to take the girls out of them. Single-sex schools are on the rise these days, particularly for middle- and high-school years, when the argument for their benefits is stronger. (See more on this topic in chapter 8.) However, single-sex education is not a good solution at this tender age. Boys and girls still interact frequently in early elementary school, in spite of their separation out on the playground. As studies of siblings have shown, boys and girls can learn a lot from each other: boys coax girls into greater physicality, which is good for both body and mind (particularly girls' spatial sense); and girls have a calming effect on boys, helping them focus and settle down to their quieter classroom tasks. Recall that highly arousable young boys are found to exhibit fewer behavioral problems the more time they spend playing with girls. Even in high school, when single-sex schooling is most fervently advocated, some schools have found that deliberately alternating boys and girls in seating can be advantageous to boys' attentiveness and work quality. So it appears that girls have a moderating effect on a group of boys, while boys have an energizing effect on girls—assuming, of course, that teachers deliberately encourage the two sexes to work together.

How then can we shape the environment to better promote boys' learning at this early stage? Several points emerge from considering boys' unique strengths and weaknesses, as well as from the tried-and-true experience of innovative teachers. (Other ideas, focusing on reading and writing skills, are presented in the next chapter.) The nice thing is, most of the boy-friendly tips are also good for girls, so everyone wins from creating a more developmentally appropriate classroom that fits the needs of *all* five- and six-year-olds.

• **More movement.** One idea everyone agrees on is that boys need more opportunity for movement in the classroom. It's hard for any six-year-old to sit still for long periods of time, but more so for boys than most girls. Giving children the chance to move from place to place in the classroom may help maintain their focus at each station. Why shouldn't kids be able to do their work while lying on a carpet, standing at a whiteboard, or sitting on stools at a high table? Many early childhood classrooms do incorporate such movement by organizing different activities in centers. The centers themselves should offer opportunities to learn beyond mere pencil and paper, such as counting or grouping with large blocks, or writing on clipboards or easels.

Recess and physical education are also key. Frequent physical breaks can recharge the mind and help kids "get the sillies out." Schools in Asia, especially Taiwan and Japan, incorporate twice as many recesses in the early grades as American schools do while managing to maintain high academic rigor. Around the world, kids are not getting as much exercise as they need: few children in industrialized nations walk or ride their bikes to school. All need more exercise during the day, and if it means extending the school day to add more outdoor recesses, longer PE classes, or simply in-class stretching breaks, then so be it. Such movement breaks will help boys focus and enjoy school more, and, like most of these tips, they are equally beneficial for girls, who are no less immune to the current obesity epidemic.

• **Less lecturing, more action.** When my son Sam was nine, we signed him up for a fencing class. He had fenced the year before and loved it; the teacher had been this gruff Eastern European guy who had basically made the kids run around for most of the hour. This next class, however, was taught by a new instructor, a young woman who was a champion fencer but who obviously knew little about children. Sam complained from the first lesson, and I could see why: instead of fencing and exercising, the fifteen boys and girls, all between the ages of seven and twelve, had to stand around in a big circle for twenty minutes listening to the instructor explain every nuance of the rules, positioning, and scoring. It was all I could do to get Sam to finish the session.

The fact is, children do not have the attention span of adults. Sure, a handful of boys can stand quietly and listen and absorb what a teacher like this is saying, but most will space out after three or four minutes, fidgeting and itching to try the real thing. Even the girls in Sam's fencing class lost it after eight or ten minutes. Kids need to get their *hands on* the fencing foils, fractions, forts, and other facts they are supposed to be learning about. They do not learn from lectures, like college students. They learn by doing. And boys, especially, tune out long monologues quickly.

Of course, children do need direct instruction to know how to perform an exercise and to correct their mistakes midstream before they fall into bad habits. Generally speaking, however, such interaction should come in small groups or one-on-one, situations in which a teacher can monitor individual children's focus and adjust the instruction to keep each child on task.

• **Training inhibitory control.** At some point, even if they're permitted to move around between lessons, kids do need to sit still and focus to get any learning done. The key to attention and self-control is inhibition, a frontal-lobe function that develops more gradually in boy than in girls. While the biological basis for this difference is not yet understood, we do know that inhibitory control can be improved through practice. Games like Simon Says; Duck, Duck, Goose; and Red Light, Green Light are fun at this age precisely because they challenge children's budding inhibitory systems. Teachers can use similar exercises requiring attention to oddball rhymes or math facts to foster children's focus and self-control. Struggling students can also be taught self-vocalization and self-monitoring techniques to improve their attention and classroom performances, hopefully before so many are diagnosed and medicated for ADHD. One curriculum, Tools of the Mind, shows this can be done by using dramatic play, private speech, partner work, and other strategies to promote inhibitory control, attention, memory, and self-regulation in five-year-olds.

• **More manipulatives.** "Hands-on" learning means children have to have something to put their hands on. Books are great, of course, but textbook publishers seem to have taken over classrooms with their fully prescribed programs for everything from phonics and counting to earth science and geography. Their content is fine, but the materials are all paper: posters, workbooks, handouts, flash cards, and punch-outs of everything from letters to tepees. Paper can only do so much. Children need to hold, weigh, move, pour, measure, and explore real objects with real mass, texture, and three-dimensional shape: plants, bugs, acorns, rocks, crystals, pennies, blocks, balls, dice, and dominoes, as well as specially designed educational manipulatives like base-ten blocks, fraction pies, and tangrams. With the right exercises, such objects and hands-on experiences will give children a quicker and deeper understanding of physical laws, mathematical relationships, and even other cultures and periods in history than they'd get from mere paper-and-pencil activities.

One curriculum that makes great use of hands-on learning is the Montessori method. Beginning in preschool, children learn reading, writing, and, particularly, math concepts through a progression of well-made wooden blocks, beaded rods, and other materials that coax a tactile understanding of counting, arithmetic, place value, fractions, and early algebra concepts. Similar manipulatives can be found in other math curricula, but their use is rarely formalized to the extent of a Montessori classroom's. While rigorous evidence is generally lacking to support Montessorians' claims about the merits of their curriculum, their materials do seem to me a wise approach for maximizing children's tendency to learn through hands-on experience.

Or, for another role model, consider Ms. Frizzle, the quirky but oh-so-dedicated teacher/driver of the Magic School Bus books. Okay, so most teachers can't shrink down their pupils and swim them through a classmate's bloodstream, but both parents and teachers can take a few tips from all the pets, specimens, collections, models, and other interesting materials in her classroom. What better way to learn about the earth, the solar system, the human body, electricity, waterworks, and so on than to simply jump in and *feel* them?

Hands-on learning is crucial for helping many boys to stay engaged in learning. It bypasses verbal challenges and merges visual and tactile senses, creating a gateway to new concepts. While girls may not need such materials as much to motivate and focus them, they can benefit equally from the spatial experience and perceptual-conceptual link that good manipulatives promote.

• **Tempt them with technology.** In one sense, computers are the ultimate manipulative. With software that can easily simulate the motion of the planets, the pouring of liquid between graduated containers, the cutting of pies into fractions, and the relationship between a circle's radius and its area, computers can vividly illustrate many important concepts and save a teacher from having to collect a bunch of messy, space-consuming materials. There's no doubt that high-quality educational software can and should be incorporated more into every aspect of the early childhood curriculum. Boys, in particular, are highly drawn to computers and may be coaxed into deeper intellectual engagement if they have the right reading, math, science, and social studies software. Girls, for their part, need to be more comfortable with technology, and an early start is essential if we are ever going to reduce the gaping technology gap I discuss further in chapter 6.

However, computers can never replace the real hands-on experience that a child gains from holding objects other than a computer mouse. Yes, children can simulate measuring liquids using software, but they cannot acquire the practical sense of what 100 milliliters looks and feels like if the volume is reduced to fit on a computer screen. Real liquids are wet! They have weight and depth and will spill if you're not careful. Computer simulations permit one kind of manipulation, but it's not the same three-dimensional experience as using a real beaker, ruler, balance, magnet, globe, protractor, or thermometer.

In sum, good educational software can be a plus for learning facts, concepts, and certain types of problem solving; it can help engage boys in learning and enhance girls' technology experience. It cannot, however, substitute for real experience in coaxing children's intuitive understanding of the world around them.

• **Practicing penmanship without penalty.** Boys' handwriting is rarely as neat or legible as girls', who even by kindergarten may have spent hundreds more hours drawing and printing. Writing is a fine motor task, so boys have that going against them. More of an issue, though, is their lack of experience with pencils and pens. Even boys who are expert LEGO builders or Game Boy players may have poor handwriting. Sad to say, these other fine hand movements don't translate into holding a pencil and forming characters any more than playing piano translates into knitting or calligraphic skill.

The only way to improve penmanship is by practicing it, which few boys find as pleasurable as girls do. Rewards may work here: points for completing rows of well-formed letters, which can be applied to some extra time at recess or stars on a penmanship chart. My own boys took a liking to number scrolls, pages of higher and higher numbers taped end to end that provided a tangible demonstration of their number printing, but they could have benefited from creating analogous letter scrolls.

While penmanship is important, it should not be the only way children express themselves in writing. In this era, when keyboards are responsible for the deterioration of many adults' handwriting, written composition need never depend on a masterful hand. Creative and expository writing should be freed from penmanship, so every child can see his or her ideas expressed in print. (More on this in the next chapter.)

• **Competition as a lure.** This one is tricky. Just as parents have largely banned competitive games such as musical chairs from five-year-olds' birthday parties, teachers have recently tended to take the competition out of classrooms. This is certainly a noble goal, an effort to boost self-esteem by not labeling individual children as losers. In the working world, groups of people are typically more successful if they cooperate with rather than compete against one another to get a job done. Why shouldn't children be taught to work together instead of against one another while mastering the skills necessary for later success?

It's an important point, and there is no question that students of all ages need to learn how to cooperate and work in teams. But classrooms are not corporations: every child needs to master lessons for him- or herself. Competition has many downsides, but it is also a natural motivator, and most children enjoy fun-spirited games and contests that allow them to stretch their minds and show off their rapidly growing knowledge and abilities. Particularly for boys, competition can be a way to engage some who may not otherwise be interested in spelling, reading, vocabulary, or memorizing their math facts.

Of course, the healthiest competition pits a student against himself, like all the computer games in which a child is striving to reach the next level of challenge. (Why in the world can't they make educational software equally addictive?) Many teachers ask children to keep track of the number of books read or the amount of time spent reading at home, tangible tables that students can feel proud of as they see their efforts add up. Low-stakes quizzes can also be a tool for self-challenge, such as mad-minute tests, in which children try to answer correctly as many arithmetic problems as possible in sixty seconds. Given repeatedly over several weeks and months, such challenges can inspire each child to attempt to beat his or her own personal-best scores.

Low-stakes competition can be very motivating, particularly for boys. It can also be good for girls, many of whom dislike or shy away from contests; it helps acclimate them to the essential fact of competition in life. By balancing competitive opportunities across domains—reading, spelling, math, science, art, music, and PE—and by organizing some contests into team efforts, teachers can create opportunities for every child to find at least one realm in which he or she is a winner.

• **Male teachers.** Last, but far from least, is a crucial way of making classrooms more boy friendly: hiring more men to teach them. In educational circles, *early childhood* refers to children from birth to age eight, or third grade. Currently, men make up just 9 percent of elementary-school teachers and less than 5 percent of teachers at the preschool level. This means that most young children are being educated in a strongly female environment, with few adult males for boys to bond with, relate to, learn from, and observe as models of mature male behavior. This is a special tragedy for the large number of children—both boys and girls—who are growing up without fathers at home.

Getting more men involved in teaching is an obvious way to

help boys feel more comfortable in school. Boys start school ready to learn but quickly discover that school is a female-dominated place, with teachers, classroom aides, librarians, secretaries, and principals who are nearly all women. Young boys may well love, obey, and want to please their female teachers, but they don't want to *be* them in the way that young girls do—coming home with their friends and taking turns emulating their teachers in an intense game of school.

Consider, by contrast, the experience of boys: except for the occasional gym teacher or custodian, many boys have only one another to look to as models of masculine behavior during the long days at school. But what if boys spent these hours under the tutelage of a man who is just as interested in literature as in sports, or in history as much as in cars? Wouldn't this example promote boys' intellectual development and desire to perform well in school?

Male teachers are more likely than females to share the interests of the boys in their classes; they can better judge the books boys will like; and they understand how to motivate them. Male teachers often have an easier time disciplining boys, although this expectation can become a burden on the token male teachers in a school who may get stuck with more of the problem children or be called upon to resolve discipline issues in colleagues' classrooms.

Nor are boys the only students to benefit from more male teachers at the elementary level. Girls also can learn a lot from the bright, caring men who are likely to go into early education, broadening their view of masculinity and possibly giving an extra boost in the spatial skills and scientific concepts that some female teachers are less attentive to.

Male teachers can provide all of these advantages, and yet their proportion has actually *declined* over the past twenty years. After increasing to 18 percent between 1961 and 1981, the proportion of men teaching at the elementary level in U.S. schools has slid to half that level.

According to Bryan Nelson, founder of a nonprofit organization called MenTeach, there are three reasons why men don't go into teaching: status, salary, and accusations of child abuse. Elementary-school teaching is not a high-paying job in the United States, a country where status and salary go hand in hand. The proportion of male teachers is much greater in countries where education is held in higher esteem. In Greece, for instance, men make up an impressive 51 percent of teachers at the primary level (grades one through six). This contrasts with the rest of Europe and North America, where males are a clear minority in early education.

Still, teaching is not the grossly underpaid profession it once was in the United States. Statistics published in 2001 by the National Education Association (NEA) show that teachers today are paid about 39 percent more than they were forty years ago, even after adjusting for the increased cost of living, with most of the gains having taken place since 1981. In other words, men started *dropping out* of the teaching profession at precisely the same time that teacher salaries made their greatest gains.

How can this be? Economic logic says that when pay increases, the demand to fill a job similarly increases. This has clearly been the case for nursing, another traditionally female profession, in which the proportion of men has steadily risen from a mere 1 percent in 1966 to about 6 percent today, with men making up 13 percent of current nursing students. Nursing, far more than teaching, was once an exclusively female domain, and yet the status issue has not prevented a rising number of men from joining its ranks as long as the salary is good and the work fulfilling.

If it's not money or status, this leaves the fear of abuse allegation as a major factor keeping men out of early education. It's probably no coincidence that the proportion of male teachers declined over the same period that the national media became fixated on several abuse scandals involving priests and child-care workers. As one early childhood teacher, Don Piburn, posted in an article called "Just Say NO, to No Touch" on the MenTeach.org website:

> Male early childhood (EC) educators face stereotypes that portray us as potential child abusers on a daily basis. Clearly, children must be protected from child abusers, yet a principal reason many men give for not entering or staying in the field of early childhood education is fear of being accused of harming young children. . . . Only at their own peril do many male EC educators offer "the intimate pats, back rubs, caresses and leisurely holding on laps and in arms that little ones need. . . ." Is it any wonder that men pass over our profession in favor of something less perilous, like working at high altitudes, driving at high speeds, or operating dangerous machinery?

Piburn continues in this article by quoting the NEA president who stated on Oprah Winfrey's show that "our slogan is, Teacher, don't touch."

As a scientist aware of the many benefits of touch for children's overall health and neurological development, I find this message deeply disturbing. Dozens of studies, in both humans and animals, have demonstrated the importance of physical contact and social touch for healthy brain and immune system development. Children crave the same hugs, pats, and handholding that our society has turned into a sin for men in any role other than father or grandfather. How can any adult effectively teach young children without the healthy, comforting bond that comes from kind words and a caring touch?

I'm especially disturbed about this policy because I've seen it close up. Julia was in second grade when she had her first male teacher, whom she still refers to as her "best teacher." At the time, I was just beginning to do some research for this book, so at one of our parent-teacher conferences, I asked Mr. E if he faced any special challenges on account of being (at that time) the only male classroom teacher in this kindergarten-to-second-grade school.

He answered, sounding as confident and noncontroversial as any teacher at such conferences, that he really enjoyed his job and that he found it more satisfying than the middle-school teaching he had recently switched from. The overall impression was that all was fine. But then he added, suddenly hesitant and not quite sure how to put it, "The only problem is that . . . well . . . the kids get pretty attached by the end of the school year and . . . well . . . I can't exactly return all of their affection."

"That's terrible," I responded, looking down to hide my sudden surge of tears. How excruciating to think that this wonderful man—smart, kind, funny, and intensely dedicated to his seven- and eight-year-old pupils—felt he couldn't even hug the children for whom he had come to care so much!

How many more such men are not even considering teaching because of presumptions like this? Boys and girls are *both* short-changed by the current culture that essentially excludes half the population from the noble and satisfying job of teaching young children.

Starting School Right

In some ways, the differences between boys and girls are more profound at the beginning of school than at any later age. Each sex has its strengths and weaknesses, but when it comes to academics, boys are the more vulnerable ones at this stage. Compared to their female counterparts, five- and six-year-old boys are less verbal, less skilled at drawing and writing, and less able to sit still and control their voices in an elementary-school classroom. Of course, there are many exceptions to this rule—boys with remarkable powers of concentration and girls who fidget and squirm throughout the day—but the overall truth is that boys tend to lag behind girls in school readiness.

The good news is that all of these skills can be learned. Indeed, education itself is the great equalizer: given the right training and a supportive environment, any healthy child can learn to write, read, pay attention, and more. In an ideal world, sex differences in school readiness would be tackled at an earlier age, particularly in preschool, when they first emerge. But the sad truth in the United States is that we are a long way from universal uniformly excellent preschool. So it is up to elementary-school teachers to make up for lost time—to help boys master the self-control, penmanship, and language skills necessary for success while simultaneously teaching a whole roomful of children how to read, print, count, and measure circle diameters.

It's not easy, but the goal is clear: if kindergarten is for five-year-olds, it must be suitable for all five-year-olds, male and female. Schools must resist the pressure to make kindergarten even more academic than it currently is and should instead focus on developing the behaviors and attitudes that will prepare every child for a lifetime of learning. Boys may, on average, mature more slowly in a few crucial areas, but with the right environment and wise, boy-loving teachers, they do not have to suffer because of it.

5

The Wonder of Words

ADAM IS A SMALL BLOND FIRST-GRADER, shy around adults but with a fondness for contact sports and a devilish smile that flashes on when he's with his playground buddies. Inside the classroom, things are a bit different. He struggles to concentrate during writing workshop and spends most of the twenty-minute free-reading period joking with a friend instead of trying to read the overly difficult Magic Tree House book he selected. Oh, Adam can read all right—simple books with just a few words per page and ample clue-filled pictures. But he doesn't yet enjoy reading by himself and rarely seeks out books for pleasure.

Adam is not unlike many boys—and girls, for that matter. Reading is hard, hard work at the beginning. Our brains evolved for spoken language, but reading and writing are brand-new inventions (as far as evolution is concerned) and not skills most children pick up instinctively. Of course, there is the occasional child who—immersed in books from an early age and with the proper calm, concentrated temperament—manages to figure it out for him- or herself even before entering school. For most kids, though, reading is the first academic skill that requires real effort—systematic, gradual, and repeated practice at sounding out and recognizing small familiar words.

Reading does, however, build on verbal communication, and this is where girls' early advantages begin paying off. Recall that even as infants, girls begin pointing and saying their first words a few weeks earlier than boys (chapter 2). Girls retain this lead throughout the preschool years, speaking in longer and more complex sentences than boys. This difference surges to the fore when children get to school, and their earlier verbal practice sets the stage for reading, writing, and all the other learning that depends on solid literacy.

Still, while the average six-year-old girl is a few months ahead of the average six-year-old boy in her skills, the difference is not nearly as large as many people currently believe. When Michael Gurian warns parents that because of their "biological tendencies," boys are as much as a year and a half behind girls in reading and writing, not only is he focusing on the most extreme measures, he's also setting up enormously different expectations for their performance.

This is where the flood of books and articles proclaiming a newfound "boy crisis" seems to be backfiring. Boys are not collectively illiterate, and people do them a great disservice by exaggerating their verbal deficiencies. I'm always struck when I hear parents (and not a few teachers) write off a child's development with comments like "Jack doesn't read nearly as well as Melissa did in first grade, but of course, he's a boy." Boys may indeed score lower than girls on reading, speaking, and writing tests, but they're not deficient in all verbal skills: their vocabularies, for instance, are as large as girls'. What's more, most language and literacy differences are small and *largely fixable* if parents and teachers focus on these abilities in an early and consistent manner. Average sex differences simply can't be an excuse to let boys slip through the cracks, growing past these critical years without the intensive literacy immersion that we know is essential to making any child—male or female—a proficient reader and writer.

In this chapter, I'll explain what's known about the language and literacy differences between boys and girls, men and women. Although the issues follow chronologically from previous chapters, my approach shifts gears at this point. Instead of proceeding by age alone, in this and the following two chapters I focus on specific areas of brain function, tracking how they evolve in both boys and girls from infancy to adulthood. While the gaps in reading, math (chapter 6), and "emotional intelligence" (chapter 7) may appear more prominent in later childhood and adolescence, all have their roots in children's earliest development, so it is essential to follow each domain from inception to maturity. Only in this way can we harness children's enormous plasticity to help overcome them.

The Numbers

So what *are* the differences between male and female language skills? While pop psychology has painted a picture of an enormous communication gulf between men and women—chatty Venus versus stoic Mars—the real data are not nearly so dramatic.

First off, let's dispense with the myth that "a woman uses about 20,000 words per day, while a man uses about 7,000." The numbers vary from telling to telling, but this urban legend was debunked in 2006 by Mark Liberman, a professor of phonetics at the University of Pennsylvania. In 2007, it was further disproven by researchers who recorded and meticulously counted the number of words college-age men and women spoke in a day. The real numbers? An average of 16,215 for women and 15,669 for men, a statistically insignificant difference.

Nonetheless, there are some real differences between male and female communication. While women don't necessarily talk *more* than men, they do speak faster. The most reliable sex difference is the verbal fluency I introduced in the last chapter. In a given minute, the average female can spit out more words with more accurate pronunciation than the average male can. However, this difference is on the small size, with a d value of 0.33, meaning that only two-thirds of women are more verbally fluent than the average man. Women are often denigrated for talking too much, but the reality is that in mixed company, men outtalk women. Among fifty-six different studies of mixed-sex conversations, women were found to talk more than men in just two studies while men dominated in twenty-four others. (There was no sex difference in the remaining thirty studies.) So conversational dominance is determined more by a speaker's power or status than by speed of articulation.

Sex differences are even more modest for other adult verbal skills. Except for spelling ability, where women have a medium-size advantage (d = 0.45), most differences in adult communication skills fall into the statistically small range. Women tend to use speech in more affiliative ways—to bond and make others feel supported—while men are slightly more prone to use speech in an assertive manner. Women are also slightly better at writing. But importantly, there are no differences between men and women in reading ability or vocabulary size, and when you put all verbal skills together, women's advantage turns out to be a mere one-tenth of a standard deviation (meaning the average woman outperforms 54 percent of men).

The gap in verbal skills is not much different in children, as we've seen in earlier chapters. But this small verbal advantage combined with girls' other valuable habits—like inhibitory control and fine motor skill—add up to a significant advantage when children get to school and begin learning to read, write, and spell.

Most attention has focused on the reading gap. According to the 2007 Nation's Report Card, otherwise known as the National Assessment of Educational Progress (NAEP), fourth-grade girls score about seven points higher than boys, which translates to some 20 percent more girls than boys who pass the proficient mark. By eighth grade, the gap grows to ten points, or about 38 percent more girls than boys who are proficient readers. It is larger still at twelfth grade, a thirteen-point spread, according to 2005 data* that amounts to 47 percent more girls than boys who finish high school as proficient readers.

So the literacy gap is substantial, especially among older students. Boys start out a little behind, but their disadvantage grows as they progress through school, which tells us that they are not getting the most out of their education in two of the three R's.

No question, the reading gap is disturbing and something that parents and teachers must pay greater heed to. But it is not—as most people assume—a new crisis. Boys have long lagged behind girls in reading, as some thirty-five years of NAEP data reveals (see Figure 5.1).

The reading gap is not unique to the United States. According to the results of an international standardized test known as PISA (Program for International Student Assessment), fifteen-year-old girls exhibit higher literacy rates than boys do in *all* of the forty-two industrialized countries sampled.† The difference, once again, is substantial, amounting to about half a level on a four-level proficiency scale. It is larger in some nations (Iceland, Norway, Finland, Thailand) than in others (Korea, Mexico, Japan), with the United States falling about midway in the size of its gender gap. PISA results from 2003 show a slight narrowing of the international gender gap in reading compared with the year 2000, but still, it's clear that around the globe, girls are better readers than boys.

---

* Twelfth-grade data are not available for more recent years.

† While females win in the industrialized world, it's crucial to point out that they are not universally advantaged in reading. Among the world's one billion illiterate adults, 70 percent are women. The cause is simple: fewer girls than boys go to school in many parts of Africa, the Middle East, South Asia, and Central America, where the lack of education is also associated with greater poverty and higher pregnancy rates.

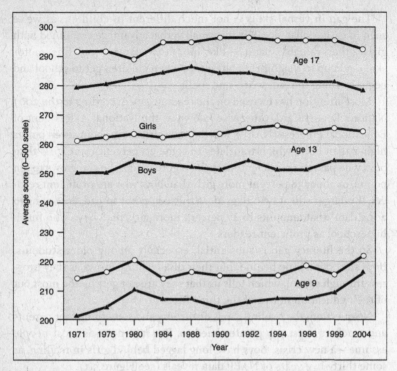

Figure 5.1. Girls outscore boys on the reading portion of the NAEP at all ages and in all years. The gap is smaller among younger children but grows as they progress through middle and high school. Data are plotted only through 2004 because they are not available for seventeen-year-olds in later years. (Source: National Center for Education Statistics, U.S. Department of Education.)

International comparisons can be useful when searching for the reasons behind educational achievement, and one factor that surprisingly does *not* correlate with the gender gap is single-sex education. According to the PISA data, boys do no better in Ireland, where 44 percent of children attend single-sex schools, than they do in Denmark, where no children are educated in single-sex schools. If anything predicts the magnitude of the reading gap, it is—very simply—the amount of reading children do outside of school. Not surprisingly, kids who read for pleasure more than others score higher on literacy tests. Worldwide, girls spend more time reading than boys, and the larger this gap in individual countries, the larger the gender gap on reading-test scores in high school.

So the reading gap between boys and girls is significant, even if it is nothing new. By all accounts, girls have always outperformed boys in measures of reading, writing, spelling, capitalization, punctuation, language usage, and using reference materials. But this doesn't mean the gaps are excusable or immutable. As we'll see in the next chapter, girls have managed to close certain long-standing gaps in math and science, and there's no reason to believe that boys can't do the same for reading and writing. There's even some encouraging evidence, as the NAEP data for nine-year-olds reveal, that the gap is beginning to narrow at the lower elementary level. But there obviously remains much room for improvement, especially for boys in middle and high school.

Reading is crucial for all learning, so this gap likely contributes to the lower grades that boys earn throughout high school and college in all subjects, even math and science. The difference amounts to about two-tenths of a point on a 4.0 scale and is most pronounced in more verbally oriented classes, such as literature, English composition, and Spanish. In college, female foreign-language majors outnumber male nearly three to one.

What's unclear is whether the language and literacy gap persists into adulthood. Current data do not reveal a meaningful difference between men's and women's verbal or reading skills, and there are two possible reasons for the discrepancy between children and adults. One is maturation: men may finally catch up to women by their early to middle twenties. Another possibility, however, is that females have indeed surpassed males in language and literacy skills but the trend is so new that it hasn't yet shown up in studies of adults; most of the data we have now were obtained from adults of an earlier generation, when women were not better educated than men.

Verbal SAT Exam

Here's another monkey wrench in our attempts to understand sex differences in verbal skills: in spite of their lower grades and test scores on focused reading and writing exams, boys actually score higher than girls on the verbal portion of the SAT college admission test. Once again, the difference is small—about six out of the famed eight hundred maximum points over the past decade (see Figure 5.2). But it is very consistent. Boys have outscored girls on the verbal SAT every year for nearly four decades. (They have also uniformly outscored women on the math portion of the SAT—as I discuss in the next chapter—only in that case, the difference is some forty points.)

Why don't females do better on the SAT, especially in verbal areas, where they decisively outperform males in the classroom? It is not, as some critics have charged, because the test questions are biased. The Educational Testing Service (ETS), which writes and administers the SAT, has been sensitive to this issue for many years and tasks a carefully chosen panel of high-school and college faculty with screening out language, symbols, and examples (such as sports or car repair) that might be more easily understood by one gender. The ETS also deliberately culls individual questions that produce a substantial sex or racial difference after thousands of students have taken the exam. Nonetheless, there does seem to be an issue of the *type* of questions on the exam. The verbal SAT obviously cannot test verbal fluency, a female strength, but it does strongly emphasize vocabulary, a skill that shows no difference between the sexes. And until 2005, when the exam was significantly revised, it included a large number of verbal analogies (such as "Fox is to den as bird is to _____?"), which is a male strength.

Further speculation about females' disadvantage on the SAT and other high-stakes exams has to do with test-taking strategy, which I discuss more in chapter 7. However, the ETS has another explanation for the gender gap in SAT scores based on demographic differences between the boys and girls who take the exam. Every year over the past three decades, the proportion of females taking the SAT (and going to college) has been rising, but much of this growth has been among lower-income girls. As a result, girls who take the SAT are, on average, less advantaged than the boys who take it. In 2005, for example, girls represented 50 percent of all test-takers whose parents earned between $80,000 and $100,000 but a striking 64 percent of test-takers from the poorest families (those earning less than $10,000). Socioeconomic status is one of the best predictors of SAT score, so when the ETS factored in this difference in family income, the gender gap in the verbal SAT actually reversed to a two-point female advantage, while the math SAT gap was reduced by a whopping twenty-five points.

Even so, many people object to the SAT producing any sex difference,

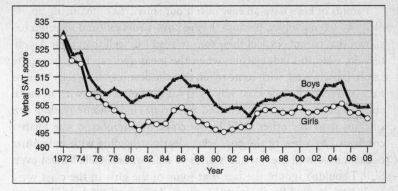

Figure 5.2. Verbal SAT scores by year and gender. (Source: College Board 2008.)

particularly in light of the higher grades females earn in most high-school classes. And in situations where SAT scores are used as a cutoff for university admission or monetary scholarships, the gender gap does look discriminatory. The College Board made some dramatic changes to the SAT exam in 2005, including the addition of a major writing component (which favors women by about eleven points), but even on this new exam, males continue to hold a large advantage on the math portion and a small advantage on the verbal portion.

Learning to Read

One of the better parent-volunteer duties at my kids' school is guided reading, or listening to children as they read aloud. First- and second-graders need a lot of help at this stage, when they're just starting to recognize and sound out familiar words. Teachers seem glad for the extra pair of ears and eyes to follow along as children read, promptly correct their mistakes, and help ensure that children actually understand all the words they are painstakingly decoding.

It's a wonderful stage, and as parents, we're eager to help—not just our children but all the other precious six- and seven-year-olds who are puzzling earnestly over unfamiliar words and the daunting stream of text on each page. The dark side, of course, is the temptation to compare your own child to the others: Is Toby reading as well as the other first-graders? How are his friends doing? Who is the best reader in the class?

Parents of boys have an easy out: if our sons aren't measuring up, it's okay, we tell ourselves, because they're boys. Cross-cultural research has shown that parents expect their sons to read less well than their daughters do, even in kindergarten, when the measured gap is in fact quite small. I discovered, with some relief, that the best reader in Toby's mixed first- and second-grade class was a girl—a tomboy at that, with so much ability and drive that she'd received three times more home-reading certificates than any other child in the class. My stereotype was further confirmed by a few of the boys who, clearly struggling with unfamiliar books, chose well-memorized stories to read when I called them over. Still, I couldn't ignore the fact that some of the girls in the class were having the same difficulties, and I was glad to see that Toby and several of his male buddies were holding their own quite well as readers.

How do children learn how to read? For a fortunate few, the process is effortless: immersed in books and surrounded by adults (or older siblings) to read to them, they intuitively learn the associations between letters and sounds while also memorizing enough sight words to bootstrap themselves into independent reading.

For most children, however, the process takes hard work. Reading has been around for a few dozen centuries at most—far too brief a time for evolution to have shaped the neural structures that carry it out. Instead, the human brain must piggyback on the circuits for spoken language to painstakingly learn the association between some arbitrary visual squiggles—letters (for alphabetic languages) or pictures (for languages like Chinese)—and the rapid-fire blending of sounds in spoken words. This retrofitting explains why dyslexia, a specific difficulty in learning how to read, is far more common than speech delay. With rare exception, every normally developing child learns how to speak—early, automatically, and with no formal instruction. By the time children are six years old, they understand an impressive thirteen thousand words. Their reading vocabulary, by contrast, grows much more slowly and with great effort, especially in this early stage.

Educators have argued long and loudly about whether reading is best acquired through a top-down (whole-language) or bottom-up (phonics) approach. This so-called reading war has finally found its armistice in the realization that both strategies are necessary. Children must have the motivation to learn to read, which comes from a love of stories, thirst for knowledge, or any other passion that drives them to the printed word for fun and fulfillment. At the same time, decoding those printed words requires several crucial technical skills, the equivalent of learning to balance, steer, and pedal a bike simultaneously (which many children master, perhaps not coincidentally, at around the same age).

At the core of the bottom-up skill set is phonological processing. Phonemes are the individual sounds of human speech. The number in each language varies, but English is fairly typical with its forty different phonemes that children unconsciously learn to distinguish well before the sixth birthday. However, the hurdle in learning how to read is making this awareness conscious. In a sense, children have to step outside themselves, dissect each tiny sound in a word like *B-A-T* to learn, all over again, the name of a flying mammal.

There are many pieces of this learning-to-read puzzle, all of them now known to correlate with a child's transition to independent reading. One is phonological sensitivity, the ability to hear and separate the individual sounds within a word. Children who can identify the first and last sounds of words; who can recognize and create rhymes; and who can pick apart each vowel, consonant, and consonant blend within a word stand a much greater chance of figuring out the relationships of these sounds to printed letters and progressing easily to literacy. The awareness of beginning sounds is best honed through alphabet books, where each page illustrates familiar nouns and adjectives all beginning with the same letter. Another trick is alliteration, where parent and child together try to make up sentences in which every word begins with the same sound. ("My mother makes me mind my manners.") Middle and ending sounds are together emphasized in rhyming songs and stories. It's no wonder that Mother Goose and Dr. Seuss remain perennially popular; rhyming verse is universally appealing, memorable, and fun to pronounce. Young children should be encouraged to sing, recite, and play with rhymes both at home and in school.

Two other skills that figure importantly into learning to read are phonological memory and phonological naming. Phonological memory measures a child's short-term recall of sounds and words, which is crucial for keeping in mind earlier parts of a word or sentence while trying to decode the remaining bits. Phonological naming is a long-term memory skill—the ability to rapidly retrieve words to express an idea, name a picture, answer a question, and so on. As adults, many of us experience word-finding difficulties, or tip-of-the-tongue syndrome: the inability, which begins as depressingly early as our forties, to come up with words we know that we know. Young children have the same difficulty, only in their case, it's not a retrieval issue so much as the still-flimsy storage of words they've only recently learned.

It's a lot of skills to pull together—many more than when learning how to ride a bike—so it's not surprising that a lot of kids have trouble learning how to read, with some 5 percent falling below the cutoff clinicians call dyslexic. Boys, as we'll see, are about twice as likely as girls to be diagnosed with dyslexia and to have trouble learning how to read. One reason is the close link between spoken and written language, all of these phonological skills that begin with early speech and hearing.

Girls, as we've seen, are ahead of boys in their verbal skills at the time of school entry. According to one large study of children four to five years old, girls score about 12 percent higher than boys on tests of phonological awareness. Their early language advantage, honed now through several years more of verbally intensive play with their parents, friends, dolls, and siblings, sets up girls for an easier transition to reading and writing: the more words you use in your own speech and the better you understand rhyming, letter sounds, and grammar, the easier you'll be able to make sense of the scramble of ABCs that constitute a written page of text. Add to this girls' fine motor skill—all that drawing and coloring is finally paying off—and boys have a substantial gap to bridge in learning to read and write.

Like most of these gender gaps, the chasm grows wider as skills take root and influence the way children spend their time. By elementary school, girls read for pleasure more than boys, though it should be noted that neither sex spends anywhere near as much time reading as watching TV. Reading fits many girls' temperaments better than boys': it is a quieter, less active pastime, not the kind of thing that appeals to high-energy young boys. Even among gifted children, boys do not have as positive an attitude toward reading as girls do.

This is a problem, since reading is just about the fastest, surest, and most proven way to get smarter: it builds vocabulary, increases one's knowledge base, and is crucial for developing one's own skills as a writer. Of course, there are other cognitive domains in which boys excel, but given a Saturday-morning choice between reading a book or playing Need for Speed on the computer, it's a no-brainer. (Pun intended.) Why opt for the slow, laborious black-and-white string of words when you can jump straight to the colorful, interactive action with a few clicks and an extra splash of adrenaline?

This, I admit, is what we used to face in our house: Julia reading stacks of books while the boys fought over who got to use the keyboard or Xbox controller. We do limit TV and computer time, however, and I'm happy to say that both boys have now become avid readers (at least when those other options are turned off).

Reading is strongly influenced by a child's environment, and there are many ways to increase boys' reading time and spark their early interest in the printed word. (See the tips at the end of this chapter.)

The Writing Gap

It's time for writing workshop in my son's first-grade class. Twenty-three heads are bent over their papers, most children concentrating hard on printing the next words in their budding stories. I circulate to help with spelling and capitalization as the teacher meets one-on-one with students to guide their revisions.

The first things I notice, as always, are the segregated tables. Given a choice (and they are, for this activity), the boys and girls invariably separate themselves. I plant myself first at Toby's table (as is the custom for parent volunteers). He is adding some detail to the fantastic thirty-minute voyage through the solar system he dreamed up for our family in this end-of-the-year fiction piece. I decipher his handwriting as best I can, proud that he's gotten over the bout of writer's block he had earlier in the year and trying to focus on the story's plot rather than on the multitude of creatively spelled words. Nearby sits Austin, struggling hard to add another word to his terse tale about mustard and "hot sos [sauce]."

I eventually make my way over to a girls' table, where Charlotte is reading her newly completed story to her classmate Hannah. This is an amazing work! Eight pages of an intricate interpersonal tale about the narrator and her two friends, Heidi and Catherine, complete with dialogue and quotation marks and with nary a misspelled word. (Okay, so at one point she wrote *to* instead of *too*, but who's keeping track?)

I move on to Meghan, Olivia, Sandeep, and William. They span quite a range at this age. Not every girl is a budding J. K. Rowling, and some boys, like Alex, surprise me with the depth (and legibility!) of their stories. Still, a pattern emerges: many girls seem to relish this creative opportunity while many of the boys seem to be struggling—wiggling, fiddling, and just hanging in there until it's time for gym class.

Writing is the one academic skill that most separates girls and boys. According to NAEP data from 2002 and 2007 (the most recent available), girls outscore boys by some twenty points in all grades tested (fourth, eighth, and twelfth). This translates to nearly twice as many girls who write at or above the proficient level throughout their schooling. In fact, writing scores differ more than scores in reading, math, and even science, with a d value of around -0.6, meaning nearly three-quarters of girls write better than the average boy. It's still not a statistically large difference (unlike, say, preschoolers' toy preference), but it is nonetheless "alarming," to use the word of University of Chicago researcher Larry Hedges, who has done some of the most thorough analyses of sex differences in cognitive abilities.

Writing emerges from reading, so it is not too surprising that boys' literacy gap hits them here. But writing demands even more: planning, organization, and penmanship—none of them a male strength at this age. Of these, it is probably the fine motor hurdle that makes writing most challenging to young boys. Planning and organization are skills of the frontal lobe, which may mature earlier in girls (see chapter 4), but it is the gaping lag between idea and printing that seems to derail boys the most in early elementary school.

A simple solution is to divorce composition from penmanship. Instead of making boys compose and print all at once, why not let them dictate their stories, personal narratives, or observations to an adult scribe, who can type up a draft for the child to edit and illustrate? Of course, boys still need to work on their penmanship, daily if possible, but separating letter formation from written expression may be especially helpful to boys at this stage.

Writing is an incredibly important art. (I think I'd say this even if I weren't writing this book.) As one writer and educator comments:

> Wordiness and confusing grammar are crippling business communication. Managers disregard the problem, and then wonder why they must re-explain themselves to employees and customers. Police exams now have a language and writing component. Chiefs are tired of losing cases when defense attorneys spotlight ambiguous police reports.

Any task that increases the number of words children put to paper and the amount of critical feedback they receive is helpful. Of the three R's, this one is arguably most urgent to fix.

Sex Differences and the Linguistic Brain

Scientists have said that boys are born with smaller language centers in their brains—and larger spatial centers—than girls and that boys develop language abilities at a slower rate, though they eventually catch up.

—Valerie Strauss, *Washington Post*

Why are boys around the world less adept than girls at reading, writing, and verbal skills? Are girls—as this article suggests—automatically better at language because of the way their brains are organized? Is there something about the male makeup that's constitutionally less adept at hearing, seeing, and producing words?

As I neurobiologist, I had high hopes for understanding sex differences by studying the brain. Unfortunately, the data just do not add up to anything like the headlines that regularly crop up in the *Washington Post, Newsweek,* and various parenting magazines. Yes, some scientists have noted differences in the way adult men and women process speech and writing. However, the data are far from conclusive and, more important, do not tell us anything about the source of these differences. Are they innate or learned? Particularly regarding children, scientists have uncovered scant evidence for the kind of inborn language bias that many teachers and parents have been led to accept as gospel truth.

Here's the data: Language is processed primarily by the left cerebral hemisphere. This left bias—or laterality—occurs in virtually all right-handed people and even in a large majority of left-handers. It is also present in both sexes, though there is some evidence that it is less pronounced among women. An initial 1995 study published in the high-profile journal *Nature* made the biggest splash in support of this theory. The study was carried out by the husband-wife team of Bennett and Sally Shaywitz and their colleagues at Yale; it used functional MRI to measure brain activity in nineteen men and nineteen women as they performed three different language tasks.

The Shaywitzes found that during one of the tasks—identifying rhymes—men exhibited strong activation of the lower portion of the left frontal lobe, while women tended to activate the same frontal area but on *both* sides of the brain. Of the nineteen women, eleven exhibited this bilateral pattern and eight activated just the left hemisphere (like men).. So the results of this study seemed to indicate that in processing language, or at least during this particular rhyming task, women were more likely to use both hemispheres while men used exclusively the left hemisphere. As one of the first reports to find a sex difference by using functional MRI, this study got a lot of press. An article in the *New York Times* Science section promptly declared: "Men and Women Use Brain Differently, Study Discovers," and the findings continue to be highlighted even in recent popular works.

The Shaywitzes' finding was appealing because it jibed with some earlier lower-tech evidence that had suggested that women process language more bilaterally than men do. Studies from the 1970s and early 1980s had indicated that men were more vulnerable than women to aphasia, or loss of language, after suffering a stroke in the left hemisphere, perhaps because women could compensate better by using the right hemisphere. Other evidence that women perceived language more bilaterally than men came from ear preference, or dichotic listening, studies.* So there was a consensus around the time of the Shaywitzes' study that females' verbal advantage was due to a greater tendency to process language using both hemispheres rather than just one, as males did.

It's such a beautiful theory! I even lectured on it to medical students when I first started teaching. Too bad, then, that little of this evidence has really held up. In spite of its lingering life in the popular press, the idea that females process language more bilaterally than males is no longer taken on faith and has even been edited out of the most recent versions of the textbooks I use in my courses.

Like any good research, the Shaywitzes' study inspired many attempts at replication. By 2008, twenty-six comparable brain-imaging studies were available for Iris Sommer and her colleagues at the University of Utrecht in the Netherlands to synthesize using meta-analysis.

Their overall conclusion: there's no sex difference in language processing. While some studies reported results similar to the Shaywitzes', others did not. Some even found that *women* processed language more

---

* In this method, a subject hears different words played into each ear simultaneously to discern which brain hemisphere is better at detecting speech.

strongly on the left side. When you put all the findings together, it's a wash: there is no significant difference in the way men's and women's right and left hemispheres are activated by language. Nor, upon closer analysis, do recent stroke and dichotic listening studies strongly support the idea that women process language more bilaterally than men.

The issue is far from resolved. Researchers are continuing their quest for sex differences in lateralization. But it seems, for now, that any differences that are found in the way men's and women's two hemispheres process language will be subtle and pertain to very specific language tasks only.

But if lateralization doesn't explain the difference between men's and women's verbal abilities, what does? Some researchers believe the answer lies at a more microscopic level, in the structure of individual neurons, which can be studied only from tissue collected at autopsy. This painstaking approach isn't for everyone, but a couple of research teams have published some provocative findings. One group is Sandra Witelson's lab at McMaster University in Ontario. In 1995, Dr. Witelson and colleagues reported that when compared to men's brains, women's brains contain 11 percent greater density of neurons in a crucial language area, the planum temporale (located at the top and back end of the temporal lobe). It's an intriguing result and suggests that despite females' smaller brains, more neurons are packed in where it counts. However, as the authors themselves emphasize, their findings are based on just nine brains (four men's and five women's) and "must be verified by other studies before they can be generally accepted." Thus far, they have not been.

A slightly earlier study by Arnold Scheibel and colleagues at UCLA found another sex difference in neurons from the same brain area. Scheibel's group was studying dendrites—the fine branches of neurons that receive synaptic input from other neurons. Women, according to this study, have slightly longer dendrites in this language area than men, suggesting greater connectivity. But interestingly, this study also found that *education* was associated with greater dendritic length: university graduates exhibited longer dendrites than did high-school graduates, whose dendrites were in turn longer than those of high-school dropouts. And guess what: the women in this study were considerably more educated than the men. So who's to know if their longer dendrites were due to their sex or their education?

This raises the most crucial conundrum about any sex difference in the adult brain: what does it really tell us anyway? If there's a behavioral difference between adult men and women, there *has* to be biological difference. We don't read, write, and talk out of thin air. Presumably, scientists will one day figure out which small differences in brain structure, function, or microscopic organization account for females' small advantage in verbal skills. But once we get there, what will it mean? That men and women think differently because our brains work differently? No surprise there.

The problem is that such data cannot tell us the *cause* of our differences. Perhaps women are born with denser neurons in the planum temporale area and therefore end up talking more. But it's just as likely that their lifetime of talking is what makes their neurons denser. Then again, maybe longer, denser neurons are somehow due to experience with nail polish, or child rearing, or grocery shopping and have nothing to do with verbal skills, making women's dendrite length a mere correlation, and not in any way causally related to verbal fluency, rhyme judgments, reading, writing, and so on.

Studies of adults can never resolve such nature/nurture questions. Only babies can tell us for sure.

Language in a Child's Brain

What about boys and girls? Is there anything in their brains that leads us closer to understanding their different verbal, reading, and writing abilities? Are the language areas of the brain really, as the *Washington Post* tells us, larger in girls?

I wish I could figure out where this one comes from. As we've seen, boys are the ones with the larger brains: an extra 9 percent of tissue, which appears to be distributed pretty evenly throughout the brain. In fact, studies of the planum temporale—the area responsible for understanding, or receptive, language—have found no difference in size or symmetry between boys and girls, from newborns all the way through adolescents.

Nor have studies of brain function revealed any clear-cut sex differences in children's language processing. As we saw in chapter 2, the handful of studies comparing newborn boys' and girls' language processing is somewhat conflicted, while a large fMRI study of older children (ages five to eighteen) found no lateralization difference between boys and girls. This and another recent study did find slightly more mature brain activation in girls while they performed language tasks, but this is hardly surprising, considering girls' faster physical maturation and stronger reading and writing skills over this age range.

So once again, there is no obvious neurological basis for girls' stronger verbal skills, even though the skill gap itself is reliable. Most likely, boys start out with slightly less mature circuits for processing words, and language experience widens this gap as boys and girls start paying attention to different features of their environment. In fact, research by Debra Mills and her colleagues at Emory University has shown that verbal experience, more than age or maturity level, best predicts how efficiently children's left hemispheres respond to speech. So if boys are somehow biased from birth to handle speech less efficiently, this is all the more reason to talk, read, and sing a lot to them, to perhaps lengthen those dendrites and stimulate their left hemispheres in a way that girls' brains may seek out more on their own.

Hormones and Verbal Skills

If the answer can't be found in brain lateralization, let's look at the other popular explanation for sex differences in language skills: hormones. We saw earlier how the surge of testosterone before birth shapes boys' later interest in rough-and-tumble play. Does this or any other sex hormone similarly shape children's skill with words?

A recent German study mentioned in chapter 2 does provide some evidence for such an effect: EEG measures showed that boys with higher levels of blood testosterone at one month of age had weaker speech processing than boys with lower levels of testosterone. Further evidence comes from a study of girls with twin brothers; they were found to exhibit slightly greater asymmetry (a masculine shift?) in language processing when compared with girls with female twins, perhaps reflecting prenatal exposure to their brothers' testosterone or other hormones. But these are just subtle electrical differences. When it comes to actual verbal skill, there's little evidence for an enduring influence of prenatal testosterone. Girls with CAH, for instance, exhibit normal verbal ability in spite of their exposure to high levels of prenatal androgens; so do females exposed in the womb to DES, a synthetic estrogen given to pregnant women between the 1940s and 1960s and known to affect their babies in other ways. So while prenatal hormones may indeed bias language circuits during early development, it appears that children's social environment largely overrides their influence, with the result that girls end up talking and reading like girls, regardless of their early hormone exposure.

There is, however, another phase in which sex hormones could influence verbal skills. This begins at puberty, when testosterone and ovarian-hormone levels surge for the second time in life. In this case, it is a female hormone, estrogen, that is suspected to have the greater effect on verbal ability.

Canadian psychologist Elizabeth Hampson found the first provocative evidence that verbal articulation changes as estrogen levels go up and down during the menstrual cycle. Specifically, women in her studies were some 5 to 10 percent faster at reading aloud while near their peak monthly estrogen and progesterone levels (during the midluteal phase of their cycles, which is midway between ovulation and menstruation), as compared to the menstrual phase, which is when estrogen and progesterone levels are at their monthly low.

Needless to say, this raging-hormones idea got a lot of press at the time, especially as it coincided with the growing acceptance of PMS as a form of mental illness. Maybe women really do think differently at different times of the months? Perhaps such hormones explain some of the cognitive differences between men and women or, at the very least, explain why the differences seem to vary so much from study to study?

Unfortunately—and like many other sexy reports about sex differences—the fluctuation in verbal skills over the menstrual cycle has not been a very reliable finding. Some recent studies have indeed replicated Hampson's results, while some brain-imaging studies have documented changes in activation of linguistic circuits during the menstrual cycle. However, other studies have been unable to detect changes in verbal fluency, verbal memory, or other measures of language ability, while a recent British study found no relationship between verbal fluency and the blood levels of six different sex hormones, including estrogen, testosterone, and progesterone. The findings are equally mixed for studies of spatial skills, as we'll see in the next chapter. So the bottom line is that cognitive skills are affected little, if at all, by women's monthly hormone changes.

The same conclusion emerges from studies of postmenopausal women, another group scientists have pursued in search of a link between hormones and verbal skills. Estrogen levels decline famously after menopause, and the large number of women who take estrogen supplements in later life (usually in combination with progesterone and known as hormone replacement therapy, or HRT) has provided researchers with a huge database for testing the possible influence of this hormone on various aspects of cognition, including verbal skills.

Neuroscientists had good reasons for thinking that estrogen would help cognitive function in elderly women. Research on both rats and monkeys shows it has all kinds of neuroprotective effects on the brain, such as increasing the number of synapses in the hippocampus, a brain structure crucial to memory storage. In rats whose ovaries have been removed (simulating menopause), estrogen treatment improves certain types of memory, and similar effects have been reported in elderly female monkeys.

But while early studies of postmenopausal women were encouraging, the evidence that HRT improves women's verbal ability, memory, or any other cognitive skill imploded in 2003 when results of the nationwide Women's Health Initiative Memory Study (WHIMS) went public. This was part of the same clinical trial that in 2002 destroyed the idea that estrogen might protect older women against heart attacks and cardiovascular disease. Much to everyone's surprise, the WHIMS trial actually revealed an *increased* risk of cognitive decline and dementia from HRT use.

The research continues, with HRT advocates holding out hope for some cognitive benefit if postmenopausal women limit their use of estrogen to the period shortly after menopause. But any objective look at the evidence tells us that estrogen isn't the magic bullet behind females' stronger verbal skills. Particularly when considering children, whose hormone levels are far lower than adults' and do not differ between boys and girls (after the first few months of life), sex differences in language and literacy skills cannot be pinned on circulating levels of estrogen, progesterone, or any hormone.

Nature or Nurture?

For all the uncertainties about the brain and hormones, biologists do have another way of quantifying the relative roles of nature and nurture in shaping language skills. And this issue is crucial for figuring out how best to close the reading and writing gaps between boys and girls.

The fact that sex differences in verbal skills emerge so early does seem to point to something innate. As we saw in chapter 2, girls gesture and say their first words about a month earlier than boys do and are one to two months ahead in vocabulary growth throughout the toddler years. In another large study, this difference amounted to about a 20 percent larger expressive vocabulary in two-year-old girls than in same-age boys. Among preschoolers, girls tend to speak in longer, more complex sentences and by four or five years of age are some 15 percent more verbally fluent than boys. It's important to recognize, however, that boys don't falter in all language skills. There is no sex difference in receptive vocabulary or word knowledge at five or any later age. Vocabulary is a key skill for learning how to read and so a cause for celebration in boys' verbal development.

Nonetheless, as we've seen repeatedly, just because a difference shows up in early childhood doesn't necessarily mean it is innate. If you count the last three months in the womb (when fetal hearing is good enough to begin to perceive a mother's speech), most babies are immersed in language for some fifteen months before they utter their own first identifiable words. Parents react differently to boys and girls, so it's certainly possible that we also talk to them in different ways. Some home-observation studies offer encouraging evidence against this idea, finding that parents talk just as much to their sons as their daughters during the first year or two of life. However, other studies find the opposite—that mothers talk more to their daughters than to their sons, and that girls talk more with one another than boys do, especially during the toddler years. So it appears that by at least the second year, social factors have begun contributing to the differences in boys' and girls' verbal production.

Other evidence for social learning comes from the study of vocal pitch. Most of us can tell a male speaker from a female one—even if it's a thickly accented customer service agent on the phone from India. Men obviously have lower voices than women, but what's interesting is that this difference emerges even before boys go through puberty, which is when they are exposed to the androgens that elongate their vocal cords, lowering their fundamental frequency.

Why do boys speak in a lower pitch than girls, even before their vocal apparatus forces them to? The reason is social mimicry: boys unconsciously speak lower than they need to in an effort to copy older males. Similarly, girls talk in a higher register than they are biologically wired for because they have unconsciously learned that this is more feminine.

Even as adults, men and women tend to speak at the lower or upper ends, respectively, of their natural frequency ranges, exaggerating the underlying biological differences between male and female.

I certainly noticed my son Sam's voice dropping a register around the time he started kindergarten, when he was well into the gender-constancy stage and seemed to be trying to express himself with greater authority. Toby does the same thing when he's trying to fit in with Sam and his friends, all three years older than him.

Disentangling nature and nurture is tricky. But there is one reliable approach, known as behavioral genetics, that can help us put real numbers on verbal ability.* What behavioral geneticists do is recruit subjects with a known genetic relationship to each other and then measure the degree to which some behavioral trait, such as reading skill, is similar or correlates between them. The best groups to study are twins, and the best design involves comparing large numbers of identical twins (who share 100 percent of their DNA) with large numbers of fraternal twins (who share 50 percent of their DNA).

Consider verbal fluency—the speed at which a subject can think of and produce words—which is one of the more reliable differences between the sexes. Research on four-and-a-half-year-old twins has shown that verbal fluency is about 40 percent heritable. Other verbal skills, such as word knowledge, verbal memory, and articulation, showed roughly similar heritability in this large study, which was conducted by Robert Plomin and his colleagues in London. However, when Plomin's group analyzed male-female differences in verbal skill, they found that gender accounted for just a tiny fraction, less than 3 percent, of the total variance, or range, of language skills. In other words, sex does not play nearly as dramatic a role in verbal skills as many parents perceive it to, whereas children's environment has a very large impact—accounting for some 60 percent of the range of verbal skills for both boys and girls.

A closer look at Plomin's data reveals one environmental variable that may influence sex differences: the sex of one's sibling. When the group of twins was two years old, Plomin's team uncovered about an 18 percent gender gap on a widely used test of toddler vocabulary, the MacArthur Communicative Development Inventory (MCDI). Once again, the data

---

* This is the same method used to estimate the heritability of gender-typical play described in chapter 3.

showed wide variation within both sexes; only some 3 percent of the total range could be accounted for by sex. But look at the graph in Figure 5.3, which shows the MCDI scores of the various types of fraternal twins.

It's important to realize that for all these groups, the twins in each pair share the same amount of DNA: 50 percent. The only difference is the sex of the twin. Girls growing up with nonidentical twin sisters show the highest verbal skills, a bit higher than girls growing up with twin brothers. Male twins are lower overall than girls, as we've seen, but those growing up with twin sisters are more verbal than those growing up with nonidentical twin brothers.

In other words—big surprise!—your speaking partner influences your verbal development. Girls get more verbal from hanging out with other girls, boys get less verbal from hanging out with other boys, and boys who hang out with girls benefit, at least at two years of age, from having a more verbally inclined conversational partner. It's possible that some other factor, such as prenatal testosterone, is contributing here (lowering the verbal skill of girls with twin brothers). But this wouldn't explain the elevated verbal scores of boys who have twin sisters as compared to twin brothers, nor is it supported by girls with CAH, whose verbal skills do not appear to be harmed by their excess prenatal androgen exposure.

Of course, siblings are not the only environmental influence on children's language development. Parents, teachers, peers, preschool attendance, the number of books in the home, educational toys, and every

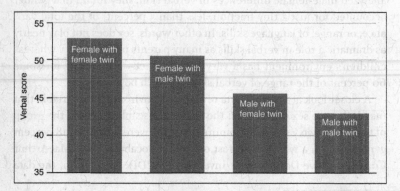

Figure 5.3. MCDI scores of two-year-old fraternal twins according to sex of the twin sibling.

other source of language stimulation are also important. It's hard to study any of these factors in isolation, but most can be lumped under the familiar variable socioeconomic status (SES) that's inevitably one of the best proxies for environment when trying to figure out how nurture affects children's intellectual growth. SES is really researchers' way of saying *social class*, and as teachers know, it has a strong effect on language and literacy skills as well as every other measure of academic achievement. (This is why teachers in the most impoverished school districts face such challenges in getting their pupils to pass mandatory achievement tests.)

According to one widely publicized study from the University of Kansas, toddlers from higher-SES homes, whose parents tend to be professionals and college educated, hear some three times more words addressed to them in a given day than children from low-SES, or more impoverished, homes. This quantity of language immersion, along with the size of vocabulary and complexity of sentences addressed to them, was found to correlate with the children's vocabulary, language development, reading comprehension, and school achievement in the preschool period and as late as nine to ten years of age.

The influence of SES on language, reading, and school achievement is pervasive. It affects every word skill and has a far greater impact than a child's sex. Depending on the particular language or literacy skill, a child's environment accounts for somewhere between 30 and 70 percent of the variance across the population. Compare this to the 3 percent of variance that's attributable to sex and you can see there's ample room to improve all children's language skills, male and female.

Dyslexia

Here's a conundrum: if biological sex contributes relatively little to reading and other language skills, then why do so many more boys than girls suffer from dyslexia, which is defined as the specific inability to learn how to read in the absence of any general intellectual or psychological problems? As one famous male describes his struggle with dyslexia:

> I'd try to concentrate on what I was reading, then I'd get to the end of the page and have very little memory of anything I'd read. I would go blank, feel anxious, nervous, bored, frustrated, dumb. I would get angry. My legs would actually hurt when I was studying. My head

ached. All through school and well into my career, I felt like I had a secret. When I'd go to a new school, I wouldn't want the other kids to know about my learning disability, but then I'd be sent off to remedial reading.

Hard to believe, but that's Tom Cruise talking, in a 2003 *People* magazine interview. Dyslexia obviously didn't hold this kid back. Still, Cruise's painful recollections of graduating as a "functional illiterate" resonate with the surprisingly large number of people who have struggled with a reading disability. Dyslexia obviously affects scholastic achievement, but it also has profound implications for a child's classroom behavior, social adjustment, and self-esteem.

Dyslexia affects at least 5 percent and possibly as much as 12 percent of Americans. The fact that so many people have difficulty learning to read reminds us that this is not a skill the brain has been programmed for through evolution. It is an add-on, a retrofit, that works better in some than in others depending on the way their left hemispheres are organized.

Boys, in particular, are diagnosed with dyslexia about twice as often as girls. Some studies put the ratio as high as four to one, while others argue that any difference is due to a referral bias by teachers—that there is no genuine sex difference in true dyslexia rates.* However, the latest data from several large, well-controlled studies in both the United States and the United Kingdom seem to be settling on a ratio of about two to one.

Why is dyslexia twice as common in boys? Are their brains, as some suggest, really less capable of processing written words? Is it a byproduct of maturity—in other words, since reading benchmarks are established without regard to sex, do more boys end up on the disabled tail of the

---

* The idea that the sex difference in dyslexia is a myth was promoted most prominently by Sally Shaywitz, the same pediatrician and neuroscientist who coauthored the high-profile report on sex differences in adult language lateralization. In 1990, she and colleagues measured the reading ability of a few hundred second-grade children in a representative sample of the Connecticut public schools. They found a small, statistically insignificant difference: about 9 percent of the boys tested in the dyslexic range, as compared to about 7 percent of the girls. Among the same students, however, teachers had identified nearly 14 percent of the boys, as opposed to 3 percent of the girls, as reading disabled. Shaywitz's team found that teachers' diagnoses of reading disability in boys appeared to be biased by other factors, particularly their higher activity levels and general misbehavior.

curve simply by virtue of their slower cognitive development? Or is it a matter of learning—are our instruction methods better suited to girls than to boys, making more boys miss the boat and fail to master the requisite skills?

Fortunately, a flurry of recent research on dyslexia and the neural basis of reading is beginning to shed light on the matter. Reading clearly grows out of spoken language, and, as we've seen, boys are typically slower to begin talking and to achieve the phonological awareness that is crucial to learning how to read. While the difference in verbal skill is small for the majority of children, the slight shift in the curve may mean that about twice as many boys as girls end up below the cutoff and in the delayed area of the language curve.

If this is the case, a simple solution would be to apply different standards to boys and to girls when determining where the threshold is. This would prevent labeling children as having reading disorders when all they really need is more time to develop their skills. However, the NAEP data (Figure 5.1) show that the reading gap tends to grow, not shrink, as children progress from elementary to high school—after boys should have caught up from their maturational delay. So establishing different criteria for boys and for girls could end up exacerbating rather than reducing the reading gap, because it would exclude many boys from the early intervention that can make a great difference in their reading proficiency.

Reading disorders are well known to run in families: parents of dyslexic children often discover that they too suffer residual signs of reading impairment, such as poor comprehension or a lifelong difficulty with spelling. Dyslexia is estimated to be as much as 80 percent heritable. In recent years, geneticists have made great strides in identifying specific chromosomal regions linked to the disorder. And yet they have not found any evidence that sex influences the heritability of dyslexia, nor have they described any differences in the way several dyslexia candidate genes are expressed in males as compared to females.

Other research has focused on how the brain processes written words and what goes wrong when dyslexic individuals try to read. This important work has isolated the problem to a region of the brain at the junction of the temporal and parietal lobes that includes the planum temporale (see Figure 5.4). This part of the brain is uniquely positioned to integrate the sight of written words (processed by the occipital lobe) with their more familiar sounds (processed by the superior temporal lobe) and shows strong activation when normal readers are shown written text. Dyslexic individuals, however, show abnormally low activity in this temporal-parietal junction during reading tasks, suggesting that their brains are not as efficient as normal readers' brains at joining the image of a letter with its memorized speech sound, or phoneme.

Inevitably (given the sex difference in dyslexia), such imaging studies tend to include more male subjects, but thus far none has reported differences between male and female dyslexics in the way they process language. However, imaging studies have found fascinating evidence that this same underactive area of the brain can be activated in children who undergo intensive training to overcome their dyslexia.

How does a child overcome dyslexia? While earlier theories posited all sorts of causes for dyslexia, including visual problems, memory defi-

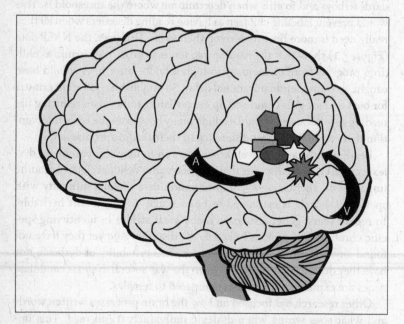

Figure 5.4. Results from several fMRI studies of dyslexia agree that the site of abnormal function is located at the junction of the temporal and parietal lobes in the left hemisphere. Each geometric symbol shows a hot spot for reading dysfunction that has been identified in a different research study. Curved arrows show how the perception of spoken and written words converge at this site from both auditory (A) and visual (V) sensory brain areas.

cits, and left-right confusion, recent research has converged on the idea that dyslexia is, at its core, a phonological problem. Though memory and other issues can cloud the effort to learn to read, the primary hurdle in dyslexia is children's inability to accurately hear, separate, and manipulate the individual sounds of speech.

As we've seen, such phonological skills are also the main bottleneck for normal children as they learn how to read. A substantial minority of children—more boys than girls—struggle with such skills; this can be revealed through tests of children's ability to manipulate individual letters and word sounds. For example, when you ask a five-year-old to omit the *k* sound from *cup*, the child should be able to say *up* and should also be able to rapidly name any letter of the alphabet he or she is shown. Other tests involve asking a child to remember a string of nonsense syllables or to pick out the rhyming words from a song. There are many ways to test phonological awareness, all of them requiring children to identify or alter individual letter sounds, letter blends, or syllables.

Children who fall well below the average for their age on such tests are at high risk for developing reading problems. But the great news is that such problems can be overwhelmingly avoided if vulnerable children are identified and given the appropriate phonological instruction at an early age. Training must involve phonemic awareness (identifying and manipulating syllables and speech sounds), phonics (the association between sounds and written letters), and fluency (developing speed in reading letters, words, and longer bits of text), along with lots of shared reading and exposure to books, which expands children's vocabulary and their interest in the written word.

Many studies have supported this early-identification, early-treatment approach. Catching at-risk children and giving them the appropriate training has been found to result in significantly higher levels of reading ability and fewer emotional problems if it's done by kindergarten or first grade, rather than delayed until third grade. In fact, some researchers argue that the high rate of dyslexia itself is really just fallout from the fact that many children are *not* receiving this crucial training in their ABCs, speech sounds, and phoneme knowledge by early kindergarten.

This argument is especially pertinent to boys. While most learn to read perfectly well with regular classroom instruction, the fact that more boys than girls are eventually diagnosed with dyslexia suggests that many could benefit from more intensive preliteracy training at home and in early schooling (preschool to first grade). In one large study from North Vancouver, Canada, twice as many boys as girls tested into the reading-disability category in kindergarten, but by second grade there was no sex difference. As the authors note, this particular school district "has an excellent reading instruction program." And that may be all it takes to help many children avoid the debilitating effects of dyslexia.

The Critical Period for Language Skills

It should come as no surprise that earlier is better for mastering the skills underlying reading. The same rule applies to learning language itself. Scientists call it the critical period and cite overwhelming evidence that most language skills, especially grammar and pronunciation,* are mastered far more easily in early life. Children, as is well known, can learn foreign languages more rapidly and speak them more fluently than adults can; their brains are also more resistant than adults' are to a language deficit, or aphasia, should they suffer an injury to the left hemisphere.

Any effort to compensate for the different language and literacy abilities of boys and girls must take into account this greater plasticity of the young brain. The critical period for language is most acute between birth and age six, meaning this is the phase when children can most flawlessly acquire a first or second language. The plasticity then gradually declines during middle childhood and adolescence. Adults obviously *can* learn new languages, but they rarely speak them as fluently as their native tongues. Languages—and, presumably, literacy as well—are best acquired early on, while the brain is still at its most malleable.

The reason for this critical period is that language experience itself wires the brain for lifelong dedication to this skill. Regardless of sex, children's brains show increasingly specific left-hemisphere activation to speech and other verbal tasks as their proficiency with language improves. In other words, it is the actual word learning, and not some kind of preset brain maturation, that is responsible for shaping the left hemisphere into the primary language-processing area. Girls may go through

---

* One notable exception to the list of skills acquired during the critical period is vocabulary, which increases throughout one's life, largely as a result of the amount a person reads. When an adult learns a foreign language, he or she can often communicate quite effectively using the new vocabulary but typically speaks in a conspicuous foreign accent, a residue of the fact that the adult brain is not as capable of mastering the pronunciation and grammatical rules of a new language.

this shift somewhat earlier than boys, but the same plasticity occurs in both sexes provided they have ample exposure to language in early life. In fact, a recent study revealed a potent relationship between activation of the left-hemisphere speech zone and children's SES, again confirming the overriding influence of environment on both brain and language development.

The point is: language and literacy are learned skills. We're not born talking, reading, or writing; we must be immersed and taught these crucial skills through interaction with other people, books, letters, words, pencils, paper, and so on. If boys are lagging in these skills, the solution lies in more effective training, not in redshirting or in single-sex classrooms. And if you still think the problem is "hard-wired," consider places like Afghanistan, where the number of female illiterates far exceeds the number of males. Education, not biology, is both the cause and the answer to sex differences in reading skill. If we are truly concerned about improving the verbal, reading, and writing skills of boys, then we have to teach them better and earlier.

Putting Nurture to Work: Tips for Promoting Boys' Language and Literacy Skills

There is no denying that the language and literacy gap is a real and important issue for both parents and teachers. Boys are not learning to read and write as well as girls, a factor that is contributing to their poorer school performance from kindergarten to college. Much as people would like to better understand this gap, scientists have no simple neurological or hormonal explanation for it. However, we do know a lot about how a child's environment and education influence the acquisition of language and literacy skills, and this is where parents and teachers *can* make a difference for boys who are struggling in these areas.

The equation is quite simple, as we saw in chapter 2: Language in = Language out. The more you talk, or read, or write, the more facile you become at each of these skills. Boys may not, for a variety of maturational, temperamental, and social reasons, be as focused on language and communication skills as girls are, but this initial bias is no reason to let them slip through the cracks and miss out on the crucial immersion that can inoculate them against verbal, reading, and writing deficits.

In previous chapters, I offered several strategies for enriching language and literacy in young children, especially boys. An early start is crucial, so please refer back to chapters 2 and 3 for ideas about talking, reading, and singing with your children from birth. Here are several more tips for developing boys' skills in the various language arts:

• **Focus on phonics.** Reading to children is crucial for fostering their love of books and motivation to learn to read. But, surprisingly, it is not sufficient to make them literate. Children also need the various phonological skills I described above—recognizing their ABCs, detecting rhymes and syllable breaks, and being aware of beginning and ending sounds in words. Preschoolers can practice some of these skills using electronic toys and computer games that focus on letter knowledge and sounds. Manipulatives also work great for boys—magnetic or wooden letters that they can handle and move around to blend sounds or explore rhymes. As always, though, adult interaction provides the best teaching: research has shown that parents who engage in such formal literacy training facilitate their children's transition to reading better than those who rely solely on informal experiences, like reading storybooks or taking trips to the library. So read them ABC and rhyming books, and pause to ask your child, "What sound does this letter make?" or "What rhymes with kite?" Be careful, though, to avoid flash cards and other drill-and-kill methods, which will make phonics a bore and turn your child off of instead of on to reading.

• **Continue reading with your child, even when he is able to read himself.** Reading to children is an outstanding way to enhance their vocabulary, promote the pleasure of literacy, and teach reading as a habit. It's easy for parents to equalize boys' and girls' exposure to books while they're still young and depend on you to read to them at bedtime. Once they begin reading independently, however, girls tend to clock more hours, due to their stronger skill and a temperament that is often better suited to sitting still and focusing on many pages of text.

One solution is to continue reading aloud to your sons to get them absorbed in a good novel or chapter book that they will then be inspired to finish on their own. Alternate reading pages with your child, if necessary, or plead fatigue after you've read for a while to get them to read aloud to you. The great thing about this activity is that you get to continue to enjoy cozy reading time even when your boys no longer need you to read to them, and you can form your own private book group to discuss the latest story you are sharing.

• **Pick books (and other reading material) they like.** Boys obviously gravitate more toward books with male themes, which reflect images of themselves and of what they aspire to be. Nonfiction is a popular choice, as mentioned earlier, especially if it empowers them with knowledge about sports, technology, world events, science, and "cool stuff" they can use in conversation. Boys also enjoy satire and other books that are funny, edgy, or appeal to their sense of mischief (yes, including *Captain Underpants*). To hook boys on fiction, pick novels with a lot of action and adventure; many boys are quite passionate about fantasy and science fiction genres. Books in series can also be very motivating (such as the Harry Potter, Series of Unfortunate Events, Artemis Fowl, and Alex Rider series). Finally, parents and teachers should look beyond books for boys, who can also rack up a lot of good reading time with newspapers, comic books, graphic novels, instruction manuals, magazines, and websites.

• **Separate reading groups in school (on occasion).** Boys and girls clearly enjoy different types of reading material. Single-sex-school advocates cite this fact as a reason for separating boys and girls into different classrooms. This seems an overreaction to me, especially for young children, who are less self-conscious about interacting with opposite-sex peers. However, when teachers give children choices in their required reading, boys and girls may naturally form separate groups, which can give some boys the space to get absorbed in books without feeling outgunned by some of the stronger female readers in a class.

• **Male role models.** Fathers and other men are important role models for encouraging boys' reading. If dad comes home and watches TV all evening, it's pretty hard to persuade Joey to go in the other room and pick up a book. Female teachers should seek out male mentors for the boys in their class, preferably younger men who are engaged in verbally oriented fields like journalism, law, the clergy, or telecommunications. The mentors can read to the children, talk about books they like, and meet with boys individually to discuss schoolwork and strategies for academic success. In her book *The Trouble with Boys,* journalist Peg Tyre describes one elementary classroom where a different police officer comes and reads to the class every day: "The kids love it. And the boys . . . get to see a man with a gun in his holster and a book in his hand."

• **Encourage other arts to promote literacy and the love of words.** Language is obviously not limited to language-arts classes. Drama, music, and even the visual arts provide pleasurable breaks in the curriculum that also inspire rich dialogue, emotional expression, role-playing, and attention to words and speech. Musical lyrics can be used to analyze rhyme, metaphor, and other poetic devices. Students love making videos, a dynamic medium for developing skills in scriptwriting, storytelling, and broadcasting.

• **Teach typing earlier.** Let's face it: we live in a digital world where handwriting is less and less necessary for real-world communication. While penmanship is still important and should be encouraged in the younger grades (see chapter 4), typing can be introduced much earlier than it currently is, which is typically around fifth grade. Commercial typing programs resemble the computer games boys already love and may provide an exciting challenge as they try to raise their scores by typing ever faster and more accurately.

• **Writing.** There is no question that this skill can and must be taught more effectively to all children, especially boys. Nationwide, only 24 percent of twelfth-graders scored at or above the proficient level on the 2007 NAEP writing assessment (and two-thirds of these were girls). As with most skills, the earlier children become comfortable with writing, the better. Many girls keep diaries or enjoy writing letters to friends, relishing the chance to fill empty pages with cute, curvy penmanship. By contrast, boys often have an extraordinarily high threshold for putting pencil to paper. Two solutions mentioned earlier involve divorcing penmanship and composition, either by using dictation or by letting boys compose on a computer (or better still, by using voice-recognition software, which converts dictation directly into a text file). Once the words are in digital form, boys can take advantage of automatic grammar checking to help hone their prose.

Of course, there's a lot more to writing than getting words on a page. Children also need to learn the craft of composing—outlining and organizing their thoughts, structuring a paragraph, varying sentence structure, avoiding wordiness and redundancy. Graphical outlining tools can help here—visual templates where students fill in themes and evidence, sketch out characters and plot, or otherwise diagram what they're going to write about before they have to stare down an intimidating blank screen or sheet of paper. Various mind-mapping tools are now available, both in hard copy and software, to help students bridge the gap between visual and written expression as they brainstorm, draft, and revise their prose.

• **Limit screen time.** It's hard to see how we are going to get boys to read more without unplugging them from their various electronic entertainments: TV, DVDs, computer games, Game Boys, iPods, and endlessly multiplying gaming systems. This has become the dominant pastime for boys, largely replacing sports, and that's had effects on both their waistlines and reading time. Most video games involve only very limited reading. And while they may be great for improving strategy and spatial skills, most boys already get plenty such practice. Although there is surprisingly little research linking screen time to academic performance, it seems wise for parents to set firm limits on when and how long children engage in electronic entertainment. (The American Academy of Pediatrics recommends no more than two hours of leisure screen time per day.)

Reading, writing, and talking are obviously crucial skills for everyone, male and female. They are not, however, pre-wired in our genes or innately programmed into our brains. All are learned, through conversation, shared reading, schooling, and practice, practice, practice. Along with high expectations and an early start, these are the keys to improving such skills—in both boys and girls.

6

Sex, Math, and Science

IT'S BRAIN DAY AT JULIA'S MIDDLE SCHOOL. One hundred twelve-year-olds, brimming with attitude, manage to sit through my little talk about neurons and synapses, perking up when it comes time to hold the specimens I've brought along.

Most are pretty game. Nervous, excited, they extend their gloved hands to accept what I consider a remarkable gift: a human brain, some-one's former self, generously donated for the purposes of research or, on this occasion, inspiring more young people to join the fight against its various diseases.

Inevitably, some make gross jokes, asking me if I've seen *Silence of the Lambs* or if the donors were dead before their brains were removed. ("Usually," I respond to one student before I realize he is actually not joking.)

When it comes to holding the brains, though, a pattern emerges. The boys appear gamer, more eager to get their latex gloves wet. Of course, there are plenty of girls who dive right in, awed by the experience. And a few of the boys look pretty shaky, with one or two refusing outright to give it a try. But overall, the girls are wimpier; several won't bother putting gloves on. Others are hyperventilating and look on the verge of fainting. Lots of kids reflexively blurt, *"Gross!"* at first sight, but generally speaking, it is the girls who shy away more from the experience.

. . .

I'm a scientist, and like many other women in such fields, I worry about the fate of females in science, math, engineering, and other technical careers. Yes, our numbers have increased dramatically in recent decades, but the trend is patchy. While medicine, biology, economics, and accounting have seen huge influxes, engineering, computer science, and mathematics have made little progress. Even where women are well represented, the glass ceiling has barely cracked. Female physicians earn only 62 percent of what male physicians earn, women scientists remain clustered at the low end of academic and industry ranks, and fewer than one-fifth of accounting-firm partners are women, even though they've made up the majority of college accounting graduates for more than two decades.

In this chapter, we'll examine an issue that is in many ways the flip side of the last: girls' poorer performance in science, math, and related domains. While boys start out behind girls in language and literacy skills, they largely catch up by adulthood, and there's little evidence that their childhood lag prevents later achievement in word-intensive careers such as law, journalism, the ministry, and politics. Given the likes of Shakespeare, Hemingway, and Martin Luther King Jr., no one seriously claims that males are handicapped by an inability to think and communicate in words.

Girls, on the other hand, start out performing just fine in math and science but slip in adolescence and continue to slide as their education and careers progress. Yes, there have been tremendous gains in some fields, but they are not what they could have been if more talented girls had stuck it out instead of defecting to traditionally female disciplines or opting out of the work force entirely.

Why does this happen? Are females intrinsically less capable of thinking in mathematical and logical ways, as the former president of Harvard Larry Summers infamously proposed? Or are the differences more cultural, a matter of learning and social forces that make females increasingly less comfortable as they move up the hierarchy in scientific and other technical fields?

As we'll see, there are real differences between boys and girls in math, science, and related skills, but the pattern is complex and in some ways surprising. Digging deeper, we'll explore the reasons—both innate and cultural—for these gaps and discuss the ways to overcome them. Without question, parents and educators can do much more to raise girls' interest and achievement in these important fields of the future.

Who's Up? Who's Down?

Girls actually start out strong in math and science. They enter kindergarten knowing their numbers and counting as well as boys do. In fact, females of all ages outperform males in computation, the ability to rapidly add or divide a string of numbers. Eventually, however, boys outscore them in most mathematical tests, including geometry, measurement, probability, and the dreaded story problems.

Consider data from the National Assessment of Educational Progress (NAEP), which tests hundreds of thousands of U.S. students. Girls score lower than boys in both math and science in fourth and eighth grades, although the difference (two to three points) is considerably smaller than boys' disadvantage on the NAEP reading and writing exams. Girls fall farther behind by twelfth grade, but it's encouraging that both the math and science gaps at this age have declined by nearly half from a decade or two ago.

In fact, school-age girls are doing better than ever in science and math. As of the year 2000, graduating high-school girls had taken *more* science and math classes than boys had, with the exception of calculus (which was very close) and physics.* Girls also earn higher grades than boys in all high-school and college classes, including math and science.

Nonetheless, boys routinely outperform girls on standardized math and science tests, an example of the grade-test disparity that is typical of most academic gender gaps. This gap is not unique to the United States: boys test higher in math and science in most countries around the world, according to the 2003 Program for International Student Assessment (PISA), which administers the same exams to fifteen-year-olds in forty different nations. Such cross-cultural data are often used to argue that males have an instinctive advantage in math and science. However, girls in several countries have recently overtaken boys on the PISA science exam. Even in math, which shows a larger international gender gap, girls in Iceland and Thailand recently outperformed boys, challenging the idea that boys' brains are innately better wired for this subject.†

---

* Twelve percent of graduating high-school boys had taken calculus, versus 11 percent of graduating girls as of 2000; for physics the numbers were 34 percent of boys and 29 percent of girls.

† American boys outscored girls by six points on the 2003 PISA mathematics test, one of the smallest gender gaps internationally.

So in spite of stereotypes, girls are doing pretty well in science and math. Though they score a little lower than boys on standardized tests, the difference is smaller than the gap in reading and writing, and also not as universal. Girls' grades are higher, and they are taking more science and math classes with each passing year.

What's the problem then?

The real issue arises toward the end of high school, when girls—now young women—must choose colleges, pick majors, and get the training they need to compete in well-paying careers. In spite of their improvement in earlier schooling, girls are not translating their skills and knowledge into technical and quantitative fields as effectively as boys are. And in today's technological society, this is where the greatest opportunity lies. While reading and writing are important, math has a special place—what some call the "critical filter"—that determines who can advance to the higher-paying and more prestigious jobs.

In other words, math and science are where the money is. While not every girl is interested in a career in engineering, computer programming, securities trading, astrophysics, market research, financial planning, investment banking, molecular biology, geology, risk analysis, and so on, there are certainly more women who could have gone into these well-paying fields than have actually done so.

The drop-off is evident at the end of high school. While girls take just as many regular science classes as boys do, they do not take as many AP science classes. In 2002, when girls made up 56 percent of all AP test-takers, they took just 43 percent of the various AP science exams. Girls also scored some 10 percent lower than boys on science AP exams as compared to 3 percent lower on English exams and about 3 percent higher on the foreign-language AP exams.

Then there is the infamous SAT gap I mentioned in the previous chapter: males have consistently scored between thirty-five and forty points higher than females on the math SAT (much larger than the verbal gap, which also favors boys). A good share of the gap is explained by demographic differences: fewer boys than girls from lower socioeconomic levels take the exam, so the boys who do take it are more advantaged than girls overall, and family income is one of the strongest predictors of SAT scores. As a result, the large math SAT gap (Figure 6.1) does not accurately portray the real difference in math ability among seventeen-year-olds. It does, however, have important consequences for college admissions, scholarships, self-perception of math abilities, and the courses men and women choose in college.

The impact of this gap is apparent in the pattern of college majors. Although the number of women receiving bachelor's degrees in certain technical and quantitative fields has soared over the past few decades, that number remains far from parity in other fields. Accounting and veterinary science are two success stories: in both fields, the proportion of women has grown from a mere 8 or 9 percent around 1970 to a considerable majority today (61 percent in accounting and 74 percent of vet students). By contrast, women make up only 20 percent of engineering majors, and the proportion of BS degrees awarded to women in computer science has actually *declined,* from a peak of 37 percent in the mid-1980s to about 28 percent in 2002.

How can it be that with our lives now so digitally enmeshed, the proportion of women in the computer work force is actually on the downswing? Despite girls' love affair with text messaging and music downloads, they're not bridging the digital divide in technical knowledge—understanding the basis of software, hardware, and communication technologies that's crucial to such careers. The computing gap looms especially large when you consider the projected growth in such

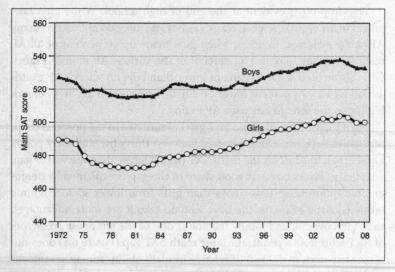

Figure 6.1. Math SAT scores by year and gender. (Source: College Board report, "Mean SAT Scores of College-Bound Seniors, 1967–2008.")

jobs over the next decades, as well as the fact that college graduates with degrees in engineering or computer science are paid some 50 percent higher than graduates in the humanities or social sciences.

Finally, we get to the glass ceiling: the fact that women who do enter technical fields don't make it nearly as high up the corporate or academic hierarchy as men. In my own field, biology, women receive nearly half of all Ph.D.'s but represent only 30 percent of entry-level faculty members and 15 percent of full professors. In mathematics, the numbers are 27 percent of Ph.D.'s, 20 percent of assistant professors, and a paltry 5 percent of full professors.

There are many, many reasons for female attrition in all careers, a discussion that goes well beyond the scope of this book about children. My aim is to examine these skills from a neurological and developmental perspective and shed light on the core questions: What is impeding women's success in math and scientific fields? Are girls trying to push higher than their brains are innately capable of? Or are there social factors that explain the leaky pipeline—the steady loss of females at all stages in technical and quantitative careers?

Fortunately, there is research addressing all these issues, and it's time to bring it to greater awareness. Certain characteristics clearly do bias boys toward better performance in math and science than girls. However, they don't add up to anything as simple as innate ability; they involve differences in interests, culture, and experience—all areas where parents and teachers can be effective in encouraging more girls to pursue these important, satisfying careers.

Males Are More Variable

In the 2004 film *Mean Girls*, Lindsay Lohan plays a mathematically talented high-school student who pretends not to understand a calculus problem so she can cozy up to the cute guy who sits in front of her in math class. Then in the climactic math competition, in which she grudgingly participates in order to avoid detention, the film depicts her, a striking beauty, as the token contrast to a bunch of nerdy male teammates. What's worse, to answer the challenge question at the end of the competition, each team selects a member of the opposite team that they consider the weak link—and Lohan's character and the sole girl on the opposing team are the ones chosen.

Of course, stereotypes are funny, and Lohan's character conquers this one by answering correctly and winning the competition for her team. Still, this was a film that's supposed to be female empowering! It does a great job of revealing the relational aggression that girls wage on one another (see next chapter), but why does Hollywood keep regurgitating the same gender stereotypes about math and science?

The answer: because the stereotypes are accurate. There really are more males than females at the highest echelons of math and the physical sciences, both on high-school teams and among tenured university faculty. But what is less commonly known is that there are also more males than females at the *lowest* end. This greater variability of male performance has been well documented in most cognitive and academic areas. It appears graphically as in Figure 6.2.

Greater variability means that more males turn up at both ends of a distribution: "More geniuses, more idiots," as psychologist Steven Pinker puts it. It's not known *why* males are more variable,* but this fact helps explain boys' higher rates of dyslexia and speech disorders. When it comes to math, however, males' slightly higher average score coupled with their greater variance mean that a significantly greater proportion of boys end up at the high end, or right tail, of the curve.

One study in particular drew attention to this high-end ratio. The Study of Mathematically Precocious Youth (SMPY) was a nationwide talent search of seventh- and eighth-graders that was begun in the early 1980s by psychologists Camilla Benbow and Julian Stanley of Johns Hopkins University. Benbow and Stanley found that within this talented pool, many more boys than girls scored at the highest level on the math SAT exam: a four-to-one ratio for scores above six hundred and a thirteen-to-one ratio for scores above seven hundred. But they made the biggest splash by speculating the high ratio was the consequence not of math education but of "endogenous," or innate, sex differences in mathematical talent. *Newsweek* seized upon their conclusion with the headline "Do Males Have a Math Gene?" while *Time* magazine declared, "a new study

---

* One theory is that females are less variable than males because of the demands of childbearing, which requires a certain threshold of physical and mental function for women to be successful. Males can take a broader range of strategies to successful fatherhood; as Pinker argues: "Natural selection favors a slightly more conservative and reliable baby-building process for females and a slightly more ambitious and error-prone process for males."

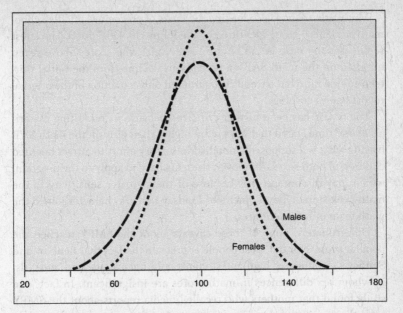

Figure 6.2. Two normal, or bell-shaped, distributions with different variance. For many traits and test scores, males (dashed line) are more variable than females (dotted line). Note how this difference can create a large ratio of males to females at both the high end (scores around 140) and low end (scores around 60) of this spectrum.

says that males may be naturally abler [in mathematics] than females."

The numbers are eye-opening and certainly agree with other statistics about high-performing math students. In the first twenty-three years of its existence, the U.S. Mathematical Olympiad team—an elite group of high-school-age mathematicians who compete in the annual International Mathematical Olympiad—did not have a single female member.

However, the tide appears to be turning. In 1998, Melanie Wood became the first girl on the U.S. Mathematical Olympiad team, and she won a silver medal at the international competition both that year and the next. In 2004, Alison Miller became the first American girl to win a gold medal in the International Mathematical Olympiad. What's more, the Chinese Mathematical Olympiad team, which nearly always wins the overall competition, has had several female members over the years, and each of them has won high medals and outperformed most male competitors from other countries around the world.

More girls are now also making the ranks of precocious middle-school math students—those identified by the Johns Hopkins group. The latest report put the ratio at 2.8 boys for every girl who scores above seven hundred on the math SAT—a precipitous decline from the initial thirteen-to-one ratio but a trend that garnered only a fraction of the original report's news coverage.

The SMPY has been further criticized because its recruiting materials at one time stated that boys score higher than girls on the math SAT, biasing what is a supposedly nondiscriminatory effort to attract talented children of both sexes. Moreover, students have to apply to the program, which may discourage girls because of their greater sensitivity to the math-geek stereotype. (Remember Lindsay Lohan's character joined the math team only under duress.)

Unfortunately, none of these caveats about the SMPY reached the popular press. Nor did most people appreciate the fact that Benbow and Stanley were studying high-end students only, not average boys and girls, in whom sex differences in math scores are insignificant. In fact, one study found that mothers who recalled media reports about the SMPY gave their own daughters' mathematical abilities lower ratings than mothers who had not heard anything about the study.

It's hard to dispute the greater variability of males and their preponderance at the highest levels of math and science achievement. Still, the fact that girls have substantially penetrated these ranks in recent years suggests these ratios are not fixed facts of biology. Even the greater variability of males is not necessarily immutable but could reflect social factors. As Steve Olson, author of a recent book about the Mathematical Olympiad, observes, "The paucity of girls at the highest levels of competition is much worse in the United States than in other countries. . . . Many cultural factors keep girls away from math. To take a more familiar example, *Rolling Stone* magazine recently published a list of history's top hundred rock guitarists, and only two women made the list, yet no one claims that women are biologically inferior as guitarists."

Math in Babies

Olson's analogy is great, but it's worth considering the evidence that males may, indeed, have an inborn edge when it comes to math and science. After all, men are now willing to grant that females have an innate advantage in verbal and relational skills, so why can't women admit that males might have a few "endogenous" talents of their own?

The usual place to begin looking for innate differences is infancy. We've already seen that girls talk earlier than boys, a small but reliable difference that seems to propel them down an easier path of reading, writing, and effective communication. Do baby boys exhibit a comparable advantage with numbers or mechanical sense that gives them a better start toward math/science/technical thinking?

Recall the highly publicized study by Jennifer Connellan and colleagues that suggested newborn boys are more object oriented than girls. This study supports a possible mechanical bias for boys. However, there are reservations about this experiment (discussed in chapter 2), while a much larger body of research has generally failed to reveal a sex difference in babies' understanding of the physical properties of objects—just the kind of thing you might expect boys to be better at. This is a fascinating, relatively new area of research, in which psychologists have tracked babies' eye movements to gauge whether they understand properties such as gravity, speed, and momentum. Remarkably enough, they do. Young infants appreciate that an object should not remain floating in midair if its support is suddenly removed; they can predict how soon a moving object should reach its target; and by about six months of age, they even know that an object travels farther when it's hit by a bowling ball than when it's hit by a golf ball. All pretty sophisticated mechanical reasoning, yet none of these groundbreaking studies has found a male advantage, and some have even found that girls are *ahead* of boys in physical understanding.

Other research has focused on infants' math skills. It may surprise you to learn that babies can do math, but it's true. Young infants can tell the difference between a picture of two frogs and a picture of three identical frogs; they know that if two acorns are placed behind a screen and three more acorns are added, when the screen is removed there will be five acorns; and they even appreciate the difference between larger quantities of items, such as a cluster of eight shiny CDs versus a collection of sixteen of them. In none of these studies, however, have researchers found differences between the performances of baby boys and baby girls. So the idea that boys are "born for math" is not supported by actual research on babies.

Of course, it's always possible that there are differences too subtle to be detected in the small number of newborns generally recruited for any individual study. And while existing research offers little evidence that baby boys and girls differ in their grasping of concepts such as quantity, momentum, and speed, other studies are beginning to hint at an early male advantage in spatial ability, which in turn may shape the way mathematical and mechanical abilities develop down the line.

Spatial Skills, or the Trouble with Parallel Parking

For a sex difference to be innate, it must have arisen through evolution. As we saw earlier, reading isn't old enough to fit this description, and neither is formal mathematics. Both are perhaps a few thousand years old—much too recent for slow selective pressures to have separately shaped the brains of cavemen and cavewomen.

Spatial reasoning, however, is one domain in which sex differences can more plausibly be attributed to evolution. Spatial skills include the ability to visualize, manipulate, and remember object locations, directions, motion, and trajectories. They are crucial to many endeavors, from the prosaic (such as picture-hanging) to the profound (such as Watson and Crick's Nobel Prize–winning discovery that DNA forms a double helix). Spatial skills are so important that Howard Gardner, the Harvard psychologist who pioneered the idea that intelligence is not a single entity but a collection of six or more intelligences, gives spatial intelligence a slot all its own. As Francis Crick described his knack for spatial visualization: "I soon found I could see the answer to many of these mathematical problems by a combination of imagery and logic, without first having to slog through the mathematics."

I'm a bit touchy about the Watson and Crick discovery, since my own university is named for Rosalind Franklin, a brilliant x-ray crystallographer who many feel should have been recognized as codiscoverer of the double helix. Nonetheless, females, as we'll see, do not perform as well as males on many tests of spatial ability, a difference that some argue is related to our ancestors' division of labor into hunters and gatherers. Was Rosalind Franklin limited by an inability to visualize in three dimensions? Are women constitutionally incapable of parallel parking and picturing macromolecular structures?

Sex differences in spatial abilities have been extensively studied, and the answer, as usual, is nothing as simple as a hard-wired difference but involves a complex interaction of both nature and nurture.

The classic test of spatial ability is mental rotation, which requires a subject to identify an object on paper or on a computer screen when it's turned in two or three directions. It's a little like opening a box from IKEA and trying to envision how the unassembled pieces will fit together to make a bookshelf. One example of a mental rotation task is shown in Figure 6.3.

In cultures around the world, males are faster and more accurate at answering questions like this, and the sex difference is one of the largest of any cognitive skill. Estimates of the d statistic for mental rotation range from 0.56 to 0.90, which is considerably larger than the female advantage in verbal fluency. Brain-imaging studies confirm that men and women activate somewhat different brain areas while carrying out mental rotation and other spatial tasks, though the neurologic differences are not as impressive as the differences in actual spatial ability.

We saw in chapter 3 that boys start to surpass girls at mental rotation at four and a half years old. In that study, four-year-old boys and girls performed equivalently, but from the age of four and a half onward, the boys scored significantly higher. Until recently, this was the earliest age at which a sex difference had been identified. Studies of infants had looked for but failed to find a difference between boys and girls in mental rotation and similar spatial skills. However, two independent research teams reported in late 2008 that baby boys between three and five months old outperformed girls on a nonverbal mental rotation task. These two studies, together with the large and universal sex difference in mental rotation found in adults, strengthen the argument that males do indeed have an innate advantage for this type of spatial ability, just as females seem to have an inborn edge in certain verbal skills.

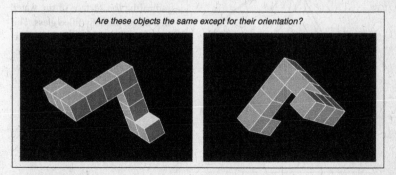

Figure 6.3. Example of a mental rotation task. (The answer is yes.)

Whether it emerges in infancy or late preschool, the sex difference in mental rotation is known to grow larger through childhood and adolescence. Spatial skills, like all mental tasks, improve with practice, so it's likely that boys' mental rotation ability is honed by the many hours they clock in spatial pursuits. Boys begin with their trucks and blocks in preschool and continue with their games of catch, hoops, and all manner of virtual driving and shooting games, so it's not too surprising that the sex that spends most of its leisure time watching objects flip, rotate, and fly through space ends up a lot better at mental rotation.

Then again, there's one test of spatial ability in which the sex difference appears to be declining. It's called the water-level task. Originally designed by Swiss child development expert Jean Piaget, the problem is deceptively simple: draw a line in a tilted glass depicting where the surface of the water should be (Figure 6.4).

Of course, the water line must remain parallel to the ground, but boys catch on to this rule earlier than girls do. Males are more accurate, drawing the line within a few degrees of the horizontal (solid line), whereas females (dotted line) may misestimate the horizontal by twenty degrees or more. Boys' advantage emerges as early as seven years of age. However, it's heartening to learn that this sex difference seems to be diminishing with younger generations. Unlike mental rotation, where males' sizable advantage has held up or even grown throughout the past several decades, sex differences in the water-level task have steadily declined among younger and younger subjects, suggesting that changes in girls'

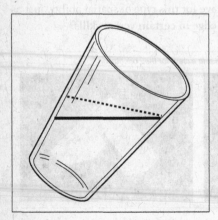

Figure 6.4. Piaget's water-level task: Subjects are asked to draw a line indicating the surface of the water in a tilted, partially filled glass. The correct answer is a perfect horizontal (solid line). Females are less accurate, tending to draw the line at an angle closer to parallel with the top of the glass (dotted line).

education and experiences may be gradually improving at least some of their spatial skills.

Sex Differences in Vision

Spatial perception is largely visual, so a reasonable question is whether the male advantage in spatial ability is due to a fundamental sex difference in visual function.

Vision is indeed the one sensory domain in which adult men have an advantage over women. Men's vision is generally sharper, particularly for detecting slight motion within the visual field, such as a car moving in the distance. This dynamic acuity may contribute to the surprising fact that adult males have a lower rate of driving accidents than females do.* Males also outperform females on tests of flicker fusion—how fast a light can flash on and off before you no longer see it changing. Men can detect the blinking of lights that women see as a single blur.

Males' visual advantage is also reflected in the fact that at any given age, fewer men than women need corrective lenses. What's more, women's distance acuity ages earlier, undergoing a marked decline between the ages of thirty-five and forty-four, whereas men's acuity declines a full decade later, between the ages of forty-five and fifty-four.

But in spite of males' advantage in adulthood—and much popular mythology about boys being more "visual"—boys do not have better vision than girls. As we saw in chapter 2, female infants develop stereovision (important for depth perception) earlier than boys. Even as adults, women perceive the 3-D effect more rapidly and accurately than men. EEG studies show that infant girls' brains process high-contrast images, such as a flashing checkerboard, more rapidly than infant boys' brains. And even though we tend to think of it as their signature ability, boys' motion detection is actually poorer than girls' until about nine years of age.

So there's little evidence that boys are innately more visual than girls, even though adult men tend to have better vision and outperform women on certain visual tasks. It's possible that hormones or genes influ-

---

* Importantly, this is not true for teenage drivers. Boys get into more accidents than girls do. And while adult males' *rate* of getting into accidents is lower (that is, number of accidents per mile driven), women have fewer accidents overall because they drive fewer miles.

ence the male visual system only after puberty, but the existing evidence suggests that, if anything, testosterone actually delays the development of male stereovision (during infancy). Neither testosterone nor estrogen has been linked to visual acuity at any later point in the life span.

Over the past sixty years, neuroscientists have learned more about the visual system than about any other part of the brain. And if there's one fact that's emerged, it is that *visual experience,* beginning at birth and continuing for most of childhood, is crucial for wiring the circuits of visual perception: children with visual problems, such as crossed eyes, can permanently lose visual acuity or stereo perception if the issues are not corrected early in life.

In other words, when it comes to detecting visual motion, men's advantage is likely influenced by their unique childhood interests in throwing, catching, batting, driving, and shooting with their various darts, Nerf guns, Super Soakers, BB guns, and virtual weapons. Similarly, females' advantage at 3-D perception may be related to their typically closer-distance visual pursuits, such as sewing, cat's cradle, drawing, and closing those impossibly small snaps on Barbie clothes. Such tasks are heavily dependent on stereovision, as you can prove to yourself by trying to thread a needle or chop vegetables with one eye closed. (But do be careful with the vegetables!)

Maps Versus Landmarks: Sex Differences in Navigation Strategies

It was a simple enough question: "Is your house east or west of Glen Avenue?" I was trying to figure out where to pick up my first-grader from a new friend's house. Amazingly (to me), his mom couldn't answer my question. She had given me detailed directions ("Our house is in the middle of the second block, gray shingles with a red door, on the left side of the street"), but I had missed something, and instead of asking her to repeat it all, I figured her answer would give me a quick picture of their location.

The difference between us? She navigates by landmarks; I prefer maps. While she was providing detailed information about driveways and door colors, I was picturing the map of our little town and where the dot for their house would go.

I'm the odd one here. It's a cultural cliché that women can't read maps, and sure enough, research has found significant sex differences in navigation. Whether it's following a map, knowing north from west, or successfully traversing a virtual maze on a computer, men really do outperform women at rapidly and accurately figuring out how to get from place to place. Figure 6.5 shows an example of one such computer-based test.

Of course, most women get around quite well in the world, even if they have trouble with this featureless maze. That's because they rely on landmarks, like Sam's friend's mom. Males also use landmarks, but they benefit from a stronger sense of direction and distance (as the maze test demonstrates) to navigate better over unfamiliar terrain. It is this *geometric* sense (awareness of north-south, up-down, right-left) that is common to both mental-rotation and virtual-navigation tasks and is said to give males the advantage in fast, successful route finding. Functional MRI studies confirm that men and women use fundamentally different brain areas in navigation: men show stronger activation of the hippocampus (a structure important in spatial awareness) while women show greater engagement of their right frontal and parietal lobes, areas critical for short-term memory storage. However, researchers have found that when men and women are forced to use one or the other strategy, both sexes

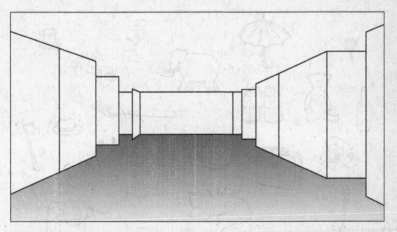

Figure 6.5. One portion of a virtual-maze task, in which subjects must learn through trial and error the correct way out of a featureless series of doorways and alleys. The male college students who participated in this experiment were faster and more accurate than the female students, according to Scott Moffatt and his colleagues at the University of Western Ontario.

are capable of using either landmarks or geometric cues, and the same parts of their brains are activated while doing so.

Women's better memory for landmarks and the spatial location of objects has been confirmed in various studies using complex arrays of items, similar to the one in Figure 6.6.

In this type of experiment, subjects are asked to examine the picture for one minute. Then they are shown another picture where several of the items have been moved to different positions. The task is to circle the items still in their original place and cross out the items that had been moved.

A recent summary of more than three dozen such studies shows that women have a small-to-moderate advantage in object-location memory, but the difference depends in part on the *type* of objects subjects are asked to remember. Women do better with most objects except for stereotypically masculine ones, like a necktie, golf ball, trophy, suit coat, and aftershave (men are better at remembering these objects' locations). And in contrast to the sex difference in mental rotation, this difference does not show up before puberty. Girls are no better than boys at remembering where they left their shin guards (especially if they can all rely on mom to dig them up).

Figure 6.6. Array similar to those used to test object-location memory.

Still, what's interesting about this sex difference in object-location memory is that it runs opposite to most other spatial abilities. Just as males excel in select language skills, females have certain spatial talents all their own, which shows that there is no such thing as a purely male- or female-dominated cognitive domain. Men and women have different spatial strengths, and although men may have the advantage in reading maps and estimating distances, women's object-location memory is pretty handy when it comes to finding those shin guards or navigating using landmarks.

Evolutionary psychologists love these differences in spatial skills and have woven various tales to explain them. One version boils down to the presumed hunter-gatherer division of labor among our ancient forebears—the Fred and Wilma scenario, as critic Richard Francis calls it: Fred and his buddy Barney were the hunters, a pursuit requiring superior targeting, navigation, and motion perception, so males with those abilities would have survived and reproduced more than males who were weaker at those skills. Wilma and Betty stayed closer to the cave, where they had less need for navigation skills and more need for object-location memory so they could forage for food and monitor the campsite for disruption by predators.

It's a wonderful theory, even if there's no physical evidence to support it—no changes in brain structure or fossil tools hinting at how male and female spatial skills may have diverged during our prehistoric evolution. And while it makes sense that targeting and mental rotation strengths could have been selected for as a consequence of hunting success, it's harder to figure out why Pebbles (Fred's daughter) would not have benefited from Fred's spatial genes just as much as her male playmate Bam-Bam* would from his father's.

Given the lack of paleological evidence, evolutionary psychologists turn to other animals, especially to some wild rodents called voles, to explain the sex difference. With their brown fur and short tails, voles

---

* Generally in evolution, traits end up differing only if what's good for one sex is actually *bad* for the other: antlers fit this description, because they're a heavy burden for female deer when they don't need them, as males do, for rutting. Other traits, like nipples, do not fit this bill, and so male primates inherit nipples, even though they serve no purpose, because they are not a burden. Spatial skills are more like nipples than like antlers, because it's hard to see how navigation and targeting ability would be a handicap for Wilma and Betty trying to scrape out a living on the ancient savanna, where recent evidence shows that some women also engaged in hunting.

are a lot cuter than lab rats, but the reason biologists study them is because of the wildly different lifestyles of two species, pine voles and
meadow voles.

Pine voles are the boring monogamous breed that bond upon mating
and raise their young together. Meadow voles are their racier polygynous* cousins whose males compete to mate with as many females as
possible. Guess which species shows the sex difference in navigational
skills? Meadow voles! In their tireless search for females, male meadow
voles cover about four times as much territory as female meadow voles
and outperform them on tests of spatial ability. Pine voles, by contrast,
show no sex difference in navigational skills. What's more, the hippocampus (the same brain area active during human spatial navigation)
is larger in the male meadow vole than in the female but does not differ
between male and female pine voles.

The contrast is fascinating and often cited as evidence that sex differences in spatial ability evolved through sexual selection: that generations of polygynous mating favored male meadow voles with larger hippocampi and better spatial skills, whereas the more monogamous male
pine voles didn't need such keen navigational skills to mate successfully
and so did not evolve larger hippocampi than females did.

Does mating style similarly explain the sex difference in human spatial skills? Anthropologists agree that, on average, humans are modestly
polygynous—that male promiscuity is more acceptable and in some
cultures perfectly permissible, whereas only a tiny handful of cultures
permit a woman to take multiple husbands. Still, humans are a lot closer
to pine voles than meadow voles in our mating strategy and are certainly more monogamous than most other primates, suggesting that
men didn't need superior navigational skills to find enough women to
mate with.

A bigger blow to the mating-style theory, however, emerges from recent brain-imaging studies that report the human hippocampus is no
larger in men than in women. MRIs show that, if anything, it is propor-

---

* The term *polygyny* is used by biologists to describe a single male mating with many
females. It is the opposite of polyandry, a rarer mating system found among certain
birds and primates in which a single female mates with many males. The more
familiar term *polygamy* is reserved for humans and refers to the marriage of one
person, male or female, to more than one member of the opposite sex.

tionally larger in females. This raises the possibility that the larger hippocampi in male meadow voles is actually the *consequence*, not the cause, of their racy lifestyle. In fact, the hippocampus is famous among neuroscientists as the most plastic part of the mammalian brain, forever modifying its circuitry to adapt to changing environments and experiences. For example, a study of London taxi drivers (who need extensive navigational ability) found that their right posterior hippocampi were larger than those of age-matched control subjects. What's more, the longer the taxi driver had been at the job, the larger this part of his hippocampus appeared.

In short, sex differences in navigational strategies are real, but the evolutionary explanations for them are not well supported. As fun as it is to speculate about the ancient origins of our differences, most theories are just that: speculation. There is no archaeological evidence behind the various evolutionary theories, nor have researchers identified any differences in genes or gene expression corresponding to sex differences in spatial skills.

Does Prenatal Testosterone Enhance Spatial Skills?

But while evolutionary theories are difficult to prove, there is stronger evidence that spatial skills are shaped by a more immediate biological influence: our old friend prenatal testosterone. The recent finding that baby boys outperform girls at mental rotation suggests that something very early in development is creating this sex difference. Does testosterone exposure before birth somehow shape the ability to navigate or to rotate objects in one's mind?

Animal studies suggest that it does. As it turns out, meadow voles aren't the only critters that exhibit a sex difference in navigation. The trusty albino lab rat shows the same difference, with males learning mazes faster and more accurately than females do. And remarkably, male rats rely on geometric cues (like the size and dimensions of the room), whereas female rats depend, like human women, on local landmarks (objects placed within the maze).

Here's where it gets interesting: each sex's strategy can be reversed by switching their hormones at birth. Male rats deprived of testosterone by neonatal castration behave like normal females as adults, navigating by landmark cues. Conversely, females treated with masculine hormones at birth grow up to navigate like males, using geometric cues like distance and direction to find their way in the maze.

But as fascinating as these findings are, the evidence linking prenatal testosterone to spatial skills in humans is weaker. A number of studies find no such link, including two that measured either amniotic or umbilical-blood levels of testosterone at birth and compared them to girls' later spatial skills. Similarly, girls with male twins, who are exposed to slightly elevated levels of testosterone before birth, perform no differently than other girls in math or visuospatial skills. Nor do studies of digit ratios offer clear evidence of a link between spatial skills and this supposed measure of prenatal testosterone exposure.

However, studies of girls with CAH, who are exposed to much higher levels of androgens before birth than typical girls are, tell a different story. A review of nine such studies found overall significantly better spatial-skill performance in CAH girls, while a recent test of virtual-navigation skills in young adults found that women with CAH performed as well as healthy men.

Where does this leave us? Early exposure to high levels of testosterone does seem to bias individuals toward better spatial skills, whereas small increases in testosterone—as in girls with twin brothers or even natural variations between individual girls or boys—do not seem to have an impact. So one scenario is that testosterone does not act directly on the brain circuits controlling spatial cognition but in other, less direct ways. For example, girls with CAH may develop better spatial abilities because they (like boys) are more active or because they play more with vehicles and with building toys that train such circuits. In other words, prenatal testosterone may be the initial spark, but it is not the full flame fueling the development of males' stronger spatial skills.

Sex Hormones After Puberty

While spatial skills begin diverging in early childhood, sex differences in math, spatial, and science performance generally grow more dramatic during the teen years. If prenatal hormones play only a modest role, what about the hormones that begin circulating at puberty? Does the rise of testosterone and estrogen in adolescence explain males' advantage on the math SAT and the lack of females on Mathematical Olympiad teams?

This question has been tackled from many angles, but the results have been disappointing for those who wish to blame it on hormones.

There are, to be sure, some fascinating reports of improved spatial-skill performance with testosterone treatment. The most titillating came from early studies of transsexual individuals who, as a first step in their sex changes, began hormone treatment with either testosterone or estrogen. A 1995 Dutch study found that women who were changing into men did indeed improve at spatial skills such as mental rotation after they started androgen therapy. These individuals also declined in verbal fluency, a hallmark female strength. Men switching to become women showed the opposite cognitive changes when they took androgen blockers combined with estrogen therapy. It was all quite fascinating—until researchers were unable to replicate these findings. In fact, the same Dutch group concluded in a 1999 follow-up study that "it is remarkable that our study found only relatively small cognitive effects despite the very strong hormonal manipulations."

Measures of testosterone in the blood have proven similarly equivocal. Some studies show higher testosterone levels are associated with better spatial performance; some find *moderate* levels are the best; some report spatial skills *worsen* the higher the testosterone level; and several find *no* relationship between testosterone levels and various cognitive skills.

The theory really falls apart when you consider a 2002 study measuring cognitive skills in boys given testosterone to treat delayed puberty. This research, led by Lynn Liben at Penn State University, is the only one that's looked at adolescents. Contrary to their expectations, Liben and colleagues found no change in the boys' performance on various spatial tasks when they were given testosterone injections.

What about female hormones? If circulating testosterone does not improve spatial skills, does estrogen suppress them? Since estrogen levels vary widely in women during the month (peaking just before ovulation, then plummeting at menstruation), it's possible to compare women's skills at different phases as a way of analyzing the effects of these hormones on female cognition.

Since 1988, several studies have reported that women perform better on mental-rotation and other spatial tasks during their menstrual period than they do during more estrogen-rich times of the month. Once again, however, other well-controlled studies have failed to find a difference. What's more, studies of rats support the exact opposite hypothesis—

estrogen *improves* female rats' spatial skills during cycle days when it's elevated, a boost that appears to be due to an increase in the number of synaptic connections in the hippocampus.

There are still other problems with the estrogen theory. First, women with the disorder known as Turner's syndrome have very low estrogen levels because their ovaries do not develop normally, but they are notoriously poor at visuospatial tasks while their verbal skills are normal—just the opposite of what you'd expect if estrogen boosted female-typical skills and suppressed male-typical skills. Second, contrary to early reports, male-to-female transsexuals do not lose any spatial ability when they begin heavy doses of estrogen treatment. Third, studies of postmenopausal women taking estrogen and progesterone treatments have shown no change in visuospatial skills with the hormone replacement therapy. Last—but of greatest interest to parents—is Lynn Liben's study, which looked not only at boys but also at girls who were given hormones to accelerate puberty. Estrogen produced no sudden change in spatial skills in these girls, just as testosterone treatment had no effect on the cognitive skills of boys.

So as much as we'd like to blame hormones for every behavioral difference between the sexes, the evidence that they affect thinking ability is weak, at best. Diane Halpern, whose own research supports a link between menstrual phase and cognitive skills, has commented that such effects are small enough that they "probably would not be detected outside the laboratory." In other words, high-school girls have no reason to postpone their geometry tests if they happen to be ovulating during finals week. Nor is circulating testosterone the reason why males consistently outperform females in mental rotation, targeting skills, math, or any other cognitive ability.

The Importance of Spatial Skills

Don't get me wrong. Just because we haven't identified a clear biological cause for sex differences in spatial skills doesn't mean the differences don't exist or that they aren't important in boys' and girls' development.

I was reminded of this in an odd place: a sunny Fourth of July at a beach in the Pacific Northwest. A bunch of young men showed up, in their late teens or early twenties—the age when they'd usually plunk down a cooler of beer and toss a football around. But this gang was far more industrious. They promptly started hauling and raising a few dozen enormous driftwood logs into a massive, fifteen-foot-high tepee. They were grunting and sweating but fast and purposeful, eagerly following the lead of one strapping dude who had both the vision and brawn to see the job through.

It was an impressive burst of energy for a holiday afternoon. It got my attention, as well as the attention of several bikini-clad young women who were stretched out and otherwise immobile on their oversize beach towels. At the time, I was just starting to think about this book and was more focused on sex differences in early childhood, when girls seem to hold the advantage. But this display was a refreshing reminder of males' wonderful strength, ambition, and keen ability to build things.

Spatial skill is important. Whether it's used to complete trigonometry homework, park the car, or calculate the angle needed to hoist a three-hundred-pound log, the understanding of shape, distance, motion, and direction factors importantly into much that we do in life.

Take geography, for example. Since the National Geography Bee began in 1989, seventeen of its twenty winners have been boys. Sponsored by the National Geographic Society, this contest for eleven- to fourteen-year-olds begins with individual school bees, progresses to the state level, and ends with a nationally televised championship. Girls do okay in the initial competitions, winning about one-quarter of their school contests, but they only rarely make it to the national level. According to researchers, spatial skill is one factor in boys' success, helping them understand and memorize large amounts of map detail. Even at the non-elite level, boys as a group score significantly higher than girls on standardized social studies tests, where such spatial and geographic abilities come into play.

Of greater concern is the link between spatial skills and math achievement, which has wide impact on many careers. Research by Beth Casey and her colleagues at Boston College has shown that young women's spatial abilities—specifically, their scores on mental rotation tests—strongly predict their scores on the math SAT. In this study, as in most, men significantly outperformed women on the mental rotation task. However, when Casey's group factored spatial ability into their comparison of SAT scores, they could completely eliminate the infamous math SAT gap between men and women.

In other words, spatial skills are a major reason why males outperform females on the math SAT exam and probably explains how they beat them on other math and science tests as well. This makes sense, considering that as many as one-third of the questions on the math SAT are geometry items. However, Casey believes that the correlation between mental rotation and math scores extends beyond geometry, reflecting an approach to problem solving that is fundamentally visual—as opposed to verbal—and more efficient for understanding mathematical concepts such as fractions, proportionality, measurement, estimation, topology, trigonometry, and calculus. Indeed, a disproportionate number of women who *do* go into professional mathematics end up in fields like algebra and statistics, the least spatial of the various areas of math.

Casey uses the following example to explain how spatial visualization can help even young children understand mathematical concepts: "The idea that one-fourth is larger than one-eighth is a very difficult idea for young students, because the number eight is larger than the number four. But if a student can visualize what a pizza slice would look like if you had one-fourth or one-eighth of that pizza, that would help. People who don't have that ability can have problems with fractions their whole lives."

The ability to visualize objects and space in three dimensions is crucial to many pursuits, both in work and play. Beyond geometry, geography, and most areas of science, the list includes engineering, architecture, carpentry, car repair, surgery, sculpture, ball sports, map reading, driving, and even a number of traditionally female occupations, such as interior decorating and fashion design. In fact, a recent study of university students in London found that female fashion-design majors performed even better than male engineering and computing majors on a test of mental rotation. The truth is, many women have an excellent spatial sense in spite of the average difference between males and females, but more need to develop this skill to keep the door open to a wide range of opportunities.

If spatial ability is, as Howard Gardner argues, one of the cardinal intelligences, it is also one that gets the least amount of teaching in school. Every other ability—verbal, math, music, interpersonal, and even kinesthetic (athletic)—has its own slot in the curriculum or its own box for teachers to check off on report cards. Not so for spatial skills, for which the only real training comes from the video games, building toys, and targeting sports that are almost exclusively the preoccupations of boys. This recreational training combined with a lack of formal training in school explain why skills such as mental rotation and map reading show the largest sex differences of any cognitive ability.

Spatial Skills Can Be Learned!

I'm a scientist and I have pretty good spatial abilities. I don't normally boast, but the reason I bring it up is that I had many experiences as a child that I believe helped promote my sense of direction, orientation, and distance. For one thing, I had three older brothers, and my memories of childhood mostly revolve around playing with them. I was always tagging along, trying to join their games of Wiffle Ball, football, or basketball, or to worm my way into their vast Matchbox car collections or Lincoln Log creations. So, compared with many girls, I had a lot of experience with the kind of play that helps develop spatial skills.

Another opportunity was provided by a wonderful fifth-grade teacher, Mr. Pohn. He let me skip the regular math lessons and challenged me to spend that time completing a set of *Motion Geometry* workbooks. These introduced concepts like congruence, symmetry, flips, translations, rotations, and construction. I remember a lot of tracing paper and drawing—which was great fun and gave me a year's immersion in visuospatial thinking.

Motion Geometry has been out of print for thirty years, but I suspect there are plenty of computer programs, games, and even existing math curricula that could give children—girls especially—comparable experience with spatial transformations. Beth Casey is developing such curricula based on her findings about the gender gap in spatial skills and math performance. As Dr. Casey puts it:

> We believe that strategies that children choose for solving mathematics problems develop right at the outset of schooling, and that we need to start at this point in time to promote spatial as well as analytical strategies for solving mathematics problems. This is particularly critical for girls, since boys appear to be more likely to acquire these strategies on their own during the early years.

Casey and colleagues have designed a series of books and manipulatives called 'Round the Rug Math to teach spatial concepts to preschoolers and kindergartners. With engaging, people-oriented adventure stories, the books and activities appeal to both boys and girls and give young children hands-on practice with visuospatial manipulation such as measuring, graphing, drawing patterns, and piecing shapes together. The early start seems wise, considering the size of the spatial-skill gap and its importance to later math achievement. As Casey laments, spatial skills are neither a recognized goal of elementary education nor something we routinely test in children. If we did, she argues, we might identify just as many girls who could benefit from intensive spatial training as we do boys who need early remediation for their reading and writing skills.

Why have schools been so slow to recognize the importance of spatial skills? One reason is that early math curricula tend to focus on counting, arithmetic, and number sense—some of the least spatially oriented aspects of mathematics. These are important skills, of course (and girls master them as well as or better than boys), but by focusing exclusively on number skills, schools may be missing a crucial early window to inoculate children against later difficulty with spatial and mechanical thinking.

Then there is the attitude that spatial abilities are hard-wired: boys just *are* better at them because evolution has programmed them this way. As we've seen, recent evidence does hint that boys have an innate edge in mental rotation, but studies of babies' other mechanical understanding is far from unanimous. Boys *do* like moving toys more than girls do, and this bias may be enough (in the absence of formal training) to set them on a surer path of spatial learning. But whatever one's initial tendencies, there's no escaping that the skills themselves—mental rotation, line orientation, distance estimation, direction sense—are a matter of experience, or "practice, practice, practice." Just like Carnegie Hall.

One reason to suspect that the sex difference in spatial skills is influenced by experience is that it is not absolutely universal. Studying Eskimo living on the Baffin Islands of eastern Canada, anthropologists found no difference in spatial skills between men and women. As it turns out, these Eskimo women engage in as much hunting as the men do, requiring them to navigate seas and sparse landscapes with few of the features that women normally rely on to find their way. Lacking landmarks, Eskimo women navigate using the same directional and distance cues that males typically employ. And guess what! They develop spatial skills that are just as strong as men's.

Even boys need the opportunity to practice, practice, practice. In a widely cited 2005 paper, Janellen Huttenlocher and her colleagues at the

University of Chicago administered spatial tests to children from three different socioeconomic groups. While boys from wealthier homes outperformed their female classmates, boys from poorer families did not. In other words, boys are not predestined to trounce girls in visuospatial tasks but need opportunity and practice at sports, video games, building toys, and the like to hone such skills. Some families are better equipped to provide such experiences than others.

Each of these activities has, in fact, been linked to spatial ability. Many studies have found that the more children play with building toys—boys and girls—the better their performance on standard pencil-and-paper spatial-skill tests.

Sports are another influence. Considering the highly spatial nature of most athletic games, whether they involve hand-eye coordination (like tennis, basketball, baseball), foot-eye coordination (soccer), or whole-body coordination (swimming, tumbling, running, dancing)—you might expect that more athletic children would do better on spatial tasks than less athletic ones. It's true, at least based on studies of adults: compared to nonathletic college students, student athletes have been found to exhibit superior visuospatial skills, such as better mental rotation, more accurate visual orienting, and quicker reaction time to a visual stimulus. One Canadian study even found that for this last skill, there was no sex difference between men and women who were serious athletes.

Of course, it's possible that women who are more athletic also happen to be naturally better at spatial skills—that sports are not the *cause* of their improved skills. But research on video games, which are about as visuospatial as you can get, provides more direct evidence that it is the experience itself that's critical. In one study published in the journal *Nature*, psychologists first demonstrated that college-age men who played a lot of video games had better visuospatial awareness than men who played few such games. Then, to test whether this relationship was causal (as opposed to merely correlational), the researchers recruited both men and women who were *not* experienced video-game players and found that just ten hours of play (at the game Medal of Honor) dramatically improved both sexes' attention to visual space.

This experiment is key, because it shows that it's the *practice* that actually develops the skill. In fact, similar effects of computer- or video-game practice on spatial ability have now been shown by many researchers. They've tested third-graders, fifth-graders, adolescents, adults, boys, girls, men, and women. For all ages and both sexes, the results are the same: spatial-skill test scores increase after people spend a bunch of hours playing Tetris, Antz, Marble Madness, Robotron, Stellar 7, Targ, Battlezone, Zaxxon, and probably any other spatially oriented video game. In some studies, the sex difference in spatial-skill test scores was actually eliminated by video-game experience. This research confirms a cause-and-effect relationship between computer experience and visuospatial skills. Or, to put the most positive spin on it: video games are actually good for kids, especially girls.

And here's one more piece of evidence that spatial skills are shaped by experience: people who are fluent in sign language (whether deaf or hearing, male or female) were found to perform more accurately on tests of mental rotation than nonsigners did. Sign language requires close attention to the space between conversational partners, where slight shifts in hand position can signal different verb tenses or other important subtleties of meaning.

It may seem obvious that spatial skills—like every other cognitive task—are teachable. But believe it or not, this idea is rather new. In 1973, psychologists more or less threw up their hands and declared that Piaget's water-level task was "unlearnable." But by 1996, researchers had shown not only that this task could be learned, but even that the sex difference could be overcome through training. In this study, male and female college students both practiced drawing the water line in bottles of progressively more challenging shapes. Compared to untrained women, whose errors were more than twice as large as males', women who went through sixteen trials of this self-discovery exercise caught up to men who had undergone the same practice.

Cross-cultural research suggests additional ways in which the water-level puzzle and perhaps other spatial skills may be learned. Chinese children master Piaget's task at a much younger age than American children do, and virtually all pass the test by age seventeen (as compared to American college students, a sizable minority of whom—male and female—never seem to get it). Researchers believe one reason Chinese kids are so strong at both spatial and mathematic skills is that they spend so much time learning to write Chinese characters. Compared to our Roman alphabet, Chinese characters are more geometric and require greater attention to horizontal lines, which is the main point of the water-level task. In fact, psychologist Chieh Li and her colleagues compared Chinese American college students who could write Chinese with Chinese Americans who had never learned to write it, and they found that, for males, at least, those trained to write these complex characters were better at the water-level task than those who couldn't write them. Most notably, their scores on the water-level task correlated significantly with their math SAT scores, again revealing the importance of spatial ability to this high-stakes exam.

Clearly, the differences in spatial skills between the sexes can be narrowed with practice. Block play, ball play, fast-paced computer games, and probably many other activities contribute to the spatial-skill sex difference, and they can also be put to work to compensate for it. Recall that the gender gap in Piaget's water-level task has actually declined over the past few decades. However, the gender gaps for other skills, such as mental rotation, have not budged and may even be growing larger, probably because of the tremendous difference in the amount of time boys and girls spend maneuvering virtual objects through space, especially with the latest hi-def video games.

If spatial skills are learned, why do we do so little to formally teach them? Chinese children get a head start in spatial skills when they master written Mandarin, and they study geometry throughout their last six years of schooling; compare that to the one or two years geometry is studied in American schools. Spatial thinking is the key to higher math, physics, chemistry, and engineering. It's little surprise that children immersed in this way of thinking perform better in such difficult fields. What is surprising, however, is that in spite of small sex differences in their own country, Chinese girls far surpass American boys in both spatial and mathematical skills.

Without deliberate training, children can learn spatial skills only haphazardly, according to their interests and hobbies. Something so important should not be left to chance.

The Cultural Hurdle

Spatial skills are important but, unfortunately, learning them is not the only obstacle girls must surmount to reach their full potential in math, science, and engineering. There are also considerable cultural barriers, exemplified by the Teen Talk Barbie fiasco of 1992.

That was the year Mattel brought out a new version of its voluptuous doll implanted with an electronic chip whose repertoire infamously included "Math class is tough," along with "Let's go shopping," "Will we ever have enough clothes?" and other deeply provoking sentiments.

The uproar came fast and furious—from parents, math teachers, and college professors—so Mattel promptly removed the line about math from the sound chip. Not, however, before the subversive Barbie Liberation Organization had a chance to switch the chips between several hundred Barbies and G.I. Joe dolls. Imagine all the little girls that Christmas unwrapping the beloved doll only to hear her growl, "Vengeance is mine," while boys heard their musclebound warrior chirp, "Let's plan our dream wedding!"

Still, the fact that a company like Mattel, which has so much influence over young children, could so blatantly bash girls' math abilities indicates just how pervasive this stereotype is. As much effort as has gone into helping girls achieve in math and science, it all becomes futile against this persistent portrayal of math and science as unfeminine.

Enlightened parents, teachers, and textbook publishers have bent over backward to show girls and women happily doing science and math. We enroll our daughters in soccer and chess lessons, buy them their own building blocks, and seek out female pediatricians as role models. And it may be working, up to a point: in fourth grade, an equal number of boys and girls (about two-thirds) said they liked science and math, according to a 2000 report by the National Center for Education Statistics.

But there is still a long way to go. Recall that parents in the United States, China, and Japan all stated that six-year-old boys were better at math even though there were no actual differences in performance at this age. Similar attitudes have been reported in other studies, and while things have begun to change, there's little evidence we've reached an era of truly gender-blind expectations.

The message is not lost on girls. As early as first grade, girls express less certainty than boys that they can succeed in math. By eighth grade, girls are some 30 percent more likely than boys to agree with the statement "I am just not good in mathematics"—even in countries such as Hong Kong and New Zealand, where girls actually score higher than boys on standardized assessments. Many researchers, including Beth Casey, have found that girls' lower confidence in math is a significant factor in their scores on high-stakes exams, including the math SAT. (To be fair, Casey also found that math confidence is less important than spatial skill in determining math performance.)

Even girls who *are* strong in math are not immune to this lack of confidence. In contrast to boys, who tend to view their success as a matter of natural mathematic talent, girls are likelier to credit their success to "hard work." Researchers have documented this attribution gap in parents, teachers, and the children themselves, and they find that it holds for math, science, and even reading ability. Girls are simply more reluctant to think of themselves as smart, and they believe that they do well only because of greater effort, while boys think just the opposite.

Little wonder, then, that girls spend more hours doing homework than boys (with better grades to show for it) in all subjects, including math. But when it comes to important tests such as the SAT, girls' lack of confidence has a considerable impact. Boys, as we've seen, score some thirty-five points higher than girls on the math SAT, and there is evidence that at least some of this difference is attributable to confidence and anxiety (as I discuss in the next chapter).

Researchers call it "stereotype threat" and have shown repeatedly that women and minorities, when made aware of negative stereotypes about their groups, underperform white males on challenging exams. The pioneering experiments were performed by psychologist Claude Steele and his colleagues; they administered a difficult math test to two sets of undergraduates—both groups included men and women, and all were students for whom math was an area of talent and pride. Before the test, students in one group were primed with a gender stereotype: the researchers told them that the math test usually revealed a gender difference. Students in the other group were told beforehand that the exam typically showed no gender difference. Figure 6.7 graphs the striking results.

As the graph shows, women who were reminded about the stereotype performed much worse. Essentially, they stopped trying to solve the more difficult problems. But in the group that was not reminded of the negative stereotype, women performed virtually as well as men.

You might also notice the stereotype *lift* the male students received when primed with the belief that they would perform better on the test. Stereotyping affects everyone, for better and for worse. Males rise to the occasion when their masculinity is on the line, but they can also *suffer* from stereotype threat in certain circumstances. For instance, white

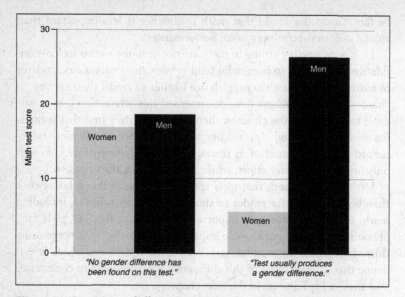

Figure 6.7. Scores on a challenging test taken by talented math students at the University of Michigan. Test questions were taken from the Graduate Record Examination; each correct answer earns one point, and wrong answers are penalized by deducting one-fifth of a point.

males performed worse on a difficult math test when they'd been told beforehand that Asian males usually did better on it, but they performed comparably when primed with an egalitarian belief. Other studies have shown that white males' performances decline on tests of athleticism when they're primed with comparisons to African Americans, and on tests of social sensitivity when they're primed with comparisons to women. So it's not just a matter of white-male dominance; any stereotype can impair people's performances in areas they are suspected to struggle with. (An interesting thought when we consider the messages young boys hear about literacy.)

For women who are good at math and science and derive much of their self-esteem from such abilities, stereotype threat looms especially large. Everyone in a demanding career reaches some point of great challenge—during academic training, an internship, or the first job. It's all too easy to succumb to a lack of confidence, especially when a woman sees male colleagues confidently climbing the career ladder without the burden of self-doubt. Stereotypes threaten by adding pressure and anxiety, distracting a person from the task at hand, and, ultimately, tugging an individual up or down based on the culture's widespread expectations about that person's abilities.

However, stereotypes do not threaten women's verbal abilities, as we saw in the last chapter. So what often happens is that bright girls and women, faced with the extra emotional burden of having to disprove a stereotype, drop off the technical track and shift into less mathematically intense fields during college or postgraduate work. As Claude Steele describes it, "Women may reduce their stereotype threat substantially by moving across the hall from math to English class."

A recent experiment showed how stereotypes really do affect women's career choices. In this study, led by social psychologist Paul Davies, two groups made up of male and female undergraduates were shown three minutes of TV commercials and then given questionnaires about their career interests. Students in one group were shown several gender-stereotyping commercials, such as a woman bouncing on her bed with joy over a new acne product and a woman "drooling" to try a new brownie mix. Students in the other group were shown gender-neutral commercials, such as ads for an insurance company, a cell phone, and a pharmacy. Although the commercials had nothing to do with math ability as such, women who viewed the ditzy ads were markedly less likely to express an interest in quantitative careers, such as engineering and computer science, and more likely to express an interest in verbal careers, such as journalism and communications. The pattern for women who viewed the gender-neutral commercials was the opposite: greater interest in the quantitative careers.

The sad truth is that stereotypes are potent role models for boys and girls of all ages. Math, science, and computer brilliance are simply not feminine, as Barbie, *Mean Girls*, and much else in popular culture tells us. Girls figure this out distressingly early. Even girls who are talented in math (like boys who are talented writers and artists) have a tough time maintaining their focus and confidence with gender identities urgently pulling the opposite way.

It was all I could do to contain myself when my daughter, Julia, then a fifth-grader, casually announced, "I hate math," one evening while doing her homework. Here she was, working way ahead of most of her classmates, a feat that up to that point had been a source of great pride. But as soon as things started getting a little too challenging, her attitude shifted, and she could find comfort in the feminine stereotype "I hate math (so I don't have to try so hard anymore)."

My response (while trying to suppress all signs of my inner heartbreak): lots of praise, extra help, and repeated attempts to show her the beauty, importance, and joy of solving math problems.

Adolescence is precisely the time when math performance starts diverging between the sexes. Girls manage to hold their own through middle school but start falling behind boys in high school. Puberty has been held to blame, though the evidence linking testosterone or estrogen to math or spatial skills is not, as we've seen, in any way convincing. Perhaps, instead, boys rise and girls fall due to the renewed gender identification that puberty brings, unconsciously coaxing children to live up to the dominant role models for their gender: pretty, talkative women and strong, logical men.

Let's face it, while parents are fixated on the three R's, teens' brainpower is far more dedicated to the three C's: clothes, crushes, and being cool. Adolescents are about as consumed by stereotypes as you can get, and it's not hard to imagine how that preoccupation colors their academic attitudes, interests, and confidence. As Bob Fischer, an early coach of Math Olympian Melanie Wood, explains, "I think it starts in upper grade school and middle school. Girls hear that math is something boys do and girls don't do. Also, in middle school, girls start to feel that many boys won't like to be around them because they're too smart, so they pull back."

Beginning in middle school, many girls "disidentify with" (as Claude Steele puts it) math and science as areas of self-worth and accomplishment. And boys, whether deliberately or unconsciously, encourage them to let go. As educational researcher Fiona McDonnell discovered while interviewing girls in high-school physics classes:

> I heard numerous stories of how being a girl does make a difference in the high school physics classroom. Almost immediately upon meeting one female student, she announced to me, "I've always dreamed of becoming an astronaut." She then proceeded to tell stories about "the division" and "the fights" between the boys and girls in her class and how she often felt "overpowered by the guys." One female explained that she did not go to the chalkboard to solve problems because of

"the joking club" and "the heckling you can get." Another . . . commented, "You know, sometimes you feel like a visitor in someone else's classroom."

Some girls do manage to fight the stereotype longer, into high-school and university education, but the pressure remains and is but one of many extra hurdles females face in pursuing careers in math and technical fields. As one microbiologist described the several causes of the leaky pipeline, it begins with "girls' fear of being unpopular with boys if they are perceived as science nerds" and includes "doubts about being able to combine family and career . . . and exclusion from informal networks." She concluded, "It takes a lot of dedication to overcome these barriers, most of which men do not have to face."

Finally, we get to discrimination, which is alive and well though operating at a largely unconscious level. The topic could fill another book, but let me just quote Ben Barres, a transgendered neuroscientist with a unique perspective on what it's like to be both a woman and a man in science. Barres recently described his experience as an undergraduate woman at MIT when she was the only student in the class to solve a difficult problem. Her professor actually accused her of cheating (by getting the answer from her boyfriend)! As Barres explains:

> If I had been a guy who had been the only one in the class to solve that problem, I am sure I would have been pointed out and given a pat on the back. I was not only not given positive feedback, I was given negative feedback. This is the kind of thing that undermines women's self-confidence.

Skeptics argue that in this era of affirmative action, women are no longer discriminated against in science or any other field. But Barres tells another story: after his sex change, he was invited back to MIT to give a lecture, where another professor was overheard saying, "Ben Barres gave a great seminar today, but then his work is much better than his sister's."

Sex differences in science and math performance are real. The good news is that they are declining, at least in some areas. Girls are now taking as many science and math classes as boys are, scoring indistinguishably from them on standardized school achievement tests, and entering some scientific and quantitative careers, such as medicine, biology, and accounting, in equal numbers as boys.

But other gaps persist—in spatial abilities like mental rotation and map reading; in school subjects such as geometry, calculus, physics, and computer science; and in women's entry into careers such as engineering and information technology. Some presume these remaining gaps represent the limit of women's abilities. Sure, women can do medicine (which requires memory and people skills much more than math), but few are up to the more hard-core technical fields such as string theory, aerospace design, and nanotechnology. Something innate, evolutionarily programmed, must wire up male brains differently to give them the talent and passion to pursue such frankly geeky stuff.

It's possible. Then again, no one was ever born knowing how to shoot pool, build driftwood tepees, or solve differential equations. Every one of these skills is learned through practice and experience. Remember that sex differences are found in a subset of math and spatial skills only, and even the largest difference—in mental rotation—is far smaller than the gaps in noncognitive traits such as height and throwing ability.

We must admit that much of the remaining career segregation—and resulting gender gap in salaries—is due to culture: to the very separate male and female universes in which we grow up, hone our skills, and fix our comfort with different types of jobs. Just as men enter careers like nursing or preschool teaching with trepidation, so do many women lack the fortitude to test their mettle as computer programmers, car mechanics, or quantum theorists. According to some research, 25 percent is the tipping point: when the proportion of women or men in a previously segregated field rises above this level, stereotyping declines and people begin to be judged by their actual abilities.

The role of culture and gender identity is illustrated in a very different way by the dramatic shift in two occupations I mentioned at the outset of this chapter—accountant and veterinarian—both of which swung from less than 10 percent female to more than 60 percent in just a few decades. The rapid swing in these math- and science-based careers proves that interests and abilities are not the only reasons for men's and women's different career decisions. Men are apparently fleeing these fields because they've tipped *too* far in the female direction, showing how gender identity continues to influence decision-making even well beyond the truck-versus-doll years.

While girls and women have made great strides in math, science, and related fields, the cultural barriers are still high, especially among adolescents, like the ones in my class on Brain Day. We must keep working hard to break those barriers down. Male or female, there simply aren't enough American-born children going into science and engineering. Graduate programs are filled with foreign students raised in countries that have better science and math education and, more important, where such careers hold higher prestige. Math and science have been the source of much of humanity's greatest creativity. We are not doing enough to harness this creative talent in American children in general, and girls in particular.

Tips

The following suggestions can help improve the odds of girls' sticking with a more rigorous, quantitative curriculum, putting them in a better position to compete for modern high-paying jobs. There's also ample room for improving the mathematical and technical education of American boys. Only about one-fifth of American twelfth-graders reach the proficient level in math, and the United States ranks twenty-second out of the forty industrialized nations in the science proficiency of our fifteen-year-olds. So, many of these tips should be valuable for girls and boys alike.

- **Spatial skills should be formally taught.** While early education focuses intensely on reading and calculation, it provides little opportunity for children to practice the visualization, estimation, measurement, navigation, and shape-manipulation skills that figure prominently in later math and physical understanding. Preschools and elementary schools should start teaching such skills using block-building exercises, maps, 3-D puzzles, and spatial disembedding (finding shapes hidden inside other designs). Girls can catch up to boys when deliberately taught skills such as mental rotation, disembedding, and Piaget's water-level task. Teachers from preschool onward should incorporate spatial-skill training into their curricula.

    One computer-based method, called ST math (for *spatiotemporal math*), looks especially promising for improving math comprehension in school-age children by using appealing visual exercises.

Developed by the nonprofit MIND Institute in Santa Ana, California, this suite of software trains students to see relationships such as fractions, proportionality, and symmetry using engaging, progressively more challenging puzzles. Thus far, the software appears beneficial in raising young students' math performance and algebra readiness.

• **Encourage the joy of building.** Girls like to build things, but they often get muscled out of the block corner in kindergarten by overzealous boys. So they turn to other pursuits, like drawing or dress-up, which develop useful skills but not particularly spatial ones. Teachers can force the issue by limiting the choice of activities to a variety of building toys during certain periods. Parents should buy building toys specifically for their daughters. (In one study, it was found that 88 percent of LEGO sets had originally been purchased for boys.) One problem is that manufacturers offer few such toys geared toward girls. But if parents viewed building toys as educational tools, manufacturers might be encouraged to go beyond Star Wars and Transformer themes and produce more building sets that appeal to girls (such as houses, stables, schools, and shops, pink and purple blocks, and a variety of animals and human figures).

Building opportunities should extend into elementary school and beyond. Science and art classes can provide opportunities to design and build model homes, bridges, windmills, and solar-powered buildings. I was fortunate to have a woodworking section in my junior-high-school art class. As I remember, we girls were just as excited as the boys to design and build our own projects. So I was dismayed recently when I suggested to Julia's high-school adviser that my daughter might enjoy the school's woodworking course and was told that "unfortunately, there aren't many girls in that class." Here's a situation (along with computer studies; see below) where single-sex classes may be beneficial.

• **Consider chess for girls.** Here's another competitive pursuit that is heavily male dominated: the ratio of male-to-female grandmasters is about ninety-nine to one. However, a recent analysis revealed no difference in skill between boys and girls in locales where equal numbers of both take up the sport. Rather, the male-to-female ratio of chess experts is a function of the number of boys and girls who enter the competitive chess circuit in the first place. Very few girls start competing, which leaves very few women in the running after the ten or more years it takes to become an expert.

But chess is a great mental workout. It challenges logical, spatial, and forward-thinking skills. While there isn't much research to prove it, many educators believe chess is especially beneficial for higher math skills, since it emphasizes spatial planning and problem solving over simple memorization.

Like most gender-stereotyped activities, chess can be a hard sell to girls. As always, an early start helps, and older girls may take a greater liking to the game if they're taught by accomplished female instructors.

• **More sports, especially for girls.** Sports are a great, healthy way to improve spatial skills. All that movement, aiming, shooting, throwing, catching, and passing (depending on the sport) are excellent for honing visual attention, spatial awareness, and reaction times. Given our current obesity epidemic, every child needs more activity, but girls in particular can benefit both in mind and body from greater participation in athletics.

The number of girls participating in sports has risen dramatically in recent decades. But as Angela Smith, past president of the American College of Sports Medicine, has noted, "Just because more girls than before are playing sports doesn't mean that *enough* girls play sports." Many girls drop out of sports in adolescence because athletics don't mesh with the feminine identity they're suddenly more interested in projecting—yet this is precisely the age when the strength and assertiveness that sports provide could be beneficial.

In her helpful book *Go Girl! Raising Healthy, Confident and Successful Girls Through Sports,* sportscaster Hannah Storm notes the many benefits of sports for girls (including stronger bones, less menstrual pain, healthier weight control, higher self-esteem, and lower risk of depression and illegal drug use) and offers tips for encouraging girls' athletic participation, including (1) attend your daughter's games whenever possible, (2) emphasize effort over winning, (3) attend women's sporting events with your daughter, (4) put up photos of your daughter playing sports, and, of course, (5) play sports with your daughter.

• **Paint technology pink.** The digital divide is very real. While many boys are spending *too much* time on the computer, many girls need more time on it—not buying music or socializing, but learning how to create and customize software and, yes, even to play fast-paced targeting and driving games (for their visuospatial skills). Girls are simply not as technologically fluent as boys. They don't understand how computers work as well as boys do and are therefore less able to solve technical problems and keep up with new advances. It's been said that girls regard computers as tools while boys treat them more like toys. So boys play on the computer more creatively than girls, which is how their digital literacy arises.

It doesn't help that most software is created with boys in mind. With a little searching, however, parents can find titles that are engaging and that encourage girls to explore on the computer. I can't comment on the various best-selling Barbie programs, but when Julia was young I did find some Madeline and Carmen Sandiego software that was both educational and appealing. More recently, Julia's been into the Sims, which has some great 3-D design opportunities—rotating buildings and see-through roofs and walls—that I'm hoping is as good for her spatial abilities as all the virtual navigation in action video games is for boys.

Another issue is female teachers, who are often slower to embrace new technologies than their male counterparts. This sets a poor example for girls but can be overcome with better training and IT support in schools.

Ultimately, however, the most important way to funnel more girls into careers in information technology will be through dedicated coursework. High-school girls don't take as many programming classes as boys do. When they do, they're often put off by the experience. To tackle the decline of women in computing, the American Association of University Women (AAUW) has proposed steps that include (1) better integration of computing into all areas of the curriculum so girls can pick up skills in the areas of their strengths; (2) enrolling more girls in computer classes, perhaps by eliminating their advanced-math prerequisites, which are not essential to many programming tasks; and (3) once the girls are there, ensuring that they make up a critical mass in any computer class so they are not relegated to observer status but get real hands-on experience. In fact, while the AAUW is generally hostile to the idea of single-sex education, this is one situation in which it has encouraged experimenting with all-girl groupings so they can be freed from self-consciousness and overt sexism about their technological abilities. And finally (4) encouraging a "hacker" mentality among girls by allowing them to take apart old computers, tinker with programming code, and conquer their fears of hurting anything, empowering them to not just consume but also control computers.

• **Hands-on science.** Just as computer mastery requires getting your hands on the hardware, so science mastery requires hands-on experience in the lab. You can't truly understand anatomy, chemistry, or optics without rolling up your sleeves and manipulating nature to see what happens. In some mixed-sex lab groups, boys take over the fiddling that's inevitably required to get an experiment to work. Girls stand back, reading the instructions or acting as scribes but less often handling the chemicals, equipment, or slimy specimens—which impairs their confidence. So while girls may understand the scientific concepts, they don't actually *do* science, a big handicap when it comes to exploring technical fields down the road.

Valerie Lee and her colleagues at the University of Michigan analyzed science performance across the United States and confirmed the value of hands-on experience for high-school-science achievement, especially for girls. Single-sex lab groups are a solution in some cases, forcing certain girls into more active roles. But some girls thrive in coed groups, and a universal policy of single-sex labs or science classes deprives both sexes of the opportunity to work together. As always, the best solution is individualized instruction—teachers remaining sensitive to who's doing what and adjusting the groupings to maximize each student's active participation. Another tactic is to reinstitute practical exams—lab tests using real specimens and equipment—to ensure that every student masters the necessary hands-on skills.

• **Check the sexism at the lab bench.** Teachers need to be aware of boys monopolizing tools or making fun of girls' mistakes in the lab. With adolescent boys seeking power and authority and girls seeking physical admiration, the combination is ripe for old-fashioned sexism. Vigilant supervision is the only way to maintain a warm climate for girls in science and technology.

• **Emphasize the human side.** Young women often report turning away from science and math because they don't find the topics relevant to their interests. Formulas and equations can seem pretty dry but, rightfully, they should be appreciated as the source of all our amazing technology, from cell phones, hybrid cars, and cancer cures to new fabrics, cosmetics, and incredible digital film and sound technologies.

It's largely an image problem, and one that can easily be solved. Teachers, textbooks, and scientists need to prove to students that science and math are relevant, that engineers don't just think about numbers all day but get to design, create, and work together to find new ways of helping people and advancing civilization. University faculty too need to educate the public about science and math, as *USA Today* contributing editor Laura Vanderkam suggests: "Math needs a Stephen Jay Gould writing about how a team cracked a spy code; engineering needs a Sylvia Earle describing how a better water system saved a Nepalese village."

• **Teach about the importance of science and math in high-paying careers.** Middle school is not too young for adolescents to begin learning some economic reality: math and science are routes to high-paying, creative jobs, not only in research but also in finance, health care, design, engineering, and telecommunications. Few children are aware of the advantages of a strong technical education for broadening their career prospects as well as for developing the logical and analytical skills prized in any occupation. A little career counseling early on may boost the motivation of *both* boys and girls to persevere with the technical side of their education, which, sadly, is all too hard to catch up on later in life.

• **Fight the stereotypes.** We may think we encourage our sons and daughters equally in math and science, but the evidence says otherwise. Parents still buy more science toys and books for boys, spend more time on such projects with them, and have higher regard for sons' math and science abilities than daughters'. Stereotyping by fathers, in particular, has been associated with lower interest in math among daughters but higher interest among sons.

But research by Joshua Aronson and his colleagues indicates that students can skirt stereotype threat using a few, easily coached strategies.

In one of Aronson's studies, college students volunteered to be mentors to seventh-graders from a rural low-income school district. The mentors met with their pupils several times throughout the school year, focusing on instilling two positive attitudes, (1) intelligence is not fixed but expands with education and study; and (2) having difficulty with a new subject is normal and does not mean a person is not smart. The mentors reinforced these messages through e-mail and by helping the students create public service websites to advocate this same view to others. (Such saying-is-believing methods have proven to be highly effective; people adopt the beliefs they advocate.) The study also included a control group of students who met just as frequently with their mentors and who also created a website but whose interactions were focused on instilling an anti-drug message.

Near the end of the school year, the students took their standardized Texas achievement exam, and guess what: Compared to the control students, boys and girls coached in the optimistic way of thinking about their intellectual growth scored higher on both math and reading tests. What's more, the sex difference in math, significant in the control group, was erased in the groups mentored in positive, nonstereotypical thinking.

Stereotypes are inevitable, but by vigilantly criticizing them whenever they arise and by teaching children to have faith in their own effort and learning potential, we can stem the self-doubt and underachievement many adolescents suffer, especially girls in science and math, and boys in writing and foreign language.

• **Role models.** It's not just our attitudes but also our examples that need some adjusting. Mothers who turn the math homework over to dads send daughters a covert message about who's good at the subject. Mothers should let their children see them balancing the checkbooks, paying the bills, or calculating square footage for some new flooring. Research has shown that Asian American mothers are more involved in helping their daughters with math homework than Caucasian mothers. The gender gap in math among Asian Americans is much smaller than among Caucasian American children, and this study suggests that maternal role modeling may be part of the reason.

Role models are also important at school. One high-profile study found that middle-school girls do indeed perform better in science classes taught by women teachers (and boys do better in language classes taught by men). The data are not as clear for math, but regardless, male science and math teachers should go out of their way to bring women scientists, economists, engineers, architects, and so forth into the classroom to talk about the value and joy of math and science.

Obviously, women have made great strides in science, math, and related fields over the past few generations, and girls are doing better than ever in these subjects at school. But there is farther to go and no reason to suspect that either sex has reached its limit in technical achievement. There is no magic way to make a child love math or science, but there is certainly a lot more we can do to help widen the pool of girls equipped to enter and compete in these important twenty-first-century careers.

7
Love and War

IT WAS A STRANGE EVENING FOR SAM, the night our school district set up a "haunted forest" for kids in a nearby park. The children were excited as they raced around in the cool mid-October twilight, all hopped up on pasta from the school fundraising dinner. I get an odd call on my cell phone—Sam has borrowed someone else's cell phone to ask if he can bring his buddy Liam over to our house with "his friends."

"Sure," I say. "Which friends?"

He hems and haws, then finally mentions some girls' names. I immediately realize why he is being so evasive (and where he got the phone).

I laugh to myself as I head home to supervise, thinking, *Sam's taking an interest in girls.*

But when the crowd assembles, I realize just how far our ten-year-old has to go. He and Liam try stirring up some interest in air hockey or Xbox while the girls are squealing about which boys are cute and who likes whom. Sam and Liam seem a little bewildered, and, mercifully, the impromptu party breaks up as a succession of parents arrive to take their kids home.

I'll say it again: boys and girls are different. But unlike the mostly subtle gaps in reading and math performance, their social and emotional differences are as clear as day. You don't have to test large numbers of fifth-graders to witness the profoundly different ways boys and girls talk to their friends, play at recess, relate to their teachers, and express their feelings. Sex differences in emotional response and social interaction are modest in early childhood but grow steadily as boys and girls segregate into their starkly different cultures. By the time they start noticing each other again, around puberty, the differences can be most striking.

In this chapter, we'll track the development of boys' and girls' emotional lives: how they communicate their feelings; their vulnerability to mood disorders; their propensity for risk taking; and their interpersonal differences in empathy, aggression, and competitiveness. Most of these gaps are amplified in adolescence, a period of gender intensification when boys and girls strive to differentiate from each other in the interest of appearing sexually attractive. Hormones are an obvious trigger, and I conclude the chapter by explaining how they launch puberty and what we know about their effects on adolescents' brains and behavior.

Compared to most of the gaps we've considered up to now, emotional differences are more dramatic, and arguably the most important. More than verbal fluency or math SAT scores, sex differences in risk taking and nurturing account for the nagging chasms between the sexes in achievement and personal fulfillment—why men are better paid but have less-intimate friendships; why women make effective managers but are still banging the glass ceiling in many workplaces.

Evolutionary psychologists have no trouble explaining such differences, pinning them on the familiar Fred-and-Wilma dichotomy. As always, such claims are hard to test, but sex differences in emotional and relational styles can more plausibly be linked to different selective pressures on male hunter/warriors and female gatherer/nurturers than most cognitive differences can. That's because differences in aggression, empathy, and competitiveness are more closely related to reproduction, the purpose of pubertal maturation and the engine that drives sexual selection. To the extent that evolution has programmed any male-female differences, it's done so through eons of mate selection, which has left its deepest mark on the way young men and women try to impress each other as worthy sexual partners.

Still, as universal as some of the differences appear, none is as hard-wired as commonly portrayed. Like every other brain function we've considered, the social and emotional differences between boys and girls begin as tiny seeds planted by evolution and nourished by hormones but blossoming only under the hot sun of our highly gendered society.

If the differences appear dramatic, it's also because emotional intelligence is not something we deliberately teach, either at home or in school. Boys and girls are left to passively learn how to love, relate, assert, challenge, compete, compromise, soothe, and support each other. And yet, as psychologists and even economists are increasingly finding, such abilities are just as important to children's futures as the academic achievements we put so much stock in.

Neuroscientists too have taken a keen interest in social-emotional phenomena. Our feelings and interpersonal lives are governed by the brain's limbic system, a collection of structures that sits deep in the middle of the brain, bridging the rational and the instinctive parts of the nervous system. Recent research has identified sex differences in various limbic structures, such as the hypothalamus, hippocampus, and amygdala. The picture is still foggy, and it's even more so when it comes to sorting out the causes of such differences. But something we do know is that the limbic system is one of the more plastic, or modifiable, areas of the brain. Regardless of their origin, there is much we can do to help reduce some of the gnawing gaps between boys and girls, paving the way toward better mental health and personal fulfillment for both sexes.

Big Boys Don't Cry

Waiting in line at the post office, I can't take my eyes off this sad little face: he's two or three years old, propped on the counter with his arms and legs wrapped tightly around his mom. She's trying to finish up her business and get him home for his nap, but it's all too slow for this grumpy toddler. His soft round face lies pouting on mom's shoulder. Gary, the amiable postal clerk, is doing his best to coax a smile out of him, but there's no melting this little guy's cranky expression.

Boys are very emotional. In stark contrast to adult men, young boys let it all hang out, and there is little evidence of significant sex difference in emotional expression in early childhood. As we saw in chapter 2, boys are, if anything, the more emotional sex: newborn males are more irritable, cry sooner when distressed, and are less easy to console than newborn females.

Things even out soon enough, but as anyone with a son knows, young boys show plenty of emotion. Like girls, their early lives are a roller coaster of soaring and sinking moods, from angry tantrums to exuberant smiles and the most heart-melting declarations of love for their parents, siblings, and pets. Boys' faces are plenty expressive (one of many reasons parents love to photograph their kids so much). Boys also cry, perhaps a bit less than girls do by age four or five, but still enough to make you want to scoop them up, cuddle them close, and do everything in your power to protect them.

But while boys most certainly *feel* all the same emotions as girls do, they quickly learn not to show them. That's the stereotype of the stoic male, and it's accurate, at least outwardly. Men really do display less facial expression, cry less, and generally disguise their feelings more than women do. But this doesn't mean men aren't experiencing the same feelings. In laboratory studies, men respond even *more* intensely than women to strong emotional stimuli such as a violent movie or an impending electrical shock. The catch is that their responses are mostly internal: compared to women, men undergo greater increases in heart rate, blood pressure, and sweating when confronted with highly emotional situations. So even though they're less visible on the surface, men's emotions are just as powerful as women's, which was confirmed by one large study that asked men and women to report anonymously on their private emotional experiences.

It's the "mask of masculinity," as Harvard psychologist William Pollack calls it, and a central fact of growing up as a boy. The feelings are there, but less and less visible to others, making it hard for their parents, teachers, and friends to know what they're experiencing. Boys learn this hiding trick early, though it takes most of their youth to perfect. By age eleven, they are about 20 percent less likely to cry than girls, and by age sixteen, 40 percent less likely. It's a smaller difference than you might expect, but a clear developmental trend that's not explained by pubertal hormones, since boys' crying declines steadily both before and after puberty. (Nor does girls' tendency to cry change with the onset of menstruation.) Rather, the sex difference in emotional expression is learned—one of the few sex differences researchers unanimously agree is a matter of socialization. Boys don't have to look far to see tough, impassive male role models all around them: their fathers, friends, and the superheroes they mimic from their toddler years on up.

Consider David, one of Toby's seven-year-old soccer pals. During the frenzy of a recent game, David spotted a pretty bird feather, stopped to pick it up, and then clutched it tightly as he resumed racing after the ball. Several parents noticed him running with the feather in his hand and commented on how cute it was. But at the next break, his dad gently wrenched it away, teasingly reminding his son to stay focused on the game. The mixture of affection, embarrassment, and concern on this father's face said it all: as sweet and sensitive as our boys may be, we feel compelled to toughen them up and teach them to maintain their game faces both on and off the playing field.

Happily, we don't hear fathers screaming, "Be a man!" at their sons too much any more. Still, boys get the message in other ways, especially from their peers. Pollack dubs it the boy code, and it gets heavily reinforced in typical boy play. As we saw in chapter 3, boys and girls segregate into separate play groups when they're quite young, and once there do, they play very differently. Boys coalesce into larger groups with a more definite hierarchy of leaders and followers, while girls mingle in pairs and trios, bonding through talk and shared feelings. Boys' bonding is less verbal, a matter of shared games and exploits without a lot of introspection. While girls get close by confessing vulnerable feelings, boys gain nothing and may even lose significant status if they display too much emotion in front of their peers.

This is not to understate the importance of boys' friendships, which can be deeply supportive even if they're rarely acknowledged (in contrast to girls, who are always declaring their best-friend status to one another). But for some boys, the constant pressure from peers and society to suppress their sadness, shame, and frustration can become an emotional straitjacket that threatens their mental health and ability to connect with others. From acting up to violence and drug and alcohol abuse, many behavioral problems have been blamed on the unforgiving boy code of emotional suppression.

The conclusion is obvious: boys need greater freedom to express their feelings. And parents and teachers need to foster their emotional intelligence. Boys who can voice their own feelings stand a better chance of understanding them in other people, which paves the way for more effective communication with their future spouses, coworkers, and, eventually, children of their own.

It's not always easy, but clinical psychologists advocate action-love as the key to bonding with boys: shared activities, preferably involving movement, seem to help boys let down their guard. It may be a game of catch, a ride in the car, or a stroll to the store that works best to open them up—and all these are probably better than the impenetrable "How was your day?" at the dinner table. So while peers, fears, and beer commercials all conspire to toughen boys up, parents and teachers need to create a safe space for them to express their feelings, solicit support, and stay connected to those who care deeply about them.

The Downside of Emotional Expression

Nancy is worried about her daughter. Maddie had a great start in middle school and was excited about her new locker, her teachers, and switching classes every forty-five minutes. But things began sliding about a month into the school year: her friends began whispering just out of earshot, she missed some crucial assignments, and her stomach started to hurt. Nancy felt Maddie growing more distant and tried to set aside some time each night to talk to her. But Nancy soon realized that these conversations were not helping. Instead of making Maddie feel better, the late-night heart-to-hearts focused her even more on her bad feelings: that she was dumb, fat, and that her friends didn't like her anymore. Nothing Nancy said seemed to lighten Maddie's mood, so she started cutting the conversations short, eventually returning to the good-night kiss and brief "sleep tight" that had been her habit before all the trouble began.

By the time they hit adolescence, boys and girls do express their emotions very differently. But while boys may be hampered by the boy code, girls could take a few lessons from it. Emotional expression is all well and good when you're surrounded by intimate friends and a supportive family, but some degree of emotional *suppression* is essential for helping children cope with adversity and prepare for the adult world. Boys get toughened in a way that girls do not, which may be a hindrance for boys when it comes to intimate relationships but can be a benefit out on the soccer field or in the workplace.

Or, as Martha Stewart said during her season on the TV show *The Apprentice*, "Women in business don't cry, my dear."

I learned this the hard way a few years into graduate school. Overwhelmed with frustration by experiments that weren't working, I let the tears flow in front of my adviser, an important scientist who had trained few female students up to that point. I guess I was looking for some sympathy, or at least some appreciation of the hard if fruitless work I had been putting in. But what I got instead was doubt: his voice changed, and though the tone was caring, I could hear the questioning in his words—maybe he was asking himself, *Does Lise have what it takes to make it in science?*

"Big mistake," as my then boyfriend (and now husband) told me. Never reveal weaknesses to a boss. Save your self-doubt for your girlfriends and pub mates, which is a lesson I now pass on to my students.

So when it comes to emotional expression, the message is different for the parents of boys than the parents of girls. While boys can benefit from greater opportunity to talk about their feelings, girls may be doing *too much* of this already. Females' tendency to admit vulnerability and negative emotion not only impairs their academic and professional confidence but also may be a factor in their mental health.

Around the world, women are about twice as likely as men to suffer from depression and anxiety disorders. At least one in ten women will experience a major depressive episode at some point in her life, which adds up to hundreds of millions of suffering women. Scientists are searching far and wide for any male-female differences in neurochemistry, genes, or brain anatomy that might explain this gender gap in depression and anxiety. The likeliest suspect is the neurotransmitter serotonin, target of the popular class of antidepressant drugs known as selective serotonin reuptake inhibitors, or SSRIs, whose brand names—Prozac, Paxil, and Zoloft, among others—have become familiar household words. Serotonin does appear to be metabolized differently in men and in women, perhaps contributing to the greater female vulnerability to these mood disorders. But these differences have not been demonstrated in children, so it's possible the serotonin system develops differently as a result of the different experiences and coping styles of males and females, rather than hard-wiring.

Hormones are another leading suspect, but once again the evidence is surprisingly weak. The main reason to consider them is that the sex difference in depression first emerges around thirteen years of age, when most girls are midway through puberty. In one large study, depression was indeed found to be more prevalent in adolescent girls with higher levels of estrogen circulating in their blood than in those with lower levels. However, the same study found that girls' testosterone levels (which also rises during puberty) correlate even better with female depression than estrogen does. This is hard to reconcile, since testosterone in males seems to buffer them from anxiety and mood disorders. Another reason to doubt that rising estrogen triggers depression in pubertal girls is the fact that irritability and a depressed mood are associated with the *fall* of estrogen during the menstrual cycle (the infamous premenstrual syndrome, or PMS). Finally, there's the study mentioned in the last chapter of adolescents treated for delayed puberty. In this research, psychologists did the more direct (and scientifically rigorous) experiment of tracking children's emotional states as they began treatment with estrogen (for girls) or testosterone (for boys) and comparing them with the same adolescents' mood when treated with placebos. They found no relationship between hormone levels and mood. So once again, the simple explanation—hormones—cannot account for the differing degrees of depression between adolescent girls and boys. Other factors are more important.

The biggest of these is self-esteem. Boys and girls are both vulnerable to depression when they feel bad about themselves, but girls' self-esteem takes a greater hit than boys' during adolescence, mostly because of girls' greater focus on appearance. In an extensive analysis of more than two hundred studies, psychologists Kristen Kling and Janet Hyde, along with their colleagues, tracked sex differences in self-esteem as a function of age. They found that while the overall difference is on the small side (d = 0.21), males' self-esteem is higher than females' at all stages of life, with the difference growing during middle school, peaking during high school (when d = 0.33), and then closing up again during college and later adulthood.

Self-esteem means, of course, feeling good about yourself. But a surprising fact is that self-esteem is more strongly predicted by the qualities associated with masculinity than those associated with femininity. That is, people feel good about themselves when they feel strong, independent, or confident but not necessarily when they feel warm, expressive, or submissive. This means that when girls enter adolescence and leave their more androgynous selves behind, the demure, feminine qualities they strive to present are the same ones that elevate their risk for depression.

Then there is body image, which is arguably the biggest factor contributing to girls' plummet in self-esteem at puberty. This decline has been documented in most Western societies and now is rapidly developing in many Asian cultures: girls enter puberty and start comparing themselves to the impossible standards of beauty in every magazine, billboard, and TV commercial. Of course, boys also become more self-conscious about their bodies at puberty, but because males' status rises with their size while females are judged by their thinness, the physical changes at puberty work to the detriment of girls' self-esteem and to the benefit of boys'. Considerable research confirms that females of all ages express more dissatisfaction with their bodies than males do, a gap that (contrary to most sex differences) has actually grown larger in recent decades.

Yet another factor in the rise of depression as girls enter adolescence is interpersonal stress—the girl wars, or relational aggression, that break out in middle school and can make some girls feel excruciatingly isolated. Again, boys experience some of this but not to the same degree. Researchers have found that these negative social experiences together with the body-image and self-esteem issues play a larger role in triggering depression than anything hormonal or otherwise purely biological.

Finally, there is coping style, or how people deal with negative emotions. Here is where all the emphasis on emotional expression may be backfiring, at least with girls. It is well known that adolescent girls and women admit to and ruminate on negative emotions more than males do, which exacerbates and prolongs depressive symptoms. Males are better at brushing off such feelings and taking an active and assertive approach to maintaining their self-esteem. They cope by playing sports, playing video games, or watching TV. And while this approach can have its downside—including denial, blaming others, and acting out through violence and risk taking—it also seems to better buffer the male psyche against the lifelong peril of periodic depression.

My husband often jokes, "We men have no feelings." He actually seems to envy women's rich emotional lives—both the ups and the downs—and the shared feelings that create such a strong network of support among females. I certainly agree with him and would never want to live on the gray side of this particular sexual divide. But there is a downside to females' more intense relational focus. As much as we want to foster the rainbow of feelings in our sons, we also need to teach our daughters greater emotional resilience, especially when they're riding the rough waves of adolescence.

Empathy

It was a breath of fresh air. After a full morning of interviewing overprivileged (and overdressed) medical school applicants, I was delighted to meet a young man from Oregon who sat down and started revealing his passion for helping people. Many interviewees struggle to conjure

260 Pink Brain, Blue Brain

moments of compassion out of their thin resumés, but David was actually doing it: He told of loyally delivering hot lunches every Tuesday to the homes of a dozen elderly and disabled people as a Meals on Wheels volunteer. He spoke movingly about his clients' chronic breathing problems, diabetes-induced blindness, and debilitating pain. It disturbed him to see how poverty, lack of health care, and social isolation compounded their physical and emotional suffering. Though a busy premed student, David took his time at each home, chatting with his clients and nourishing them as much with his company as with the food he delivered. He told me he had considered careers in both medicine and social work, but he loved science and felt he could make the greatest impact as a physician. I told David it would be hard to maintain his idealism after four grueling years and the mounting debt of medical school, but his commitment to the underserved seemed genuine and deep.

David is a very empathetic young man, one of the most important qualities we look for in our medical students. Empathy is the ability to discern, understand, and share another's feelings. It's crucial for physicians and anyone else in the caring professions, but it also binds people's closest relationships together and is key to the emotional intelligence we hope to instill in all our children.

Empathy differs between males and females, as we saw in chapter 2. Women typically appear more empathetic than men. Even among physicians, female doctors and medical students score a bit higher on a standard empathy test.

The sex difference in empathy fits with the differences seen in emotional expressiveness: the sex that shows more emotion is also better able to read and relate to it in other people. Empathy is the polar opposite of aggression, which is obviously greater in males. You can't face down a fierce opponent if you're distracted by how he might be feeling. So the sex difference in empathy is another instance where the stereotype is accurate. As usual, however, the actual size of the difference is smaller than most people believe it is. There are many men, like David, who are much more empathetic than the average woman, and these are the ones we hope to attract to medicine, teaching, and other helping professions.

Researchers have studied empathy in many ways, but to do it right, the testing must be in a controlled laboratory situation. While some psychologists have described large sex differences in empathy, these accounts are mostly based on self-report. Asking people how much they agree with statements such as "Seeing people cry upsets me" and "I tend to get emotionally involved with a friend's problem" is not the best way to gauge empathy, because women's answers tend to be colored by their desire to appear kind, and men's by their desire to look tough. Such surveys inevitably produce much larger sex differences than more objective tests of empathetic responding.

Better data come from studies in which subjects are asked to identify the specific emotion exhibited on another person's face. Compared to men, women are faster, more accurate, or both at detecting expressions such as joy, sadness, anger, or fear in a photographed face. There is also growing evidence that men's and women's brains process such faces differently, and especially that women tend to use both sides of the brain symmetrically while analyzing facial expressions whereas men tend to show more right-sided activation of various brain regions. However, the actual sizes of the differences are not large; one analysis of many studies computed a d value of 0.40, which means that a third of men perform *better* than the average woman at recognizing other people's facial emotions. And further research shows that the sex difference depends on who is looking at whom. Women do better at detecting the emotions on men's faces than on women's, while men are actually better than women at detecting anger but only when it's expressed by other men.

This raises the issue of *why* there should be sex differences in emotion perception. The pat evolutionary view is that females have been selected for greater sensitivity because of their role as nurturers. The idea is that in order to best care for young children, women needed to be acutely aware of their moods and able to read emotions from subtle nonverbal cues (since children need the most care when they're too young to talk). Recent research confirms that mothers' brains are indeed more responsive to babies' cries and facial expressions than fathers' brains. However, such studies have found that *both* types of parent are better able to discriminate babies' cries and facial expressions than are men and women who have never been parents. Similarly, the brains of parents—both male and female—process infant faces and crying sounds differently than adults who have never been parents. In other words, it is likely the *experience* of caring for babies that shapes our emotional sensitivity and neural responses to them, not the other way around. Neural plasticity wins again.

The sex difference in emotional sensitivity has another explanation that has little to do with evolution. It is based on the difference in power and status between men and women in every culture. Women, according to this logic, may *need* to be more sensitive to others' moods and behavior because they are physically and socially weaker. This would explain why *both* men and women are more sensitive to emotions expressed on men's faces than on women's, and why men's best-tuned awareness is to anger in other men (their most threatening adversaries).

The status hypothesis also explains something else: why women smile more than men, a sex difference that is similar in magnitude (d value about 0.42) to the difference in empathy. Smiling is nice, of course, but it is also a good strategy for thwarting aggression from a more powerful adversary. In fact, the fear grin is common in many other species of primates and is usually displayed to higher-ranking individuals. Boys and girls, who are about equally powerless before puberty, smile in similar amounts, according to a summary of twenty studies. However, a sex difference emerges at adolescence, when boys grow bigger and stronger and presumably gain admiration without having to smile all the time.

When it comes to empathy, the sex difference in children emerges earlier than the difference in smiling. As we saw in chapter 2, some studies from the 1970s found that newborn girls are likelier than boys to cry in response to another baby's crying, suggesting a difference from birth. But the empathy difference is considerably smaller in children than between men and women, telling us that social factors also contribute. Just as boys learn to suppress their outward expression of emotion, they lose—or fail to develop—an acute awareness of other people's emotions when they enter their male peer groups, where the emphasis is more on strength and toughness.

The truth is that boys are plenty responsive to other people and they can stay that way, given the right upbringing. What parent or teacher hasn't been touched to see a boy lovingly care for an injured animal or shower gentle affection on a younger child? Boys *are* empathetic and can learn to be even more so if we don't exaggerate stereotypes and try to focus on their emotional development as much as on their athletic and academic skills.

Scientists have documented many ways in which social experience changes the brain and hormone systems, often permanently. Whatever their instinctive origins, nurturing and empathy are strongly honed through learning, and they are learned by receiving as well as by giving. The evidence is all too clear from studies of abused and abandoned children, who often suffer deep emotional problems such as attachment and conduct disorders. When such children grow up, they often fail as parents themselves, because they never learned how to nurture during the early critical period for social-emotional development. Decades of animal research have documented widespread negative impact on the brain, stress response, bonding hormones, and mental function of young mammals, both male and female, that were deprived of normal nurturing in early life.

Happily, most boys receive plenty of loving attention, but they have fewer male role models for nurturing behavior and little opportunity to be on the giving side. Fathers who are the primary caregivers for their young children become more sensitive to their children's needs than fathers who interact less with their children. As we've just seen, men's brains, like women's, shift after they become fathers, responding to infants in a more female way. Also like mothers, men experience many of the same changes in nurturing hormones (oxytocin, vasopressin, and prolactin) when they become fathers, which likely fuel some of these shifts in their emotional brain function.

Boys' brains are every bit as plastic, if not more so, than those of the men who raise them. Family studies indicate that children who are raised in more emotionally intelligent homes—where parents talk about feelings and their causes—score higher on empathy tests than children from families who talk less about feelings, and the influence is every bit as strong for boys as for girls. Other research has focused on improving the social skills of children who have deficits in these areas. A team at Emory University found that just six sessions of structured training in identifying other people's facial expressions (they used an Uno-like card game as well as puzzles that required putting together the different parts of a facial expression) improved the abilities of eight- and nine-year-old boys and girls to recognize facial expressions and other nonverbal cues.

Empathy is the glue that holds our social world together. Clearly, it can be cultivated in boys if we don't let misguided exaggerations about their innate wiring deter us.

Aggression

Poor Toby! He was so eager to join the big kids in their evening games of Capture the Flag and Kill the Pill. It's a great tradition among the kids at the marine biology lab where we've spent many summers. But little did Toby know that he would soon become "the pill." He came home in a fury, bruised, crying, and as angry as I'd ever seen him. Luckily, his injuries weren't serious, but when I interrogated him to find out who'd done this, the real source of his rage became clear. It wasn't all the tackling and roughhousing he was upset about. It was that he'd been beaten by . . . *a girl!*

Aggression is not something we usually associate with females. Empathy may have something to do with this: with their greater sensitivity to others' feelings, girls are perhaps less inclined to let their anger spill over into a physical or even a verbal assault on another person. But it's possible—and even more likely—that cause and effect works the other way around here: boys may not develop as much empathy because it interferes with their ability to fight.

Aggression is one of the largest and most reliable of the male-female differences. Around the world, and among all ages, males are more likely than females to strike, shoot, or merely yell and curse at other people, familiar or not. Homicide statistics from thirty-one countries show that almost 90 percent of those arrested for murder are male. War-making is an exclusive male enterprise in many cultures, and even in modern armies, women are almost never placed in frontline combat.

The difference is less dramatic but equally reliable in everyday life. Psychologists have documented sex differences in aggression through hundreds of studies, ranging from questionnaires about hostile feelings, to experiments testing subjects' willingness to deliver electric shocks (which were fake, though that was unknown to the subjects) to other people, to tallies of real-world acts of aggression admitted by the subjects themselves or witnessed by peers, parents, or teachers. Importantly, this research has found little difference in the male and female threshold for *anger*. What differs instead is how each sex acts in response to hostile feelings: men aggress, and women suppress.

Overall—considering both physical and verbal forms of aggression—the sex difference is moderate (with a d statistic ranging from 0.42 to 0.63), meaning that the average man is more aggressive than about 70 percent of women. The difference is much larger, however, when you consider physical aggression alone: as high as 0.8 or 0.9 when peers (as opposed to subjects themselves) report the behavior. A difference of this magnitude means that only 20 percent of women are more likely to hit or assault another person than the average man is, a stereotype you can almost rely on.

Even verbal aggression is more common in males, though the sex difference is not as large. In spite of women's stronger language skills, females generally don't yell and scream at one another (with exceptions like my former neighbor, who unloaded on me early one Saturday morning when I asked her to refrain from using her leaf blower).

Sex differences in aggression emerge as early as two years of age. We've already seen that play fighting, wrestling, and rough-and-tumble play are more common in boys than girls from a young age (chapter 3), much like many other young mammals. Such play is *not* motivated by aggression—an important fact for parents to keep in mind. (On the contrary, growing evidence indicates that it may actually promote interpersonal skills.) Nonetheless, boys' propensity for intense physical interaction does make it more likely that a truly aggressive boy will resort to hostile blows than an aggressive girl.

The sex difference in aggression grows gradually through childhood, peaking in early adulthood. Boys don't suddenly turn aggressive at the onset of puberty. Nor do they remain that way as adults. In fact, the sex difference in aggressiveness declines dramatically after age thirty, when it is actually smaller than the difference between boys and girls. So it takes a while, but most boys eventually do learn that aggression doesn't pay and grow into the gentle men we strive to raise.

Aggression in Evolution

The fact that boys are more physically aggressive than girls, even early on, makes a strong case that the difference is innate. Of all the sex differences, aggression is the easiest to link to evolutionary causes. Just think of all those jousting bull elks or wrestling elephant seals on TV nature shows and you get the idea: males are the more aggressive sex in many species, from beetles to chimps. And the reason, of course, is reproduction. Male animals battle, first and foremost, to gain mating rights, and second, to guard females they have already mated with.

It's all about competition. Male-male competition and aggression is the rule in any species, and particularly in mammals, where females invest much more energy and time into rearing their offspring. So even if there are an equal number of males and females in the population, the time involved in pregnancy, lactation, and provisioning offspring means that at any given moment, the actual number of females available to reproduce is lower than the number of males. The result: males must compete for access.

Of course, males don't always compete through battle. Some species use colorful adornments (picture the male peacock, the male cardinal, and—within our own primate family—the brilliant blue-nosed male mandrill) to advertise their genetic superiority. Other male birds compete by singing beautiful songs. Still other species use large armaments, like those impressive elk antlers and stag beetle jaws, to impress the ladies. While those weapons have obvious value in battle, females prefer such well-endowed males even before they've proven their superiority as warriors.

This is evolution by sexual selection, which is responsible for the fact that human males are larger, stronger, and more physically aggressive than females. Prehistoric women presumably preferred males who were successful in battle or in hunting (another goal of aggression)* or who were simply bigger than their competitors. Men today still compete over size. (What else could be fueling the billion-dollar steroid industry?) And they also still compete through aggression, though the more successful do this with ritualized battles, otherwise known as sports.

Nor is male aggression limited to the physical realm. We're a social species, like many other primates, with a hierarchy and status system that is more important to mating success than the ability to outpunch competitors. In modern society, men compete as much or more over wealth, job status, and swanky cars as they do over biceps or in barroom brawls, and their aggression is more often verbal than physical—the one-upmanship or playful putdowns that raise or lower each guy's relative rank on the team or in the corporate division.

The same jostling for status is seen even among boys, whose play groups tend to be larger and more hierarchically structured than girls'. For boys, play time is a chance to test their powers, to stack themselves up against many others to see how good they are. The process begins as

---

* As we saw in the last chapter, sexual selection for hunting ability has also been proposed to account for males' superior visuospatial abilities (such as mental rotation and visual tracking), but unlike the difference in aggression, evidence from other animal species does not lend much support for this theory.

early as age three, when boys prove more adept than girls at wielding influence over children of both sexes. (Girls generally can only influence other girls.) Feeling powerless with their male playmates, girls split off into smaller groups of two or three, where—in contrast to groups of boys—they often scrupulously avoid the appearance of outdoing one another.

Testosterone and the Neurochemistry of Aggression

It's never long before a discussion of aggression comes round to the big T. While the evidence for testosterone's contribution to other sex differences, such as verbal and spatial skills, is weak, here at last is a behavior we can convincingly link to the famous steroid. Testosterone is known to increase aggressive behavior when it's injected in rats and mice. The hormone also clearly shapes aggressiveness in humans, though 'roid rage is hardly the whole story, and testosterone does not act in our species as simply as it does in rodents.

Remember that testosterone surges at two distinct times in males' lives. First, before birth, when it (and other androgens) literally creates the male body, dissolving potential female organs and molding the penis, scrotum, and internal plumbing to connect them. This prenatal surge also masculinizes the brain—or at least we presume it does, based on studies of girls with CAH. Given the link between prenatal testosterone and rough-and-tumble play in many animal species, it's well accepted that this early surge, known as the organizational phase, factors importantly in males' propensity for physical aggression. Recent studies of finger-length ratios further support this influence. As I described in chapter 1, the ratio of index and ring finger lengths is thought to be an indicator of prenatal testosterone, and this ratio has been found to correlate with aggressiveness in both sexes.

However, when most people think about aggression, they imagine the testosterone that floods through males' veins beginning at puberty. This is the activational phase of the hormone, and while it is essential for all the outward manifestations of male maturation (growth spurt, enlargement of the scrotum and penis, voice deepening, and the growth of facial, underarm, and pubic hair), it does not, surprisingly, turn boys into raging combatants.

As we've seen, the sex difference in physical aggression appears well before puberty. After its initial surge (lasting between the first trimester in utero and just a few months after birth), testosterone settles down in boys to a level barely distinguishable from girls'—even while the sex difference in physical aggression remains significant. When testosterone does rise dramatically, during male puberty, there is no pronounced jump in aggressiveness in boys, according to an analysis of many studies by British psychologist John Archer. Although there's some suggestion that men with constitutionally higher testosterone levels are more inclined to try to dominate others, researchers are increasingly finding that testosterone works the other way around in adults: it's more the *consequence* of male competition and aggression rather than the cause.

This new theory is called the challenge hypothesis, and it's based on studies of fish, birds, lizards, and monkeys and many other mammals, including humans. In virtually all instances, testosterone levels are found to increase *after* an aggressive encounter with an intruder or some other provocation.

In men, such testosterone surges have been measured during all kinds of contests, physical and otherwise: tennis, soccer, judo, wrestling, even chess and math exams. It doesn't have to be an explicit competition—even a well-placed insult that threatens another man's honor is sufficient to raise the perpetrator's blood testosterone. Nor do the men have to be competing directly: testosterone levels rise in men who are merely *watching* their favorite team play in the World Cup or the Final Four NCAA basketball tournament. Whatever the source of the competition, men's testosterone levels jump first in anticipation of the event (pumping them up for the contest) and then rise even higher if they happen to win it.

The result is that testosterone levels increase mostly as a result of achieving dominance. And importantly, the rise in testosterone with a sports contest or having a high-status job takes place in women as much as men, even though women's baseline testosterone is just a fraction of men's.

So circulating testosterone is clearly a factor in males' greater aggression and competitiveness, but it is not a simple cause-and-effect relationship. Prenatal exposure seems to prime boys' greater appetite for active physical contests, but once they're out of infancy, both males and females exploit much the same hormonal circuitry to compete and establish dominance, whether they're serving balls on the tennis court or whispering in the school lunchroom.

Testosterone is not the only neurochemical involved in aggression. Growing evidence suggests that the neurotransmitter serotonin is actually a better marker than testosterone for aggression and violence. Ironically, low serotonin levels are linked to both violent aggression and clinical depression in humans. Though seemingly opposite extremes, these two responses may simply reflect the different ways society sanctions men and women to cope with feelings of low self-esteem or stressful life events.

Then there are some fascinating hormones that have long been known to participate in maternal behavior but that are increasingly being found to temper aggression in males as well. One of these is prolactin, the milk-making hormone that surges in women after they give birth. It may surprise you to learn that prolactin levels also rise in men after they become fathers, a change that correlates with their sensitivity to babies' cries. Another pair of hormones, oxytocin and vasopressin, is also important in lowering aggression. Oxytocin is best known for triggering childbirth and milk letdown, but it is now recognized as crucial to bonding and affiliation in many social mammals, including humans—earning it the nickname "the love hormone." In one recent study, men's ability to discern others' emotional expressions was enhanced—get this!—after taking a snort of oxytocin. However, for males it seems to be the closely related hormone vasopressin that is more important in bonding and paternal care. Finally, it's known that testosterone declines when men become fathers, a finding that's been replicated in many different cultures.

Aggression and affiliation lie on the opposite extremes of social emotions, so it's not surprising that men might shed their aggressive tendencies (and hormones) when they marry and have children. And while testosterone is clearly a factor in males' greater aggressiveness, this behavior is far more plastic in humans than in other animals. Males are not slaves to their circulating steroids, nor are women immune to the aggressive impulses fostered by similar hormonal circuitry.

Relational Aggression

As much as we worry about boys, anyone who's ever been (or raised) an adolescent girl knows that competition for status is not limited to males. Girls wage war in their own special way, and it has only recently been named: relational aggression. Through whispers, ostracism, gossip, and, now, harassing text messages, girls battle under the radar, behind one another's backs, manipulating friendships and intimacies in their own contests to rise to the top of the popularity heap.

It's the *Mean Girls* syndrome, fictionalized in the Lindsay Lohan comedy I mentioned in the last chapter only not so comical to the girls who endure it. Just as some boys bully with fists and taunts, aggressive girls dominate through exclusion, cattiness, and lies. It is far less visible but arguably every bit as painful as the playground bullying teachers are trained to look out for.

This is the one type of aggression that's more common in females than males, and that's because it's covert. Girls know they will be ostracized for physical aggression, so they resort to hidden tactics to intimidate their opponents. Boys, by contrast, generally *gain* stature through overt aggression, whether it's athletic prowess, pushing another kid around, or heckling a teacher.

It begins as early as preschool with girls' jockeying for best friends. As with boys' aggression, this behavior peaks in the teen years and appears to be universal across all cultures. Primatologists have noted a similar behavior in Old-World monkeys, in which high-ranking females steadily harass lower-status females—even to the point of suppressing their ovulation or causing spontaneous abortions. So female-female aggression is a fact of nature that apparently serves the same end as male-male aggression: competition for status and reproductive success.

However, human females are unique in the lengths they'll go to hide their aggression, and here is where nurture, or culture, must be held accountable. Recall that researchers find no sex difference in feelings of *anger*. Rather, girls learn early on to suppress their expression of such feelings. Both mothers and fathers participate in this emotional training: mothers are known to accept anger and tantrums more from their sons than their daughters. Fathers, for their part, are likelier to soothe their daughters and upbraid their sons when they express fear or sadness. By grade school, girls who appear aggressive are judged less likable by their peers than nonaggressive girls, while the opposite is often true for boys. And of course, the double standard is alive and well in adults, when assertive behavior is generally heralded in men but disparaged in women.

Simply put, it's not ladylike to be mean. While boys and female monkeys can get away with hitting or yelling in one another's faces, girls cannot. So they keep their aggression under the table, behind one another's backs, and strike at the one thing girls hold most dear: their friendships. It's ironic, but girls' greater empathy is also their greatest weapon—knowing exactly how much another person will hurt when she is ignored, besmirched, or rejected by someone she thought was a close friend.

Unlike physical bullying, which parents and schools have come to take very seriously in recent years, relational aggression is difficult to stamp out. How do you write zero-tolerance policies for eye-rolling and betrayed confidences? Because it's covert, relational aggression is much harder to police than physical or verbal aggression is.

But some groups are trying. Rachel Simmons, author of an influential book on the topic (*Odd Girl Out: The Hidden Culture of Aggression in Girls*), has developed empowerment training that teaches girls—both the victims and the perpetrators—about the truly violent nature of relational aggression. When the issue is posed this way, in the language of abuse, some girls may start to get the message. Other tactics involve training high-school girls to mentor middle-school girls and show them how to recognize and prevent relational aggression.

The issue is not trivial. Researchers have found that children who are frequently targeted for relational aggression—both male and female—suffer increased rates of depression and anxiety. And surprisingly, so do the aggressors. Even though it produces no bruises or broken bones, relational aggression can be just as dangerous as physical aggression. Parents and teachers need to make a greater effort to fight this form of bullying, which—though more common in girls—actually threatens the mental health of both daughters and sons.

Learning and Unlearning to Be Aggressive

The different aggressive styles of males and females are quite familiar to anyone who's lived through adolescence. But it's important to point out that such behaviors are hardly fixed by nature. For one thing, the fact that boys and girls bully in such different ways is almost certainly influenced by social learning. Physical aggression is simply taboo for girls beyond the toddler years, so they resort to the next best thing—best-friend wars. Boys are more prone to direct physical and verbal aggression, but if those are off-limits, they are quite capable of spreading rumors to reach the same outcome.

Happily, most teenagers eventually grow out of their various styles of bullying, further evidence that social behavior can be learned. It's a slow process, but then again, the prefrontal cortex, upon which most self-control and social intelligence depends, continues maturing until at least twenty-five years of age.

A quick glance around the globe proves that aggression and violence vary dramatically by culture. Many religious groups, including the Quakers and Baha'i, advocate a devout pacifism. And consider the contrast between Nazi Germany and Switzerland in the early 1930s: the two groups had negligible genetic differences but very different tolerances for war-making. The enormous range of aggressiveness in different cultures shows that this behavior is not hard-wired but highly modifiable by environment and social teaching.

Aggression is another behavior for which we need to separate the average from the tail end of the curve. The vast majority of men are not violent. Most crime is committed by a small proportion of chronic delinquents. According to military experts, only 2 percent of soldiers are what they'd call natural killers; the other 98 percent have to be arduously trained to attack the enemy.

Cultural influence is especially obvious in the case of spousal abuse. Recent research by John Archer has shown that the amount of violence between heterosexual partners varies dramatically in different countries, largely as a function of the degree of sexual equality in a given society. In present-day America, females are actually likelier than males to strike their partners (although they're less likely to inflict harm). By contrast, wife-beating is treated as a God-given right in many other contemporary cultures.

Whatever our innate tendencies toward violence and aggression, it's clear that cultural learning plays the larger role in determining who gets assaulted by whom. Boys are not destined to be violent but learn whether and when such behavior is acceptable and even heralded, depending on social circumstances. Girls are not without their aggressive impulses and have greater leeway than ever to assert their dominance over others. Relational aggression shows no sign of abating among adolescent girls, while females' use of physical aggression has increased dramatically in the last few decades. Just think of the violent females in movies like *Tomb Raider, G.I. Jane,* and *Kill Bill,* and you'll realize how rapidly cultural views about aggression and violence can change. On a brighter note, Steven Pinker has recently argued that violence of all types has steadily declined over the long history of civilization, which offers us hope that culture can largely conquer this instinct in both males and females.

Competition

One of my greatest regrets is that I never played a sport in high school. I did try track, basketball, and volleyball in junior high, but there was one thing that turned me away: a dislike of competition. I still remember, back in seventh grade, crouching at the starting line for the 440-yard dash, a surge of total nausea washing through my guts and down to my legs, now too weak and tingly to walk, much less run a quarter-mile as fast as possible. The team sports were better, but I found an easier out—dance—which became my physical refuge throughout adolescence and into adulthood.

Apparently, I'm not alone. Consider this from Melanie Wood, the first female to represent the United States at the International Mathematical Olympiad in 1998: "I hated the idea of winning and competing. This is something I've had to work at and get over. That seems to be a difference between me and the boys, but I don't know if it's a gender thing."

It clearly is. Girls and women compete everywhere these days, but there remains a significant gender gap in competitiveness that parallels the difference in aggression. While some men are undoubtedly glad to escape the schoolyard ethos, males as a group are simply more comfortable than women are with overt contests of all sorts—sports, games, spelling bees, and, most important, vying for a prestigious job or valuable promotion.

It was just this discrepancy that inspired a pair of economists to carry out a fascinating experiment. Muriel Niederle at Stanford and Lise Vesterlund at the University of Pittsburgh were trying to understand why women make up only a tiny minority (2.5 percent) of the highest-paid corporate executives at a large number of U.S. firms. So they recruited an equal number of male and female college students and actually paid them to perform a relatively easy job—add up strings of five two-digit numbers. There was no difference in men's and women's basic ability to perform these calculations, but there was a difference, as it turned out, in how each gender chose to be compensated for the task.

The experiment involved three rounds. Initially, each participant was rewarded for correct answers using a piece rate of fifty cents each. In round two, however, the participant had to perform the same task in a four-person tournament (with two members of the opposite sex and one other member of the same sex). Whoever solved the most problems would be paid four times more than the previous round per correct answer (two dollars), but the three losers would earn nothing.

In both situations, the groups of men and women performed equally, earning virtually identical amounts of money. But something interesting happened in round three, when each participant was given a *choice* between being paid by the piece rate or competing in a tournament.

Now you might think that everyone would make this choice rationally, choosing the piece rate if round two had shown the person was likely to lose in competition and the tournament if previous experience had shown that person was likely to win. But that's not what happened. Overall, 73 percent of the men selected the tournament while only 35 percent of women did. Even among the top performers, who presumably would be most confident about their ability to win big by joining the tournament, only 50 percent of the women chose to compete as compared to 80 percent of the men. And, lest you think men were choosing rationally and women irrationally, the sex difference was similar at the lower end of performance: among this group, 38 percent of the women made the wiser choice to stick with the piece rate, while 64 percent of the men lost considerable earnings by opting for the tournament.

So there is nothing rational about this gender gap. Men and women have different tastes for competition, which ends up punishing men on the lower end of performance but also women on the higher end.

The real-world implications of this study are compelling. If talented women are avoiding competition, they're simply not going to land the highest positions and best-paying jobs. They're going to slam into the glass ceiling, not because of discrimination or lack of talent but because they lack the competitive drive to try to crash through it.

Of course, there are certain domains in which females do compete, regularly and intensely. While many avoid overt competition in school, sports, and the workplace, females can get downright nasty when it comes to clothes, hairstyles, and attracting a certain boy's attention. We wouldn't have international beauty pageants or spend billions of dollars on fashion and cosmetic surgery if women weren't interested in outdoing one another in this arena. In one British study, university students were asked to keep diaries about their competitive feelings and encounters.

Both males and females reported similar numbers of competitive interactions regarding school achievement, financial success, and popularity. However, there were big differences in two areas: males reported many more competitive interactions with other men when it came to sports, while females reported that they felt competitive when comparing their appearance to other women's.

In other words, women *do* compete, but that competition is mostly over beauty and—at least among teens and younger girls—friendships and social status. And in striking contrast to males', female competition is largely covert, just like their aggression, the perpetrator hiding behind some anonymous insult or exclusionary tactic. It's simply not nice to compete openly, so even when participating in sports, women explain their drive to win in more socially acceptable terms ("not letting the team down") while men are more comfortable asserting their simple desire to be the best.

Evolutionary psychologists struggle to explain the unique machinations of female competition. Sure, you'd expect *some* female-female competition for mates. We may not be a perfectly monogamous species, but most men do substantially contribute to child rearing, whether as stay-at-home dads or as tribal hunters who provision mom and the kids with meat. So women must compete to mate with the best among them. It's certainly the case among other primate species, where the mother's dominance rank is the most important factor influencing the number and success of her offspring.

But it's not clear why female-female competition should be so dedicated to appearance or why it should be covert. Sure, females want to look healthy to attract mates, with symmetrical features advertising good genes and untroubled development. (Facial symmetry is universally regarded as attractive in both women and men.) Female competition, along with male selection, is likely also the basis for certain physical traits such as breasts that are large even when they're not lactating and males' preference for youth and a small waist-to-hip ratio (indicators that a woman is fertile but not already pregnant). Even in cultures where men prefer decidedly plumper figures, women with smaller waist-to-hip ratios are regarded as more attractive than females with thicker middles.

Less obvious is the selective value of females' other primping efforts: makeup, jewelry, and trendy clothes. Don't get me a wrong—I love a new pair of earrings every now and then. But if we accept the evolutionists' line that women are the choosier sex, then men should be the flashier dressers and the ones most interested in touching up their facial flaws. At least, that's the way it works among other species, like peacocks, cardinals, ducks, and mandrills, whose vivid males must flaunt it to gain access to the choosier females.

And if reproduction is really the goal of female-female competition, then shouldn't we see women also advertising their strength and intelligence, attributes that would have been even more important to successful childbearing and nurturing out in the hardscrabble savanna where all this selection is presumed to have originated? Instead, we have quite the opposite, at least in modern industrialized societies, where waif is in, and girls compete—often to their deaths—to be the skinniest (not to mention least fertile) in their clique. No doubt anorexia and other eating disorders are a pathologic extreme, but they do show the lengths females will go to outcompete one another.* They're not starving themselves to appeal to men, who generally state a preference for heavier figures than what women regard as beautiful.

So female competition is a bit of a paradox. Clearly, women do it, but they don't want to be seen as competing, and the focus of their competitiveness is not necessarily in areas that will improve their reproductive fitness. Of course, male competition can turn maladaptive as well, as when an overly aggressive man ends up in jail instead of on the corporate fast track. But the very covert nature of female competition and aggression is uniquely human, and it suggests that other, purely cultural factors are responsible.

Patriarchy is an obvious cause. Men have more power and higher status in every culture. Physical strength was probably the original reason, but once male dominance was established, the best way for men to maintain that power was to create rules that systematically denied it to the opposite sex. Female aggression and assertiveness are universally stigma-

---

* There is nothing universal—or evolutionarily advantageous—about thinness as a standard of beauty. In poor countries, fat females are considered more desirable, presumably because they come from wealthier families and are likelier to be fertile than women who are literally starving. Sharon LaFraniere reports that in the West African republic of Mauritania, men "want women to be fat," so young girls are actually force-fed, like geese, by their mothers to make them more attractive for marriage.

tized as unfeminine—maybe not quite so much today as in generations past, but enough to force your average twelve-year-old underground or behind her cell phone where she can secretly assert herself without violating the feminine taboo of appearing too aggressive or competitive.

Competition in Children

Children too show a substantial sex difference in competitiveness, which would seem to favor the idea that boys are innately inclined in this direction. However, this difference is not evident much before five years of age. What's more, cross-cultural studies reveal a wide range of sex difference in competitiveness among preadolescent children. According to one large 1981 analysis, the largest gender gap in competition was found among children in India, followed by white Americans. By contrast, African American and Canadian boys and girls were not found to differ in competitiveness, while a few studies of Israeli children (including those reared on kibbutzim) actually demonstrated the opposite difference: greater competitiveness among girls.

So the sex difference in competitiveness emerges early, but it is magnified through social learning. Children—all children—are naturally competitive, just as they naturally hit and kick until it's trained out of them. (There's certainly no sex difference in sibling rivalry, as far as I've witnessed or uncovered in my research.) But girls lose their taste for overt competition, as they do for overt aggression, because it is not considered feminine and because they rarely see it modeled by older girls or women. (Imagine if Miss America said what was *really* on her mind during the pageant interview!) Boys, by contrast, see nothing but socially sanctioned competition in every sporting event and video game they occupy themselves with.

Empathy is another factor. Girls may lose their taste for competition because they don't like feeling responsible for others' losing. Competition also interferes with girls' more intimate friendships, and it contrasts with the nurturing style they experience with their mothers and teachers.

Regardless of how it comes about, the sex difference in competitiveness has important implications for children's education and later opportunities. In spite of the overall trend of reducing competition in children's lives (who plays Pin the Tail on the Donkey at birthday parties anymore?), it is neither realistic nor desirable to eliminate it completely. Competition can be very motivating, albeit more for boys than for girls. And while some girls may not relish it, they can nonetheless benefit from it.

One recent study of fourth-grade (nine- and ten-year-old) children is revealing in this regard. Economists Uri Gneezy and Aldo Rustichini set up a running contest during the children's normal physical education class. First, boys and girls were asked to run a forty-meter dash one at a time, with the teacher recording and ranking each child's speed. Next, the children were asked to run in pairs, beginning with the fastest two and proceeding down the ranking so that each child ran against a child of virtually identical speed. This procedure meant that some boys were paired with boys, some girls ran against girls, and the remainder were paired with children of the opposite sex.

What do you suppose the competition did to each child's performance?

Overall, competition improved the boys' performance but not the girls'. Even though there was no sex difference in average speed in the noncompetitive situation, boys did better and girls did slightly worse when they ran competitively.

However, children's performance also depended on who they were competing against. As you might guess, boys ran faster when they raced against other boys. Girls, by contrast, actually ran *slower* when competing against other girls. But both sexes made the greatest improvement when running against a child of the opposite sex. That is, boys ran fastest when running against girls, and girls ran fastest when running against boys (though their gains were not as large as the boys').

Sure, boys *have* to beat girls in order to save face, so we can assume their motivation was heightened in mixed-gender races. But girls also benefited from competition, at least when it was against boys. (The fact that girls did worse when racing against other girls highlights their aversion to *overt* competition with one another.)

If the results from this physical competition hold up for academic achievement, it has important implications for education. On the one hand, competitive situations do seem to benefit boys more than girls, a useful tactic for improving their performance in, say, reading or spelling. On the other hand, girls also benefit from competition, but mostly against boys.

It takes two to tango. It's pointless to compete against someone who's not competing back. To the extent that we want to cultivate competition in girls, we have to bring boys into the mix. You can see this in girls with older brothers or athletic fathers; they often adopt the competitive ethos much more comfortably than girls without such role models. This finding also has implications for single-sex schools, in which girls may not get enough experience with competition, and boys may get too much.

No doubt competition can be overdone, and there's a lot to be said for reducing children's experience of losing, especially when they're too young to have developed the skills to compete in all the sports and other contests we prematurely inflict on them. A better strategy is to promote self-competition, encouraging children to strive for personal bests in their academic, athletic, musical, and other endeavors.

But we can only shield children for so long, especially in a free-market society where competition reigns supreme. And it seems an inescapable fact that if women are going to continue their forward progress, we've got to make girls more comfortable with competing. According to Sylvia Rimm, a psychologist who interviewed a thousand successful women for her book *See Jane Win*, the one experience that was most common in all these women's childhoods was "success in competition." Dr. Rimm concludes that girls "will need to cope with competition whether or not they enjoy it."

Risk Taking, Confidence, Fear, and Conscientiousness

Sam and his buddies made a great discovery last summer: the Punch Bowl, a clear deep pond accessible by bike and perfect fun on a hot day. The pond itself is lovely, but the best thing is the sturdy knotted rope hanging from a tree on the bank and perfectly situated for swinging over the water and jumping in. It took one jump and Sam was hooked. He wanted to go every day and was eager to bring the rest of the family.

We joined him on his next trip and were thrilled to see Sam's brave exploits—he added a new flip or spin to each successive jump. When my turn finally came, though, I confess I was *very* scared. It took me the longest time to work up the nerve. I stood on the bank, rope in my hand, until the gang of boys got tired of waiting and Sam yelled, "Come on, Mom!" Not wanting to hold up the crowd, I finally went for it, sailing through the air and letting go for an exhilarating ride into the perfect water.

It was awesome. And I went a few more times, but I wasn't dying to jump a hundred times like Sam, Toby, and their pals. Neither, I might add, were Julia and most of her girlfriends, who came along on a couple of the trips but mostly sat on the bank chatting and reading.

As with competition, boys and girls have different appetites for risk. Although girls are increasingly taking up skateboarding, ice hockey, rock climbing, and other extreme hobbies, the numbers still show reliable sex differences in risk taking of all sorts. The largest differences are for physical risks as well as gambling, dangerous driving (like speeding, tailgating, and not wearing seat belts), and even intellectual risks (such as choosing difficult versus easy problems to solve). Sex differences are smaller for other risky behaviors, such as drug and alcohol use, and there's no difference in abuse rates for two substances—cigarettes and cocaine (both drugs with anorexic effects, showing that young women are willing to take risks for one thing: slimmer figures).

So the sex difference in risk taking is wide-ranging, though not absolutely universal. Nor is there any sign of it in infancy: recall the ramp-crawling experiments (chapter 2), which showed that boys and girls were equally game to crawl down slopes of comparable steepness. Rather, the difference emerges between two and three years of age and peaks during the teen and young-adult years. What's more, the overall sex difference in risk taking may be declining in recent decades: studies conducted before 1980 reveal larger d scores (averaging 0.20) than those conducted since 1981 (d = 0.13). So either girls are getting braver (which is likely) or boys are getting more cautious (also likely, in our safety-first culture), and possibly both.

Risk taking is both a blessing and curse. Boys are 73 percent more likely than girls are to die from accidents between birth and age fourteen. Males of all ages are more likely than girls to get stitched up in the ER, drown, or die in a car accident or ATV crash. They also break more rules. Boys are likelier than girls to cheat on exams, get expelled from school, drive drunk, and sell drugs. As adults, men are arrested at least four times as often as women are for every crime (except prostitution).

But risk taking also has its upside, as any successful entrepreneur will tell you. "Nothing ventured, nothing gained" is what drives people to put their money, reputation, and very lives on the line. For those who manage to stay physically and financially intact, risk taking can pay off big when they're seeking coveted jobs, investing money, or scaling Mount

Everest. Just as men in the study by Niederle and Vesterlund were more willing than women to risk competition in order to increase their earnings, so, according to economists, are male investors about twice as comfortable as women when investing in stocks using their retirement savings. In financial parlance, women are more risk averse, a trait that has historically worked to their disadvantage as investors.

So risk taking is the route to all kinds of concrete gain, but it is arguably even more important for its effect on children's psyches. Successful risk taking builds confidence, an attribute that women often lack when compared to men of similar abilities. How can you speak up in class, ask for a raise, or negotiate maternity leave without the self-assurance that comes from a long history of successful risk taking?

Differences in confidence and risk taking may even be partially responsible for the performance gap on the SAT and other high-stakes exams. Boys shine on such tests in spite of the fact that girls get better grades in both high school and college.

Here's why that's the case. Exams like the SAT are scored with a guessing penalty: you get one point for each correct answer, but you lose a quarter of a point for every wrong answer. This means there's a certain gamble in selecting an answer you're not positive about. However, if you can eliminate two of the five choices, then guessing among the remaining three becomes advantageous, giving you a 33 percent chance of gaining one point versus a 67 percent chance of losing a quarter of a point. Males are generally more comfortable with this type of gamble, according to research on both high-school and university students. By contrast, females are more likely to leave a question blank if they're not absolutely sure of the answer.

Risk taking also influences the way students manage their time on tests. If cautiousness makes girls equivocate on difficult questions, they lose valuable time that the more confident males use to move on to the next problem. In fact, the Educational Testing Service itself conducted a study that found that females' scores improved markedly when the time constraint was lifted, whereas males' scores changed very little.

Sex differences in risk taking flow from the difference in competitiveness we just considered. How can you best the rest without pushing yourself and the rules to the absolute limit? If males really do need to compete more intensely for mates, then evolution may indeed have wired their brains for more robust risk-taking.

Risk taking is related to a more basic psychological trait known as sensation seeking. This term incorporates not only the desire for adventure and thrills but also a person's susceptibility to boredom and propensity to seek ever-changing experiences, risky or not. Sensation seeking is also more prevalent in males, a trait that is associated with both attention deficit disorder and various forms of substance abuse. This trait is strongly heritable, and researchers have made great strides in identifying several genes related to it. One gene variant is for the enzyme COMT (catechol-O-methyltransferase), which breaks down dopamine, the brain's natural reward neurochemical. In women, the more active variant of COMT is associated with higher levels of sensation seeking, on par with typical males. Surprisingly, this variant does not have the same effect in men. Nonetheless, COMT activity was found in one study to be 17 percent higher in the frontal lobes of men than women. So it may be a lower level of frontal-lobe dopamine that drives men more than women to seek out new experiences. Genes do not explain males' higher COMT activity, because the different variants are equally distributed between the sexes. However, some studies hint that testosterone is related to sensation seeking and could, theoretically, interact with COMT or other enzymes to affect dopamine levels in the brain.

Then again, the sex difference in risk taking may boil down to another personality trait: fearfulness, particularly of physical injury. According to evolutionary psychologist Anne Campbell, women are innately more wary of physical competition and risks because, reproductively speaking, they have more to lose. A man can lose a leg and still father a child, but how well would a one-legged pregnant woman have managed on the savanna fifty thousand years ago? How could a one-armed woman have carried an infant and still managed to forage enough calories to stay alive, much less nurse her baby? Citing evidence from various cultures that children whose mothers die are less likely to survive than children whose fathers die, Campbell argues that evolution selected women who were more cautious because their long-term reproductive success depended on being intact and fit enough to see a child through several years of gestation and sustenance.

We all know women are more fearful than men. That's why Hollywood scream-fests always show young women being attacked by zombies, overgrown gorillas, or creepy guys jumping out of their showers. Such fearfulness is most pronounced for physical threats, as Campbell hypothesizes, but it also extends to psychological challenges, including the financial risk aversion, math anxiety, and test anxiety we considered earlier. Most troublesome is the fear of success itself, an ironic but very real handicap many women still grapple with as they try to resolve the conflict between achievement and femininity.

Fear is one emotion that neurobiologists understand well, so several recent studies have been carried out to compare how men's and women's brains process fearful stimuli. The key structure is the amygdala, buried deep in the two temporal lobes. In brain-imaging studies, the amygdala is activated, or lights up, when subjects view emotional expressions on other people's faces. Of all emotions, fear is the one that best activates the amygdala, and it can be triggered by scary pictures of snakes, spiders, or merely a terrified expression on another person's face.

It may seem counterintuitive, but the amygdala is actually larger in males, a difference that may emerge in childhood but widens during puberty, probably as a result of testosterone action. Several recent brain-imaging studies have found that the amygdala responds differently in men and in women when they're confronted with fear-inducing pictures or stories—women show stronger activation in the left amygdala, men in the right. It's not known how such activation relates to the actual experience of emotion, but the finding supports the idea that men and women do process fear differently, and there's some evidence that this difference may be influenced by prenatal testosterone.

Another area underlying sex differences in fear and risk taking is in the frontal lobe, specifically the orbital prefrontal cortex (named for its location at the base of the frontal lobe, just above the orbit of the eyes). Unlike most parts of the brain, this zone actually matures earlier in boys than in girls. Toddler-age boys outperform girls on a task called object reversal, which involves subtle risk taking. In this game, toddlers must first learn which of two objects (say, a red block and a blue ball) conceals a tasty treat (a Froot Loop). Boys and girls master this particular task equally, learning that the red block, for example, conceals the treat. However, they differ in the next round, when the experimenter moves the treat under the blue ball. Boys more quickly catch on to the switch and start selecting the blue ball. Girls, however, persist in choosing the red block, perhaps because they're unconsciously afraid of breaking the rule they learned in round one.

Research on monkeys confirms that it's the orbital prefrontal cortex that is responsible for this task. What's more, young female monkeys that are injected with testosterone shortly after birth perform as well as young males on the object-reversal task, suggesting this hormone is responsible for the more rapid maturation of the orbital prefrontal cortex in males and perhaps for their greater tendency to take risks.

These studies tell us that there may indeed be inborn differences in the orbital prefrontal cortex that bias boys toward greater risk taking and lower fear. By three years of age, girls catch on, and there is no longer a sex difference on the object-reversal task. There are, however, other gaps in risk taking that persist into adulthood and have been linked to different patterns of orbitofrontal activation between men and women.

So there likely is an innate basis to sex differences in fear, risk taking, and sensation seeking. At the same time, there's no denying that such differences are reinforced through learning. For one thing, the difference in fearfulness is very hard to detect among infant boys and girls. Some studies even find that baby boys are quicker than girls to seek solace from their mothers when confronted by an approaching stranger, and the overall difference in timidity between baby boys and girls is minuscule.

This changes in the toddler years. By fourteen months, more girls than boys exhibit what psychologists call an inhibited temperament—shy, more fearful, and more uncomfortable in new situations. Sex differences in inhibition continue to grow throughout childhood, and by the time they enter school, girls are clearly more likely than boys to express fear or anxiety. The same pattern is evident in both Western and Chinese cultures: a small difference in toddlerhood amplifies into significantly greater shyness and inhibition in girls during later childhood and adolescence.

Though few of us would admit it, parents are one wedge that exaggerates the different fearfulness of boys and girls. Recall again those infant-crawling experiments, where mothers judged their sons more capable of crawling down a steep incline than their daughters, even though both sexes were equally game when tested without such presumptions. It's no different out on the playground, where mothers of boys are slower to intervene than mothers of girls when their children climb too high or swing too fast. Other studies find that fathers and mothers are more tolerant of daughters' expressions of fear than sons', just as boys are discouraged from crying more often than girls are. And outside the family there is no end to the contrasts between brave males and more cautious females children see among their peers and in the popular media.

Cross-cultural studies also support the idea that fearfulness can be enhanced or discouraged through social learning. In one study, psychologists found that Chinese infants (both boys and girls) were more likely to exhibit a fearful or inhibited temperament than Canadian infants. This most likely reflects parental teaching (as opposed to genetic differences) and highlights the value placed on risk taking in Western societies as compared to the self-control and dutifulness that are more highly prized in Asian societies.

Which brings us back to the downside of risk taking and the fact that, for all of its disadvantages, girls can benefit in some contexts from their greater fear and cautiousness.

"Girls are teacher pleasers," as one of our own kids' elementary-school teachers once told me. They don't want to disappoint the adults who care about them, so they get their assignments done on time, print with legible handwriting, and actually study for those weekly spelling and math quizzes. Sure, girls have other strengths that help them in school, particularly their ability to sit still and better plan and organize their work (functions of a different area of the frontal lobe), but the fear of failure is arguably an important motivator for girls, and one that some boys just don't seem to experience.

Fear, in other words, makes girls more conscientious, a quality once described to me as the most important attribute in the success of the medical students we train (50 percent of whom are now female). No doubt when it comes to completing schoolwork, or just about any job out in the real world, conscientiousness is paramount. And of course, many boys, as well as many successful men, are highly conscientious. It's the boys who seem to lack all fear or concern about their teachers' opinions who are the hardest to teach and the likeliest to get into trouble.

Boys and girls clearly differ in fearfulness and risk taking, a fact that likely originated through evolution but that is also potently reinforced by parents and the masculine and feminine stereotypes all around us. Both traits have their advantages and disadvantages, with a balance between them being the optimal goal. Just as any parent would strive to challenge a fearful son or rein in a daredevil daughter, so should we attend to our anxious daughters and reckless sons. Given the substantial dangers as well as the benefits of risk taking, it's hard to argue that the old way, simply letting boys be boys and girls be girls, promotes either sex's best interests in today's world.

Puberty and Sexuality

It's one of my favorite scenes in the Harry Potter books: Harry is rummaging through his school trunk for the precious Marauder's Map and tosses out a box of chocolate cauldrons. His buddy Ron gobbles up three of the candies, thinking they're a birthday present for him. Moments later, Ron is swooning over Romilda Vane, the girl who'd secretly spiked the chocolates with a love potion meant for Harry. Ron is saved by an antidote but not before his current girlfriend hears him lusting over Romilda.

Finally, we get to the biggest sex difference of all: sexuality itself. While they start out quite innocently, these adolescent crushes soon enough grow into the full-blown sexual desire that leads to some of the riskiest behavior both boys and girls engage in.

Puberty marks the dawn of sexuality. After years of ignoring or actively detesting each other, boys and girls start noticing and, eventually, obsessing about the opposite sex. As in many things, girls lead the way. They may not be concocting love potions, but they are probably gossiping about boys, flirting with them, and shopping for the latest makeup and lacy tops that our omnipotent marketers have convinced them boys will find hot. The gap is most obvious in middle school, when the girls, midway through puberty, may spend weeks plotting and planning for an upcoming school dance. (The boys, many of them still prepubescent, see it only as an opportunity to stay out late with their buddies.) Eventually, of course, the boys catch up and are every bit as consumed by sex, which is when parents' real worrying sets in.

Most boys and girls discover the opposite sex at some point during adolescence. By adulthood, attraction to the opposite sex is one of the largest sex differences. In the range of d values, which for most behaviors is lower than 0.5, sex differences in sexual preference are very, very large. Men report being sexually attracted to women much more than women do, with a d value of 3.5, while women are more attracted to men than other men are, with a d value of 3.99.

Now *that's* a sex difference—miles larger than the measures of empathy and math skills we've been so carefully parsing up to now. But sexual attraction also plays a role in shaping certain mental and emotional differences. As we've seen, differences in aggression, risk taking, and competition all become magnified during adolescence, precisely when boys and girls start the real business of their biological lives: finding mates. While such social-emotional differences have effects that ripple well beyond romance, this is their primary purpose and the reason they are often more substantial than most of the cognitive differences we worry about so.

So what is it that triggers the onset of sexual attraction? For once, hormones provide the simple explanation. Sexual attraction is the most clear-cut psychological effect of the hormonal changes at puberty, even if the other effects are not as wide-reaching as many believe.

It is typically around the age of ten when kids come home from school reporting that "Jake's got a girlfriend!" or "Emily likes Nick." (Happily, they're still innocent enough to let parents in on the secret.) University of Chicago researcher Martha McClintock calls it "the magic age of ten" and notes that such feelings dawn in both boys and girls at this age, and equally in those who will be gay as those who end up heterosexual. Regardless of culture or sexual orientation, ten years is the average age that most adults remember first liking someone in a way that transcends platonic friendship.

Why so young? Few children show any obvious signs of puberty by fourth grade. However, this age does coincide with some initial hormonal changes, caused by an event known as adrenarche. While we usually think of the ovaries and testes as responsible for initiating sexual maturation, this earlier phase is triggered by activation of the adrenal glands, those lumpy pyramids that sit atop the kidneys and are the source of so much trouble in congenital adrenal hyperplasia (CAH, the genetic disorder leading to overproduction of male hormones). Adrenal hormones carry out many jobs, including regulating fluid balance and mobilizing the body's response to stress. For women and early pubertal children, they are also the major source of androgens, the male hormones that promote muscle growth and have increasingly been linked to sex drive.

Adrenarche begins between six and eight for both boys and girls. By age ten, both sexes are producing about ten times the amount of the hormone dehydroepiandrosterone (DHEA) than they produced during the toddler years. DHEA is an androgen that has little effect by itself but that can be converted into estrogen, testosterone, and other active androgens elsewhere in the body. The conversion to androgens is responsible for the emergence of pubic hair, the only outward sign of adrenarche, evident as some sparse, downy growth as early as age nine in girls and ten and a half in boys. It may also be responsible for those first magical feelings of interest in the opposite sex. Androgens, and testosterone in particular, are well known to trigger sexual interest and arousal in adults and adolescents. Indeed, testosterone is probably the closest thing we have to a love potion in our normal human chemistry. (Hogwarts girls, take note!)

For girls, the adrenal glands remain the major source of androgens, whose levels rise about four-fold throughout adolescence and peak between the ages of fourteen and sixteen. For boys, the rise is larger and more protracted, especially once the testes become involved. Testosterone reaches a level in boys at least ten times higher than it does in girls, peaking between sixteen and eighteen years old.

Which brings us to the next and more familiar phase of puberty, gonadarche, when the hormones start flowing out of the newly activated ovaries and testes. This phase begins around age ten in girls and leads to the familiar bodily changes—breast development, rapid height increase, underarm hair, and the onset of menstruation (menarche), which the average girl undergoes between twelve and thirteen. In boys, gonadarche starts a year or two later and begins with growth of the penis and testes, followed by a rapid height increase, ejaculatory potential (spermarche, typically around age fourteen and a half), underarm hair, voice change, and beard growth. Though the age of onset can vary by several years, once puberty begins, a child is transformed within a matter of three or four short years into a reproductively mature young man or woman.

Puberty and the Brain

But what about their minds? Do the raging hormones that so dramatically change children's bodies have as great an impact on their brains and mental abilities?

The surprising answer is no. While androgens clearly trigger both sexes' interest in sex, neither these steroids nor the various ovarian hormones act as the neuropoisons adults assume they are when we lament about teens' rebelliousness, turbulent emotions, and questionable priorities. (Other features of the teen brain, notably the very slow maturation of the prefrontal cortex, *are* important here but do not appear to be related to gonadal hormones.)

We know this based on research by Richard Udry, a biologically inclined sociologist at the University of North Carolina. Udry's group has found some of the strongest evidence linking rising testosterone levels to the emergence of sexual behavior in both boys and girls. In the same study, however, they observed no significant change in aggressiveness in boys between the ages of twelve and sixteen, in spite of their dramatically increased testosterone levels and in agreement with other research I described earlier.

Similar findings emerge from a series of studies conducted at Penn State University. Researchers there tracked psychological changes in teens who were being treated with hormones because their own natural puberty was delayed. Boys were given several testosterone treatments, girls were given estrogen treatments, and in both cases, these treatments were alternated with placebos, allowing researchers to rigorously explore psychological effects without the subjects' knowing when they were on or off the hormones.

Once again, hormone treatments definitely increased teens' interest in sex. Girls engaged in more necking, and boys reported more nocturnal emissions (wet dreams) when they were on hormones as compared to placebos. Boys and girls also both reported feeling more attractive, more competent, as well as more physically aggressive while on hormones. But there was little effect on verbal aggression, mood, or overall behavior, nor did the hormones alter spatial ability (specifically mental rotation performance; see chapter 6).

These studies agree with other long-standing research that's found no dramatic change in thinking skills or personality when children pass through puberty. Intelligence grows continuously over the adolescent years, more in accordance with children's chronological age than with their pubertal development.

This is what you'd expect, given that the brain itself does not undergo any dramatic alterations during adolescence. Despite all the recent hype about the teen brain, there is no sudden change in its growth or maturation at puberty, according to MRI studies of hundreds of children led by Jay Giedd at the National Institute of Mental Health. It's true that girls' brains mature earlier than boys', but both sexes go through the same sequence of neural changes. The gray matter switches from a net growth to a net pruning phase, when excess synaptic connections are eliminated to streamline mental processing. Girls typically reach their peak level of gray matter about a year or two before boys, in accordance with their earlier progression through physical puberty. Gray matter maturation is followed by a more prolonged phase of white matter growth, lasting until about age twenty-five, when the brain can finally be thought of as mature. (The same age, Giedd wryly notes, that Avis and Hertz finally trust them to rent cars.) But the point is, the overall trajectory is a smooth one from birth to adulthood. Neither sex undergoes any dramatic neural shift at the time of puberty.

Still, it's possible that pubertal hormones exert their effects at a more microscopic level, influencing small but crucial brain areas that don't show up on an MRI scan. In fact, the one place in which sex differences in the brain have been most convincingly demonstrated (beyond the globally 9 percent larger brain size of males) is in the tiny hypothalamus. This makes sense, considering this is the brain area that controls reproductive behavior. What's more, the hypothalamus has the highest density of sex-hormone receptors of any brain region, meaning that it is the likeliest to respond to testosterone, estrogen, and other reproductive hormones.

Two tiny clusters of cells in or near the hypothalamus have been identified that appear to differ between men and women as a result of pubertal hormones. One is the SDN-POA (sexually dimorphic nucleus of the preoptic area) we heard about in chapter 1, which is larger in men. The other is called the bed nucleus of the stria terminalis (BNST). In rats, the SDN-POA is known to be influenced by testosterone exposure around the time of birth. However, the only study that has looked at its development in humans found that the sex difference in the SDN-POA does not emerge until about ten years of age. Similarly, the one study that has investigated the development of the human BNST found no sex difference in infancy but a 39 percent larger size in adult men, a difference that appears to emerge around the age of puberty.

The BNST is closely connected to the amygdala, another structure that, as we just saw, differs between males and females. Men's amygdalae are some 26 percent larger than women's (even after correcting for the overall size difference in the brain). The difference is smaller or absent in children, emerging only during later childhood and puberty, suggesting it is influenced by pubertal hormones. This effect has been proven in rats: when adult male rats are castrated, their amygdalae shrink; when testosterone is added back, the structures grow larger. Recent research indicates that the amygdala is an important center for sexual arousal in men but not women, so if testosterone causes it to grow in adolescent boys as it does in male rats, this may explain their sudden obsession with sex.

Obviously, puberty is a time of major upheaval. Boys' and girls' bodies are rapidly transformed from child to adult, and they are beginning to grapple with mature feelings and expectations. The same hormones that drive their physical changes do reach the brain, but their effects are largely confined to reproductive behavior itself. Sex hormones bind to receptors located in the hypothalamus, the amygdala, and related limbic structures, waking up teens' sex drive and, perhaps, their competition with same-sex peers. This feedback from the gonads back to the brain makes good biological sense, aligning adolescents' minds with their bodies' newfound reproductive potential, which after all is the whole purpose of puberty.

But for all their seismic effects on reproductive behavior, pubertal hormones do not similarly affect the rest of a teen's brain. Receptors for the various sex hormones are much sparser in the cerebral cortex than in limbic areas. This difference may explain why overall brain growth is relatively unscathed by puberty and why teens' core personalities and cognitive development do not undergo the same upheaval as their sexuality. In fact, teens' minds do not catch up to their maturing bodies until many years after puberty, which is arguably the greatest source of trouble in adolescence.

So while hormones are clearly a factor in certain teen behaviors, they are not the whole story. Equally powerful are the social changes of adolescence, which stem from physical development but take on lives of their own. Girls who look like women are treated differently and see themselves differently, even if their judgment has yet to catch up. They are whistled at on the street and checked out by everyone—boys, one another, and finally themselves, in long hours of self-inspection. As their bodies mature, they sense the power but also the vulnerability of womanhood, and especially the tyranny of beauty. Body image declines in many girls during adolescence, which as we've seen is the dominant cause of low self-esteem and depression. Girls' assertiveness and confidence may also decline, especially in domains like sports and math, which are traditionally viewed as unfeminine.

Not that boys have it any easier. While their physical transformation is more welcomed—adding height and muscle mass never hurt any male's self-esteem—the social challenges can be equally daunting. Boys command a new respect during adolescence but also considerable fear, especially when they hang out in their preferred large groups, which can be intimidating even to their own parents and teachers, not to mention the little old ladies down the block. Boys seek affirmation in their gangs but also face constant challenge within them—to be brave, aggressive, and, of course, "cool," which requires proving oneself through some combination of physical risk, sexual exploit, and drug or alcohol use. Rarely is an individual boy as tough inside as he is expected to be on the outside. The changes can be exhilarating, or scary, or both, but they are certainly destabilizing and alter the way boys' brains and behavior are shaped during this still-rapid phase of cerebral maturation.

So once again, hormones are only the beginning. To the extent that pubertal hormones influence teens' emotions and thought processes, their effects are indirect, via a much larger feedback loop: hormones mature the body, which changes the social landscape, which alters one's confidence and motivation, which in turn shapes every type of learning and, thereby, how one's brain finishes developing in the critical period of adolescence. Even sexual activity—the one aspect of behavior most closely related to hormones—is crucially shaped by peer culture. Girls who go through precocious puberty don't become sexually active at nine or ten, when their hormone levels have already started to peak, but wait until their early teens, when their peers have started to catch up. In general, children who go through puberty either much earlier or much later than their peers experience more emotional problems than those who go through it at the average age, with the notable exception of early-maturing boys (who gain social stature). All of this shows that it is not the hormonal changes as such but their impact on peer relations that most sculpts the adolescent psyche. Whether the issue is sexuality, risk taking, mood swings, rebellion, or any other hallmark of adolescent angst, hormones are not the direct trigger but merely one piece of a complex interaction with social and intellectual experiences that play out in every teen, male or female, a little bit differently.

Different Hearts Versus Different Minds

Finally, in this chapter, we've gotten to the heart of the differences between boys and girls—distinctions in emotional expression and interpersonal style that essentially define masculine and feminine and have an impact on all aspects of their lives. Children understand these categories from a very young age, though puberty delivers a booster shot of gender intensification, when teen boys and girls diverge even farther emotionally in their effort to appear attractive. Hormones kick things off, both before birth and again at puberty, but sex differences in aggression, empathy, fear, and competitiveness are also clearly shaped by children's social environment and their unconscious aspirations to masculine or feminine ideals.

In quantitative terms, sex differences in emotions and interpersonal behavior fall mostly in the small-to-moderate range, not so different from the various cognitive differences we considered earlier. They're noticeable but don't absolutely separate boys and girls. There are many warm, caring boys out there, as well as some intensely competitive and aggressive girls. But small differences add up in large populations, and when you consider the tradeoff between career and family that most adults now face, it's clear that these social-emotional differences play a large part in many of our more nagging gender gaps.

Girls generally have the advantage in emotional intelligence, the expression of emotions and ability to perceive them in others. This sensitivity can be valuable in the workplace, where women's consensus-building style has been widely heralded in management circles. But it comes at the cost of lower assertiveness and discomfort with competition, which, like it or not, is an important ingredient in much achievement.

Who's to say which style is better? Early feminists pushed women to toughen up, to hide their emotions and play hardball with the big boys. The male way was the only way, so women had to emulate men to succeed. Then came the newer wave of difference feminists—launched by Carol Gilligan's book *In a Different Voice*—who celebrated women's verbal and relational strengths, proposing that even moral judgments are made differently by women and men. Girls and women are said to take a care orientation toward ethical problem solving, while boys and men rely on a more black-and-white justice orientation. But even this difference turns out to be rather small when researchers look closely at it. In one large analysis of 160 studies, sex was found to account for only 16 percent of the variance in moral reasoning across the population. In other words, both men and women are capable of making moral judgments based on relationships *and* absolute principles.

This really shouldn't be surprising. Moral reasoning, like most interpersonal skills, is not instinctive. Children don't know right from wrong at birth, just as they don't automatically understand what other people are feeling. Such abilities are learned—perhaps more easily by some children than others—but to the extent that we want to foster prosocial behavior in any child, boy or girl, we have to temper their aggressive, competitive impulses. If there is any difference between boys and girls, it may be that girls learn these positive interactions more easily, perhaps because of their faster maturation, or because they see more women in caregiving roles, or because they're more fearful of confrontation.

The good news is that there is plenty of plasticity in every child's brain to nudge them in either the empathetic or assertive direction. Women in free societies have obviously grown steadily more ambitious in recent generations, while men are increasingly comfortable with their role as nurturers. Whichever style evolution biases us toward, our big, malleable brains offer all of us the chance to be our very best selves.

Tips

Here, then, are some ideas for evening up boys' and girls' social-emotional differences:

- **More people, fewer toys in infancy.** Babies—both male and female—are instinctively social and do their best learning through interaction with those who love them. It hasn't been proven, but I believe that the proliferation of toys, bouncy seats, and electronic entertainment is actually depriving babies of social interaction at the age they need it most, perhaps even tipping a few into the growing pool of children (mostly boys) diagnosed with autistic-spectrum disorders.

- **Talk about feelings.** Empathy may look instinctive, but the evidence from children who are highly neglected or abused proves otherwise. It is learned, largely by example but also from parents who recognize, discuss, and help children identify emotions in themselves and others. There's no shortage of teachable moments here. Every sibling squabble, birthday party, soccer game, and squirt-gun battle presents an opportunity to see things from another's perspective. Parents who talk more to children about feelings raise more sensitive children, but this is an area that can go sadly neglected, especially for some boys.

• **Don't banish the boys.** Parents often direct their children, especially boys, outside or to the basement when company comes or when they are interacting with other adults. But interpersonal skills need as much practice as anything. Why not use such opportunities to teach boys to stop and say hello, make eye contact, shake hands, and try conversing with the guests?

• **Coed partners in early elementary school.** While boys and girls start segregating themselves beginning in preschool, they can still interact comfortably through most of the elementary grades. Teachers will probably have to assign such pairings for various academic projects, but, especially for children without opposite-sex siblings, this may be the only opportunity to work with a peer of the opposite sex, exposing both boys and girls to a wider range of intellectual strengths and working styles.

• **Babysitting for boys.** Parents with sons love to hire responsible boys to babysit, and it's a great opportunity for older boys to foster their nurturing and leadership abilities. Many communities offer babysitting classes to train teens in basic skills and safety, but be sure to get a group of boys to sign up together, so they won't feel out of place in a sea of girls. (Or offer separate classes for boys and girls.)

• **Coed sports.** It would be hard to argue in our sports-saturated culture that there's some better way to find and develop the best athletic talent. And given the intensity of early training, most sports are likely to stay single sex at the elite levels. But what about everyone else? Most kids these days need more physical activity and may enjoy games like soccer, volleyball, basketball, tennis, softball, field hockey, and lacrosse on coed teams. An equal blend of boys and girls can help tone down the hypercompetitiveness that drives some boys away from sports while at the same time help girls learn the value and joy of team-based competition. Psychologist William Pollack notes that coed sports emphasize wider participation and more confidence building than typical sports programs and argues they can be "wonderfully transformational" for both boys and girls.

• **School uniforms.** This idea gets trotted out every now and then as a panacea for every educational problem—academic performance, discipline, truancy, drug use, you name it. I doubt uniforms can do all that; but they certainly prevent teenage girls from turning themselves into sex objects every morning before school and help stem the consumerism and material-status issues that infect boys and girls alike. On the downside, most uniforms are made from horrible poly-cotton blend fabrics, so some improvement in comfort and design will probably help sell this idea.

• **Balance competition and cooperation.** Clearly, children need both skills to succeed in our free-market society. In recent years, schools have de-emphasized competition, judging that too many running races, spelling bees, and math contests unnecessarily discourage the majority of children who don't win anything. Instead, children today often work in groups, learning to collaborate and cooperate in ways that more realistically prepare them for adult work environments. Still, many children are highly motivated by the prospect of winning, and competition is a fact of society that everyone has to cope with at some level. So it's a delicate balance, but not an impossible one. The beauty of our two-gendered species is that boys and girls can serve as role models for one another in these opposite but equally crucial ways of interacting. For those who are turned off by competition, coed teams may coax them into the spirit of striving for excellence and recognition. And for those who seem to be motivated only by the opportunity to beat their peers, teamwork can cultivate more prosocial consensus building. Like all things, moderation is the key, and teachers should strive to keep both types of interaction in balance, even in single-sex classrooms, where the tendency may be to let girls cooperate and boys compete.

• **Teach organizational skills, especially for boys.** Girls are typically more conscientious about schoolwork, a big reason why they earn higher grades than boys. Part of their advantage comes from planning and organization skills, frontal-lobe abilities that seem to mature earlier in girls. Girls may also work harder because they're more fearful of disappointing their parents and teachers. Nonetheless, a few organizational tricks, taught early and enforced by teachers, mothers, and fathers, can help boys out. Early on, boys should be given assignment notebooks (or electronic organizers, which may be especially appealing) and taught to copy down every homework task and deadline, giving them points toward grades, if necessary, to reinforce the habit. Separate folders for each subject, with to-do and completed slots, can help them keep track of important papers. Timetables and micro-deadlines can help students stay on track with big projects. Finally, parents should think about how they model planning and accountability: if moms are the only ones keeping track of family members' schedules, this is going to look like a female task that boys will have less incentive to master.

• **More exercise.** This applies especially to girls, to ward off depression and help maintain a positive body image. Recent research keeps uncovering all kinds of benefits of exercise for the brain and mental function. Neurochemically, depression involves overactivity in the stress-hormone system, which regular exercise can help buffer. Girls, as we've seen, are about twice as vulnerable to depression as boys. Then again, many boys today also don't get enough of the exercise that is just as beneficial for their moods and attention spans. Schools can help by reinstituting daily physical education and recess and by offering more intramural and afterschool sports opportunities to non-elite athletes. Children's fitness—and not just their weight—should be assessed and targeted to certain standards, just like their academic work. It may sound like boot camp, but the obesity epidemic is a serious matter threatening children's physical and mental health.

• **Limit media exposure and teach media literacy.** One reason kids don't go outside to play is that they have all those indoor thrills accessible on TV and their various electronic devices. Many parents struggle to stem the flood of inappropriate images, language, and lyrics in children's wide-ranging media exposure. Heavy TV viewing is associated with stronger gender stereotypes in children, while exposure to violence on TV, in movies, and in video games has been shown to exert a real, albeit weak, influence on children's real-life aggression. It's hard to see how so many boys can spend so many hours gunning down virtual opponents without rewiring their brains in decidedly nonempathetic ways.

Nor is mass media doing girls many favors. Although female characters are getting more assertive, they're still a small minority of the strong, heroic decision makers in most movies and cartoons. Worse is the growing sexualization of girls, typified by the miniskirt-and-fishnet-clad Bratz dolls and TV cartoon characters that are marketed to girls as young as four. A recent report by the American Psychological Association describes "stunning" negative effects on girls' problem-solving abilities when they were distracted by their appearance and blames sexualized cultural ideals for the seventy-seven thousand cosmetic surgeries American children under eighteen (mostly girls) were submitted to in one recent year alone.

The solution is to tune out whenever possible, or at least to supervise and discuss with teens any inappropriate or stereotypical content in the movies, TV, and Internet sites they're exposed to. Schools can help too by educating children about the commercial interests behind the media and steering them toward movies, books, and websites that provide more positive images of masculinity and femininity.

• **Social service.** What better way to take adolescents outside their celebrity- and consumer-drenched culture than by focusing on the needs of those who are less fortunate? Community service should be required of all middle- and high-school students, though the projects need to be well organized and well supervised to successfully instill the values of empathy and self-sacrifice. Even disadvantaged children can learn the joy of altruism and feel empowered by participating in projects to help the sick or elderly.

• **Encourage risk taking and leadership skills in girls (and boys).** From sports to science to art, girls tend to play it safer than boys. In spite of their steady achievement gains, females as a group are still less comfortable asserting themselves physically, creatively, and intellectually. The difference is most exaggerated in adolescence, when boys start taking more risks to prove their power and courage while many girls retreat into clothes, makeup, and a more passive femininity.

Unlike boys, who gain stature by asserting themselves, girls denigrate one another for standing out and seek a kind of equality in their friendships that boys don't worry about. This is great for supportive, cooperative interactions, but not so good for selling their strengths or striving for achievement. While boys are more tolerant of one another's boasting, girls bond through self-deprecation ("I'm so stupid!" or "I'm so fat!"). The problem is, if you criticize yourself enough, you start to believe it.

To boost the confidence of girls, some middle schools have devised after-school or weekend leadership retreats. The idea is to challenge girls to assert themselves both physically and interpersonally using self-defense training, rope climbing, public speaking, and team-building exercises. Of course, many boys also struggle with their confidence and self-esteem during adolescence and can benefit from similar training, which therefore should be offered to both boys and girls.

8

Truce Time

GENDER DIFFERENCES ARE A MOVING TARGET. Even in the time I spent writing this book, the concerns, the conventional wisdom—even the gaps themselves—have all noticeably shifted. The boy crisis has ballooned, with fears about boys being kicked out of preschool, being unfit for college, or failing to launch as independent adults. Girls are surging ahead, led by a new crop of "alphas" who are bringing home the top scholastic awards and stomping over one another to get into the best colleges (which continue to reserve about half their slots for arguably less deserving boys). At my own daughter's recent middle-school graduation, with flowers wilting in the stifling gymnasium, parents fanned ourselves while listening to the six student speakers, all chosen for their perfect GPAs and every one of them, female.

But wait a minute: there were some other speakers at the steamy ceremony, all adults and every one of them—the principal, superintendent, president of the school board, and the lead eighth-grade teacher—male.

So have things changed or not?

I believe they have, and mostly for the better. In spite of all the fears about boys adrift, the truth is that their reading, writing, and math scores are on the rise, their high-school dropout rates are down (Figure 8.1), and more men are going to college (even if their percentage of the total college population has declined). While there are still many troubled schools out there, kids are learning more than ever. Everyone's getting better educated; it's just that girls' improvement has been more dramatic. Boys are losing only by comparison to girls.

But there are important lessons from girls' recent successes. First and foremost: mental and emotional abilities are not fixed. Second: they are *not* strongly determined by gender. For most traits, there is far more overlap than separation between the sexes, and plenty of room for learning or plasticity to raise achievement in any child. Expectations are crucial, and girls have done better as we've raised the bar for them.

Or rather, girls have soared as we've begun expecting them to be more like boys. Sports, math, science, and leadership training have been good for girls. We still have farther to go—to boost girls' spatial skills, technological fluency, and comfort with competition—if we want to narrow remaining gaps in fields like engineering and computer science and continue chipping away at the glass ceiling. Time will tell whether girls' newfound academic successes extend to the working world, where progress has been much slower. In some major cities, women in their twenties are actually outearning men. But the next, and more challenging, step will be for women to maintain their recent gains through the child-rearing years, when the lack of support for working mothers derails many careers.

Or, as Princeton's president Shirley Tilghman put it, encapsulating

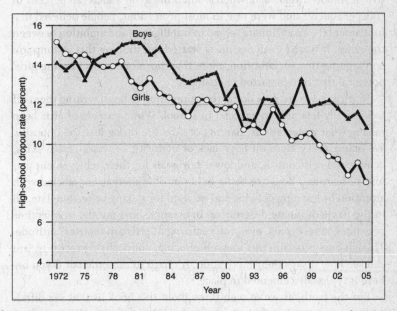

Figure 8.1. High-school dropout rates by gender. (Source: National Center for Education Statistics, Digest of Education Statistics.)

the largest remaining hurdle our daughters will face as adults: "It's the daycare, stupid."

Then again, one likely fallout from young women's surging achievement is that more men will opt to be the stay-at-home parent or primary caregiver in a family. Traditionalists will have a tough time swallowing this. Many people still can't see men who choose this role as successful. But the truth is that fathers today are much more actively engaged in parenting than those of a generation ago, and many are not even apologetic about taking breaks in their careers to care for children. What looks like a lack of ambition may simply be men who are taking a broader view of achievement, as women have long done. This shift may even bode well for our national economy, if it means that professional ability rather than gender becomes the main factor determining which parent remains in the workforce.

Of course, most families will continue to need two incomes to support themselves, and boys' current plight demands that we do more to help them develop the skills and intellectual confidence to succeed in school and beyond. Here is where girls' recent success can be instructive. A hundred years ago, when women made up a mere 20 percent of college students and were rare in most professions, people believed in a fundamental—read "innate"—gap in intelligence and ambition between the sexes. It wasn't until feminists started questioning this assumption and arguing that socialization, more than hormones, was limiting girls' potential that things started to change.

Ironically, we are now doing the same thing to boys: writing them off as naturally less able to succeed in school. While parents of girls keep raising their expectations, parents of boys are doing just the opposite. We blame every lapse on boys' lack of maturity, or lesser verbal skills, or minimal self-control, and lower our goals for their achievement and love of learning. Boys are being held out of kindergarten, labeled with attention or learning deficits, and excused for talking or reading late, all in the name of innate differences. In essence, boys are the new victims of gender stereotyping, even internalizing a "girls are smarter" attitude. Though boys proclaim this in a self-effacing, politically correct way, you can see how it saps their motivation. It's hard to raise the bar if you believe it is already cemented in place.

But the truth, as we've seen throughout this book, is that sex differences are not nearly as large or as fixed as this new wave of essentialism projects. The truly innate differences—in verbal ability, activity level, inhibition, aggression, and, perhaps, social perception—are small, mere biases that shape children's behavior but are not themselves deterministic. What matters far more is how children spend their time, how they see themselves, and what all these experiences and interactions do to their nascent neural circuits. That's why gaps in areas like speaking, reading, and math are modest compared to the differences in areas we don't formally educate, such as empathy, self-control, competitiveness, conscientiousness, spatial ability, and organizational skill. And yet, as we've seen repeatedly, there are ways to enhance every one of these, if we start early and create the opportunities for children to incrementally practice such important skills.

Clearly, we need to do more for boys at this moment, without forsaking the gains girls have made. But the solutions must come not from pigeonholing boys as hard of hearing, colorblind, immature, nonverbal, hyperactive, and disorganized, but by raising expectations and then creating an environment in which they can succeed. Beginning from birth, boys need more one-on-one verbal engagement, literary immersion, and opportunities for physical play, hands-on learning, and exploration of all types. They must be presented with small, achievable goals for concrete mastery: sitting still, printing their names, stating opinions, sorting seashells, adding fractions, reading books, and belting out songs.

Neither parents nor teachers can do it alone. Both need to hold to the same high standards coupled with firm structure and a deep love for boys. Throughout this book, I've suggested ways to keep boys more engaged and productive in school: more physical breaks, earlier reading and writing instruction, more male teachers, a broader range of literature, opportunities to compete, a blend of multiple-choice and essay exams, hands-on learning, and computer-based learning.

But we parents must also do our part: reading, talking, and singing to them; taking them to the library; organizing a quiet time and space for homework; fostering their outdoor, athletic, and musical pursuits; finding every opportunity to help them create, explore, and care for others. Everyone agrees that boys today need less medication, more campouts, and, above all, decreased time in front of the TV, computer, Nintendo, and Xbox.

Or, if we could reduce boys' issues to a simple tag: "It's the video games, stupid."

No question, this is a challenging time for boys. Parents and teachers are grasping at solutions to keep boys engaged in school, from kindergarten to college. All this attention is a good thing and has already led to boy-friendly improvements in many schools and a leveling-out, if not yet a reversal, of the college-enrollment gap.

But considering the various gender gaps that remain, should we be attempting something more radical? Around the country, more and more school districts are turning to single-sex education as the cure. Proponents argue that it's easier to cultivate boys' reading, writing, and artistic talents in classrooms where they don't routinely feel outgunned by girls, and, in the same way, that it's easier to promote girls' risk taking and math and science achievement if they're away from domineering boys. Their logic is very persuasive, but the question is, Does such gender segregation really work?

Single-sex education has been touted for girls for many years, but the movement has picked up new steam as parents and teachers begin grasping for better ways to help boys. A big boost came in 2006, when the Bush administration lifted restrictions on the Title IX ruling that had previously precluded gender segregation in U.S. public schools. Then there is Leonard Sax, who left his medical practice to direct the National Association for Single Sex Public Education (NASSPE). A fervent proponent, Dr. Sax has been very vocal with his pronouncements on the different needs of boys and girls. Unfortunately, his presentation of the actual data on single-sex schools suffers from the same sort of cherry-picking as his proclamations about the neurologic differences between boys and girls, showing little effort to fairly weigh all the evidence supporting the essential sameness of both sexes or both types of schools.

Here's a favorite Saxism we visited earlier: boys and girls should be in different classrooms because boys can't hear as well as girls. It's true—females of all ages can detect quieter sounds than males, with the notable exception of newborn infants. However, the differences are minuscule,* detectable only with the most fine-tuned auditory testing and certainly not relevant to what goes on in a classroom. Nor are dif-

---

* Sax's claims have been thoroughly deconstructed by University of Pennsylvania linguist Mark Liberman, who demonstrates on his Language Log website the absurdity of Sax's claim that a teenage girl perceives her father's voice to be "ten times louder" than her father perceives it to be.

ferences in boys' and girls' color vision as dramatic or well proven as the doctor proclaims.

Leonard Sax is not alone in preaching that boys and girls are so differently wired that they can't learn well in mixed classrooms. Psychologist Michael Gurian invokes a similar rationale for separating boys and girls in school. According to a 2005 *Newsweek* story, children as young as first and second grade were segregated by sex at Foust Elementary in Owensboro, Kentucky, where desks were removed from boys' classrooms based on Gurian's teaching that "males have less serotonin in their brains" and therefore need plenty of room to move around. Girls, by contrast, were given "a carpeted area where they sit and discuss their feelings" because their supposedly higher oxytocin levels required them to bond more. Testosterone too was used to justify differential treatment, in this case to give boys timed multiple-choice tests while girls were given extra time to complete the same tests.

Now, there are many sound reasons to advocate single-sex schooling, but sex differences in children's brains or hormones are not among them. As I've shown throughout this book, scientists have uncovered only small differences between boys' and girls' brains. Nor have measurements of testosterone, serotonin, or oxytocin revealed any appreciable difference between prepubescent boys and girls. So the argument that boys and girls need different educational experiences because "their brains are different" is patently absurd. The same goes for arguments based on cognitive abilities, which differ far more *within* groups of boys or girls than *between* the average boy and girl.

Proponents are on firmer ground when they base their arguments on some of the motivational and interpersonal differences between boys and girls—particularly, the idea that each sex may benefit from some protected time away from the other during their formative years. Perhaps boys would thrive in a more disciplined, competitive atmosphere and girls in a more supportive, nurturing environment. By eliminating the boy-girl contrasts that inevitably arise in mixed classrooms, each sex might be freer to excel in a wider range of pursuits. If teachers can manage to teach boys and girls the same curriculum in a separate but substantially equal way (as the federal statute requires), perhaps gender segregation is the best solution for addressing these current educational gaps.

It all sounds quite reasonable, except that existing research does not support it. Around the world, educational researchers have analyzed dozens of studies comparing the achievement of students in single-sex schools with those of students in coed schools. And without exception, analysts from the United States, Britain, Canada, New Zealand, and Australia have come to the same conclusion: academic achievement in single-sex schools is not convincingly superior to that in coed schools.

The most thorough analysis comes from Rosemary Salomone, a St. John's University law professor who in 2003 published the thoughtful book *Same, Different, Equal: Rethinking Single-Sex Schooling.* She herself is a graduate of a private girls' high school, which she credits "for having nurtured and validated [her] intellect at a time when society valued young women far more for how they looked than for what they knew or thought," and so Dr. Salomone is a cautious advocate of single-sex schools. Still, after her careful review of existing research, she admits that the evidence of benefit from such schools is still "tentative" and "not sufficiently weighty to yield definitive conclusions."

A similar consensus was reached by the U.S. Department of Education, which in 2005 commissioned a systematic review of dozens of studies comparing single-sex education versus coeducation. Its overall summary: the benefits of single-sex education are "equivocal." Though the report did find modest evidence that single-sex schools promote short-term academic success (in one-third of the studies), the majority of studies (53 percent) found no difference between children's achievement in single-sex versus coed schools. And the data are even weaker when you consider long-term differences in college graduation rates or graduate-school attendance; 75 percent of the studies found no difference between students who attended single-sex high schools and those who attended coed high schools.

Most recently, British researchers Alan Smithers and Pamela Robinson reviewed some fifty years of data in a report titled "The Paradox of Single-Sex Education." They similarly found "little conclusive evidence" that single-sex education creates an advantage. Instead, they reported, "The main determinants of a school's performance are the ability and social background of the pupils." It's a "paradox," they concluded, because everyone claims to know that either single-sex or coed schools are better, but no one can prove it.

How can Sax, Gurian, and other single-sex-school advocates be so wrong? *Newsweek* is not alone in publishing glowing stories about the burgeoning number of single-sex academies around the country. According to a steady stream of print and TV news reports, boys are devouring literature, and girls' confidence in math and science is soaring, thanks to the new old-fashioned arrangement of separating their classes. Further proof of the benefits of single-sex education is offered by the impressive number of female leaders, such as Hillary Clinton, Nancy Pelosi, Madeleine Albright, and Condoleezza Rice, who graduated from all-girls' high schools or all-women's colleges.

But it was precisely the study of women's colleges that first revealed the apples-versus-oranges problem of evaluating single-sex education. In 1973, a high-profile study reported that women who attended elite all-women schools, such as Wellesley, Smith, and Bryn Mawr, went on to higher-prestige jobs and higher salaries than their contemporaries who graduated from coed colleges. However, later analysis of the same data found that the women who attended such colleges—mostly in the 1950s and 1960s—were both smarter and from more privileged backgrounds than the women at the coed colleges they were compared to. These were the best and the brightest women in an era when the most prestigious colleges still refused to admit them or, at best, were struggling under the growing pains of a new coeducational structure. Once Harvard, Yale, Brown, and other top-rated colleges became well integrated, women's colleges lost their best-qualified students. In fact, applicants' SAT scores at Smith College declined during the 1970s.

The popularity of all-women colleges has fallen precipitously over recent decades. Only 3 percent of girls who take the SAT now even express an interest in applying to such schools, and many of the less prestigious ones have been forced to go coed just to stay open. But this has not deterred advocates of single-sex education from revving up the movement at the kindergarten-through-twelfth-grade level. Beyond the silliness about oxytocin and hearing differences, the strongest argument for single-sex education is that it can counteract the gender stereotyping that boys and girls impose on each other, especially during adolescence, when everyone's sorting out his or her sexual identity. Single-sex schools or classrooms are said to free up children to be themselves without trying to impress the opposite sex, or to conform to masculine or feminine stereotypes. In an all-girls' science class, girls will actually *do* the experiments, because there aren't any boys around to take over the equipment; in an all-boys' school band, boys may actually consider playing the flute, because it doesn't involve crossing some invisible divide exposing them to gay insults. And so, the argument goes, all-girls' schools should encourage more girls to stick it out in math and science as well as to assert themselves as leaders, while all-boys' schools should promote boys' language and artistic abilities and generally help them buy into school and a serious academic focus.

It's persuasive reasoning, and advocates point to many outstanding schools where this is precisely what is taking place—all-male or all-female academies where teens are thriving, learning, and expanding their horizons as they might never have by attending their local, nonselective, and probably oversize coed public high school. There are, unquestionably, many top-notch single-sex schools out there, and families who can get their children admitted to them should by all means take advantage of the opportunity.

The caveat, however, is that the advantage of such schools is *not* demonstrably related to their single-sex structure. As study after study confirms, the differences in test scores, AP course enrollment, and other objective measures of academic success between single-sex- and coed-school students melt away when researchers statistically adjust the data to correct for preexisting differences between these groups of students.*

Very often, single-sex academies are more selective than the coed schools they are compared to. The children who attend them come from more affluent, better-educated families. Their entrance requirements are stiffer, and more of the students are enrolled in college-preparatory classes than the comparison students in coed schools. And in some cases the schools themselves are actually better, in the sense that they offer newer facilities with smaller classes that are taught by more dedicated faculty—differences, in other words, that have nothing to do with sex segregation. Of course, the best research on single-sex schools takes such demographic and infrastructure differences into account when statistically analyzing the data. But even when these are factored out, there remains the pro-academic choice that parents and pupils make when choosing to attend a single-sex school over their default option. In other words, families that go to the trouble to find, enroll in, and commute to single-sex private or charter schools probably care more about school

---

* The problem is analogous to the healthy-user bias that made estrogen therapy look so beneficial for postmenopausal women. In the final analysis, the women who were taking hormones were already healthier than the control group, and we now know that sustained hormone use is not beneficial to elderly women.

achievement than demographically comparable families who do not seek out such options. As several researchers argue, it may be this difference in attitude alone that is sufficient to boost achievement and test scores.

In medical research, we have randomized, placebo-controlled, and double-blind studies that are the gold standard for determining whether a new treatment works. You can't possibly do a blind study of coed versus single-sex schools, in which neither the students nor the researchers would know what kind of school they were enrolled at. But the State of California did attempt something close to this in the late 1990s, when it established a pilot program to test the value of single-sex education in a dozen different public middle and high schools.

The experiment was a failure. Six school districts participated, each establishing one all-male and one all-female school. In spite of a $500,000 grant to each district, four of six districts closed their single-sex schools after two years, and a fifth closed its schools after three years. Researchers who studied the pilot schools did find a few advantages (less distraction by the opposite sex) but also some disturbing disadvantages and no evidence for academic improvement. Segregation tended to reinforce stereotypes, especially in some of the dual academies where the same teacher taught the same classes separately to boys and girls. (The same kind of stereotyping is evident in the *Newsweek* description of Foust Elementary School, where boys have their desks removed to better move around the classroom while girls are encouraged to sit on the rug and discuss their feelings.) Although some students were heartbroken when their academies were closed, interviews revealed that their single-sex classes were not exactly a haven; some students missed the buffering effect of the opposite sex, the absence of which had resulted in more cattiness among girls and teasing among boys.

When all was said and done, everyone agreed the California pilot program was ill-conceived. Teachers were not specially trained in how to teach to each sex without sending a mixed message about gender roles—that is, to address each sex's special needs without reinforcing gender stereotypes. Despite the noble goals of this project, its failure serves cautionary notice that successful school reform is going to take a lot more than simply separating boys and girls.

As we've seen, girls are doing better and better in school, taking more math and science classes (with the exception of computer science and engineering) and getting better grades in them than boys do. And the vast majority have done this in coeducational schools and classrooms. As girls ascend in academic achievement, the argument that they need protecting in all-female schools grows weaker. No longer do they need to opt out of the male world to be successful. Or as one market analyst, hired to figure out how to attract applicants to all-women's colleges, put it: "High-performing young women tend to see themselves as high-performing students, and not as students in need of some kind of special care."

Now, the focus has shifted to boys, who some educators hope can be saved by classrooms that are geared exclusively toward their needs. Never mind that the existing data on single-sex schools find few if any advantages for boys.* The reality is that many schools are failing and in need of drastic change. Single-sex campuses or classrooms seem like as good a weapon as any in the armament of school reform.

In fact, while the overall research has proven few benefits of single-sex schooling, some researchers find it is precisely the more disadvantaged and at-risk children who benefit from such groupings. This stands to reason, if by setting up single-sex classrooms or campuses, school districts are creating the pro-academic ethos that was lacking in their struggling schools. And this seems to be exactly what's happening in a number of all-boys', largely African American, schools around the country, such as the Thurgood Marshall Elementary School in Seattle, the Bluford Drew Jemison Science Technology Engineering Math Academy in Baltimore, and the Urban Prep Charter Academy for Young Men in Chicago. Neatly uniformed children have signed on for long days of intensive instruction, some running as late as 7 P.M. With dedicated teachers and a strong emphasis on discipline and high expectations, these smaller magnet and charter schools are bucking the odds, inspiring boys as young as kindergarten to work hard and aim for college—essentially changing the culture overnight from one in which school achievement is ridiculed to one in which it is held in the highest esteem.

Thanks to the new ruling on Title IX, we now have a vast experiment going on in the United States: hundreds of public school districts opening dual single-sex academies or, at the very least, offering single-sex

---

* While the overall findings are equivocal, those studies that have found an advantage of single-sex education tend to find it only for girls. According to this existing research, based mostly on middle-class children, boys tend to do better in coed schools. This leads to the impossible conclusion that both sexes fare better with girls as classmates.

classrooms within their coed campuses. This movement may even yield valuable new data for researchers, though the fact that the statute requires the choice to be purely voluntary means there will always be an apples-versus-oranges, or self-selection, problem comparing students in single-sex with those in coed classrooms.

Still it's hard to see how school districts can meet the legal challenge of providing equal educational experiences in both single-sex and coed classrooms. It's also hard to see the rationale in early elementary school (kindergarten through third grade), when boys and girls are still optimistic about learning, can interact effectively, and have much to learn from each other. The argument is stronger at the middle- and high-school level, when gender differentiation and peer opinion become overriding influences for some students. Single-sex schools don't eliminate the social distraction of adolescence—there's plenty of hazing and cliquishness to make up for the lack of sexual attraction. But they do automatically expand the leadership opportunities available to both boys and girls, and they may increase the odds that each sex will enter nontraditional disciplines. Though the evidence for this from studies of single-sex schools is weak, other research suggests that any time the ratio gets above three to one, the minority gender tends to disengage and may be evaluated less fairly. So even in coed schools, I agree that there should be some provision for single-sex classrooms in subjects such as computer science and studio art, and preferably they should be taught by teachers of the same sex as the students.

Which leads us to perhaps the greatest asset of successful single-sex schools: the gender composition of their faculty. At all-girls' schools, one often finds strong, dedicated women serving as role models in math and science. Ditto for all-boys' schools, which manage to employ a strikingly larger proportion of men than usual teaching in every subject (not just math and PE) and providing a conspicuous, pro-academic contrast to the other male role models in many boys' lives. The importance of teacher gender is supported by several studies, including one from Belgium, where sixth-grade boys were found to suffer from the preponderance of female teachers, and another from Nigeria, where girls' favorable attitude toward math depended on their having a female teacher in the subject. Even in coed middle schools, there is evidence that girls and boys each do slightly better in science, English, and social studies classes led by teachers of the same sex as themselves.

Of course, you don't have to open a single-sex school to hire more male elementary-school teachers or women high-school math instructors. The same focus on individual needs and gender empowerment can and should be a focus of any curriculum, coed or single-sex. Just as successful single-sex schools need to balance their premise of gender differences against the goal of gender equity, so coed schools need to remove their neutral blinders and accept that gender is an important basis of children's individual needs.

In the end, the strongest argument against single-gender education lies in the reality of adulthood: girls and boys ultimately need to learn to work together, respect each other, and also compete against each other. While single-sex schools may be a good Band-Aid for some students, they are never going to close the real gender gaps that still trouble us. As I've tried to show throughout this book, each gender has much to learn from the other. Girls in the past few decades have obviously benefited from emulating boys' more assertive and ambitious ways, while boys have always gained from the studious example and calming influence of girls in classrooms. While single-sex classrooms may reduce boys' and girls' attempts to differentiate themselves from each other, coed classes are ultimately a better environment for deflating stereotypes of the opposite sex, just as any stereotype withers when you get to know people as individuals.

It sounds trite but it really is true: every child is an individual. While single-sex schools or classrooms may be a temporary fix that works for some boys and girls, I believe that the greater risk—of gender stereotyping and the loss of mutual understanding—makes such segregation a step in the wrong direction.

. . .

I'll confess that this has been a difficult book to write. As a biologist, I set out naively thinking I could explain the differences between boys and girls simply by extending what scientists have learned about lab animals. But in searching for the fuller explanation of boy-girl differences, I've had to grapple with the true manifestation of our remarkably plastic brains—human culture, which puts gender squarely at the front of individual identity and is at least as important to sex differences as the genes and hormones we biologists are so fixated on. Happily, a new field of cultural neuroscience—comparing brain function in people from different cultures—is pushing the research on this frontier and promises to bring much-needed balance to the issue of sex differences in the brain and mind.

No question, gender does matter, in education and everything else we do as humans. But if we're going to pay attention to sex differences, we have to do it carefully. After eons of unequal rights and opportunities, no one wants to undo recent progress with distorted stereotypes about either boys or girls. The challenge is to acknowledge and use our understanding of sex differences to help children, without turning this knowledge into self-fulfilling prophecies.

So we've got to do this right, scientifically and accurately, which explains the depth and detail of this book. I've gone to some lengths to present the real data on sex differences in children in response to the many glib generalizations that I believe are leading parents and teachers astray. Even in adults, most sex differences in cognitive and personality traits are smaller than commonly believed. For children, they are smaller still; just as their bodies have yet to fully diverge, so will their minds take time to grow into male and female styles of thinking. As we've seen for all their attributes—cognitive skills, school performance, motivation, emotions, and relational styles—boys and girls do differ somewhat, but they are hardly worlds apart.

Yet just because the differences are small doesn't mean we can ignore them. Small gaps can end up having large consequences depending on how we let them fester. Adults need to be aware of boy-girl differences so we can help children compensate for them early on, allowing every child to flourish in his or her strengths, but always looking for opportunities to stretch them in ways that their peers and the broader culture will not.

Obviously, we can't be gender blind; parents and teachers must recognize and accept boys' and girls' different needs and interests. But the trick is to shed our gender blinders without turning everything into a Mars/Venus dichotomy. We need to be aware of gender but also of the imprecision of stereotypes. Above all, we need to assiduously avoid prejudging any boy or girl. Presuming that girls will be less interested in science or boys will not enjoy writing virtually defeats the purpose of education. What's the point of teaching children a broad range of skills if you convince yourself they're hard-wired for some tasks but not others? Girls and their parents have increasingly rejected this notion, but not the parents of boys, who find it easier to blame testosterone or brain maturation for every misbehavior and lagging achievement rather than examine their own nurturing, discipline, and expectations for diligence and self-control.

Admitting that sex differences *are* plastic puts an unquestionably greater burden on parents and teachers. I'll confess that my husband and I struggle with these issues all the time: how to keep our boys on task, reading instead of playing computer games, conscientious about their schoolwork and chores, and thoughtful of the feelings of others. The goals are different for our daughter: teaching her to assert herself, take more intellectual and athletic risks, and cultivate interests beyond her social life and appearance. Such gender worries are part and parcel of our larger parenting challenges—balancing discipline with self-esteem, fending off the superficialities of youth culture, and shielding our children from the deluge of sex and violence in the media.

None of it is easy, and in some ways, it's even harder in modern society, where all of this plasticity becomes a double-edged sword. Given the broad freedom to pursue their different interests, along with a culture that gives them every extreme of choice, boys and girls naturally end up exercising different neural circuits, promoting the gender-specific strengths we're so familiar with—athletic and mechanically adept males, verbal and socially more sensitive females. As girls expand their range of interests, boys are fleeing them—from school choruses, student councils, newspaper staffs—maintaining their male identity in sports alone. And while girls have greater latitude in their various pursuits, they are as constrained as ever by the dictates of beauty and femininity.

But plasticity also offers the best chance to reduce some of the gaps between boys and girls, if parents and teachers can rein in some of these choices and ensure that each gender exercises a wider range of skills while growing up. Throughout the book, I've offered concrete suggestions for doing just this—for keeping boys reading, communicating, and creating, practicing their leadership, nurturing, and organizational skills; for exercising girls' spatial, math, and technical abilities and easing them into greater competitiveness, risk taking, and confidence in their own leadership abilities.

The human brain is an amazing organ, but none of its remarkable faculties is limited to one sex or the other. Every boy and girl deserves the chance to develop the full range of his or her potential, unhindered by gender, race, or other societal assumptions. If we can pull it off—break down remaining barriers by taking a more proactive approach to sex differences—we all stand to gain. Boys and girls will both lead richer lives, and our society will benefit from the more complete development of their diverse and balanced talents.

I was reminded of this balance just recently, at Julia's first high-school choral concert. It was a lovely evening of music performed by several different choirs, but the best was the upper-class chorus, which had managed to maintain equal numbers of boys and girls. With its lush mix of pitches from bass to soprano, its sound was more resonant than the mostly girls' choruses that had preceded it. Together, these young men and women were creating far more than the sum of their individual parts—a rich and complex harmony that comes only from blending the full range of human voices.

Notes

Illustration Credits

Bibliography

Index

Notes

1   *68 percent of expectant parents learn the sex*: According to a Web-based survey
      of 10,648 parents (Sharp 2004).

3   *boys more likely to die in accidents*: UNICEF 2001.

      boys more often victims of violent crimes: Wordes and Nunez 2002.

      girls more likely to attempt suicide: Centers for Disease Control (CDC) 2002.

      boys more likely to commit suicide: CDC 1997.

      57 percent of college students are women: This figure is based on 2005 data,
      the most recent available (U.S. Dept. of Education 2007).

      boys score twenty-five points higher on SAT: This figure is the average gap
      between boys and girls since 2006, when a mandatory writing section was
      introduced (College Board 2008).

      college engineering degrees: National Science Foundation 2006.

      women earn less than eighty cents for every dollar men earn: U.S. Government
      Accounting Agency 2003.

4   *hormone receptors present at an early age*: Although there is ample evidence
      of it in other species, there is surprisingly little proof in humans that an-
      drogen receptors (which bind testosterone and are the primary trigger for
      sexual differentiation in primates) are present during the first half of prena-
      tal gestation. The one study of human fetuses that looked failed to find them
      (Abramovich et al. 1987), while a study of male rhesus monkeys found that
      androgen receptors begin to appear at the end of the first trimester but only
      in a restricted area of the hypothalamus (Choate et al. 1998).

5   *brain size*: Newborn boys' brains are 8 percent larger than newborn girls'
      brains, comparable to their 9 percent higher birth weight, according to

Gilmore et al. 2007. The difference in brain volume at age thirteen is closer to 11 percent, according to Lenroot et al. 2007.

brain versus physical growth: CDC growth charts for boys and girls between two and twenty years of age, accessed at http://www.cdc.gov/growthcharts. The brain-size difference is based on average cerebral volumes in nine-year-olds reported in Lenroot et al. 2007.

EEG sex differences: Clarke et al. 2001; Barry et al. 2004; Matthis et al. 1980; Benninger et al. 1984.

7   *heritability of weight:* One recent meta-analysis found that 77 percent of the variance in body mass index (BMI) is attributable to genes. Cited in Speakman 2004.

8   *males and females* share *roughly 99.8 percent* (footnote): Ross et al. 2006.

girls' brains are wired for communication and boys' for aggression: Brizendine 2006, p. 14.

serotonin and oxytocin: Claims about sex differences in the levels of these neurotransmitters are found in the writings of Michael Gurian and have been widely influential in his "brain-based" teachers' training. However, the only studies I've found report no significant differences in these hormones between boys and girls: For oxytocin, see Fries et al. 2005; for serotonin, Flachaire et al. 1990.

boys do math using the hippocampus while girls use the cerebral cortex: Sax 2005, pp. 101–2. This claim is based on differences in *spatial* processing by adult men and women, which I discuss in chapter 6. However, this difference has not been shown in children, nor has it been shown between males and females during an actual *mathematical* task.

belief that girls are left-brain dominant while boys are right-brain dominant: Gurian 2001, p. 21.

9   *Michael Gurian stresses the nature side:* Gurian and Stevens 2005, p. 42.

trumpets biologically programmed differences: Sax 2005. Quote is from the flap copy.

"The female brain": Brizendine 2006, p. 8.

10   *file-drawer effect:* Halpern 2000, pp. 66–68.

corpus callosum and women's intuition: Begley 1995.

review of fifty such studies: Bishop and Whalsten (1997) also detail the various popular media reports about corpus callosum dimorphism that appeared in *Time* and *Newsweek* magazines as well as in Phil Donahue's 1985 book *The Human Animal* (Simon and Schuster).

no sex difference in child or fetal corpus callosum: Bell and Variend 1985; Clarke et al. 1989; Koshi et al. 1997; Giedd et al. 1999b; Ng et al. 2005.

persistent claims about the corpus callosum: Gurian and Stevens (2005, p. 48) write that "a boy's corpus callosum . . . is a different size than a girl's (some studies show up to a 25 percent difference)." Leonard Sax ignores the consensus data and continues to cite selective studies about the corpus callosum on his webpage about "brain differences" between boys and girls: http://www.singlesexschools.org/brain.html.

11   *good research on sex differences:* Cahill 2005.

difference between the sexes in adult height (data and figure): Lippa 2002, pp. 5–6.

most psychological sex differences are small: Hyde 2005.

13   *"Men are from North Dakota":* This line appears to have originated with Dr. Kathryn Dindia, professor of communications at the University of Wisconsin in Milwaukee.

15   *sex-roles survey of college students:* Lueptow 2005.

16   *gifted teenagers:* Lippa 1998.

18   *both sexes are earning higher grades:* Corbett et al. 2008.

1. PINK AND BLUE IN THE WOMB

20   *absolute earliest for gender to be evident:* Efrat et al. 1999.

Sex determination not as cut-and-dried (footnote): Nelson 2000, pp. 122–25

22   *MicroSort method:* Fugger et al. 1998.

MicroSort success rate: Schulman and Karabinus 2005; Matken et al. 2003.

23   *maternal blood testing for Y chromosome genes:* Kaiser 2005; Kaplan 2008.

Shettles method: Shettles and Rorvik 2006.

24   *not the quicker, wilier critters:* Han et al. 1993. In fact, recent research suggests that in seminal fluid, female sperm swim faster than male sperm (Maligaya et al. 2006).

timing of conception has no bearing on baby's sex: Wilcox et al. 1995.

dramatically skewed sex ratio in Asia: In China, where couples are limited to one child and most girls end up leaving their families after marriage to take care of their in-laws, a recent estimate shows 117 boys born to every 100 girls; in more prosperous regions, it's as high as 135 boys for every 100 girls. Similar ratios are reported in parts of India (Parliamentary Office of Science and Technology, UK, 2003).

IVF and PGD discouraged for family balancing: Robertson 2002.

excess of second- and third-born boys: Almond and Edlund 2008.

25   *improving the status and equality of women in Asia:* There is good news on this front from South Korea: as educational and job opportunities for females have dramatically improved, the ratio has gone from nearly 117 boys born for every 100 girls in 1990 to about 108 boys for every 100 girls in 2007 (Sang-Hun 2007).

"being a boy or girl is a medical handicap": Dr. E. Scott Sills, as quoted in Kranz 2001.

SRY discovery: Simpson et al. 1987.

27   *undescended testicles:* Fujimoto and Soules 1998.

testosterone levels: Nagamani et al. 1979; Reyes et al. 1974.

AIS individuals are unequivocally female: Hines et al. 2003.

28   *DAX1:* Vilain 2000.

role of ovarian hormones: In rats, early postnatal estrogen is known to decrease the size of a female's corpus callosum, the massive bundle of white
matter that allows the two sides of the brain to communicate with each
other. It is not known whether estrogen acts similarly in human infant girls,
but this seems unlikely, since the human corpus callosum is, if anything,
slightly larger in women than men. Nonetheless, these findings in rats have
led to a new theory, a revision of the classic dogma: the female brain is no
longer considered merely the default version of the human brain, the one
without testosterone, but also depends on ovarian hormones for normal
development (Fitch and Denenberg 1998).

alpha-fetoprotein: Bakker et al. 2006.

29   *Turner's syndrome:* LeVay and Valente 2002, p. 130; Ross et al. 2002.

30   *male rats castrated within the first six days of life:* Beatty 1992.

rat ovarian hormones never begin cycling: Kelly et al. 1999.

31   *masculine behaviors in testosterone-treated female rats:* Meaney and Stewart
1981.

hypothalamic sex differences: Kelly et al. 1999.

testosterone receptors as early as seven days before birth: Vito and Fox 1981.
Paradoxically, testosterone exerts many of its effects on the rat brain via
estrogen, into which it is converted by an enzyme called aromatase. In primates, however, testosterone exerts its masculinizing effects mainly through
the androgen receptor.

32   *prenatal testosterone doesn't lessen females' interest in babies or increase mounting:* Herman et al. 2003; Wallen 2005.

infant vocalizations and spatial navigation: Wallen 2005; Herman and Wallen
2007.

sex differences in human hypothalamus: Four different research groups have
looked for sex differences in the human hypothalamus, identifying three
different tiny zones that are larger in men than women. It's agreed by three
American research groups (Allen et al. 1989; LeVay 1991; Byne et al. 2001)
that one of these areas, known as INAH-3 for the "third interstitial nucleus
of the anterior hypothalamus," is larger in men. However, Allen et al. also
found a difference in the second interstitial nucleus (INAH-2), which neither LeVay nor Byne et al. detected. A Dutch team initially failed to find
a difference in either INAH-3 or INAH-2, but based on a larger sample of
brains, it reported about a two-fold difference between the sexes in the size
of the first interstitial nucleus, INAH-1 (Swaab et al. 2002). Very recently,

however, Swaab and a colleague confirmed that a portion of INAH-3 is about twice as large in men than in women (Garcia-Falgueras and Swaab 2008). Nonetheless, this difference is considerably more modest than the sex difference in rats and located in a startlingly small area of the brain.

macaque hypothalamus: Vasey and Pfaus 2005.

33 *female psychological profile in AIS:* Hines et al. 2003.
effect of SRY on rat and bird brains (footnote): Arnold 2003.

35 *male identity mirrored in twin:* Iervolino et al. 2005. This large study found that same-sex twins exert stronger influence over preschoolers' gender-typed behavior than regular siblings do.

less famous case: Bradley et al. 1998.

preferred way to rear children with ambiguous genitalia (footnote): Reiner and Gearhart 2004; Meyer-Bahlburg 2005.

36 *CAH girls more aggressive, more physically active:* Pasterski et al. 2007.
more fond of rough play: Servin et al. 2003.
preference for traditional boy toys: Berenbaum and Snyder 1995.
real human infants: Hines 2004, p. 152.
career interests in CAH women: Berenbaum 1999.

37 *sexual orientation in CAH women:* Zucker et al. 1996.
parents of CAH girls do not treat them differently: Pasterski et al. 2005.
current medical practice dictates CAH individuals raised as girls: Joint LWPES/ESPE CAH working group 2002.

38 *testosterone and effect of uterine position on female rodent behavior:* vom Saal and Bronson 1980; Vandenbergh 2003.

39 *larger (more male-size) teeth:* Dempsey et al. 1999.
girls with twin brothers are more lateralized: Cohen-Bendahan et al. 2004. On the other hand (so to speak), a large study of girls with twin brothers failed to find consistent evidence for masculinization of their finger lengths (Medland et al. 2008a).

otoacoustic emissions in girls with male twins: McFadden 1993.
lesbian and bisexual women also have more male-range OAEs: McFadden and Pasanen 1998.

40 *androgens do not cause female homosexuality:* Gooren 2006.
girls with twin brothers are no more inclined toward lesbian: Bearman and Brückner 2002.
aggression in opposite-sex twins: Cohen-Bendahan et al. 2005a.
risk taking in opposite-sex twins: Resnick et al. 1993.
spatial skills in opposite-sex twins: Cole-Harding et al. 1988. This preliminary finding was never published in a peer-reviewed journal.
toy preference in opposite-sex twins: Rodgers et al. 1998. Similarly, a more recent large study found no evidence of altered play interests between girls with male twins and girls with female twins (Iervolino et al. 2005).

appear just as feminine: Loehlin and Martin 2000.

math in opposite-sex twins: Kovas et al. 2007.

cognitive skills in opposite-sex twins: Luciano et al. 2004.

no difference in age of puberty: Rose et al. 2002.

prenatal testosterone's effects and contemporary reproduction (footnote): Lummaa et al. 2007. The large study of modern Finns is published in Rose et al. 2002. The recent study of Dutch, Australian, and American twins is in Medland et al. 2008b.

41   *older-brother effects on play interests of younger sisters:* Henderson and Berenbaum 1997.

42   *how much testosterone boys pass on to their twin sisters:* Cohen-Bendahan et al. 2005b. This study rules out the possibility that testosterone reaches female twins via maternal blood, but it remains possible that hormones cross directly between the fetal membranes of twins.

subtle sex difference in the lengths of specific fingers: Manning et al. 1998.

finger ratio found in fetuses as early as nine weeks: Malas et al. 2006.

43   *2D:4D ratios and amniotic testosterone:* Lutchmaya et al. 2004.

the more crucial test: Without this test, the correlation between digit ratio and prenatal testosterone is virtually inevitable in a mixed population and could be due to any maleness factor, genetic, hormonal, biochemical, or otherwise.

three studies of CAH girls' digit ratio: The two studies that found masculine finger ratios in girls with CAH were Brown et al. 2002 and Okten et al. 2002. A third, very large study disagreed with the others, finding that girls with CAH had typical female digit ratios, but that study was based on measurements of the left hand (Buck et al. 2003).

two out of three studies of girls with male twins: Van Anders et al. 2006; Voracek and Dressler 2007. The more convincing study that failed to find masculine digit ratios in girls with male twins is Medland et al. 2008a.

every possible behavioral sex difference: See, for example, Hampson et al. 2008.

2D:4D and homosexuality: Lippa 2003; Manning et al. 2007; Collaer et al. 2007; McFadden et al. 2005.

prenatal sex hormones exert only "a modest predisposing influence": Collaer et al. 2007.

45   *no sex differences in fetal activity or heart rate:* Eaton and Enns 1986; Robles de Medina et al. 2003.

cell division in male embryos: Pergament et al. 2002.

faster development protects male embryos from estrogen: Ingemarsson 2003.

46   *risk for serious respiratory problems:* Perelman et al. 1986.

fetal lung maturity: Lazarus 2001.

risk of newborn death due to respiratory problems: Mage and Donner 2004.

fetal habituation: Leader et al. 1982.

sex difference in fetal mouth movements: Hepper et al. 1997; Miller et al. 2006.

preterm girls are known to suck harder: Lundqvist and Hafström 1999.

47 *170 males for every 100 females conceived:* Pergament et al. 2002.

boys 7 percent more likely to be born prematurely: Cooperstock and Campbell 1996.

sex difference in infant mortality rates: Ingemarsson 2003; Mizuno 2000.

declining sex difference in infant mortality: Bhaumik et al. 2004.

48 *fit mothers raise more males:* Trivers and Willard 1973.

Trivers-Willard effect in human societies: Hrdy 1999, pp. 318–50; Cronk 2007; Gaulin and Robbins 1991; Koziel and Ulijaszek 1991.

1 percent higher chance: Almond and Edlund 2006.

male vulnerability and natural selection: Wells 2000.

H-Y antigen: Gualtieri and Hicks 1985.

androgens suppress immune function: Muehlenbein and Bribiescas 2005.

boys' vulnerability: Geary 1998, pp. 216–17.

49 *fetal sex and a mother's abdomen shape:* Perry et al. 1999.

10 percent more calories: Tamimi et al. 2003.

50 *Swedish study of hyperemesis gravidarum:* Askling et al. 1999.

UK study of hyperemesis gravidarum: del Mar Melero-Montes and Jick 2000.

women pregnant with boys actually scored higher: Vanston and Watson 2005.

51 *women's intuition:* Perry et al. 1999.

52 *maternal blood testosterone and fetal sex:* Klinga et al. 1978; Meulenberg and Hofman 1991.

2. UNDER THE PINK OR BLUE BLANKIE

56 *labor twenty-four minutes longer with boys:* Eogan et al. 2003.

use of analgesic medications: Boatella-Costa et al. 2007.

fetal distress: Bekedam et al. 2002.

boys and C-sections: Eogan et al. 2003; Lieberman et al. 1997; Harlow et al. 1995.

Apgar score: Sheiner et al. 2004.

57 *Brazelton scale:* Boatella-Costa et al. 2007; Lundqvist and Sabel 2000.

boys' higher metabolic rate and less mature lung function: Rosen and Bateman 2004.

9 percent larger brains: Lenroot et al. 2007.

larger kidneys in boys: Schmidt et al. 2001.

58 *girls' skeletal system:* Flory 1935; Tanner 1990, pp. 56, 175.

Swedish EEG study: Thordstein et al. 2006.

EEGs at three and six months: Jing et al. 2008.

more mature EEG patterns in boys: Barry et al. 2004; Clarke et al. 2001; Benninger et al. 1984; Matthis et al. 1980.

59   *all the existing studies of newborn touch:* Maccoby and Jacklin 1974.

pain tolerance versus threshold in adults: Baker 1987.

60   *sex difference in pain response in two-day-old infants:* Guinsburg et al. 2000.

newborn cortical activation by pain: Bartocci et al. 2006.

analgesia recommendation for circumcision: American Academy of Pediatrics, Task Force on Circumcision 1999.

many older doctors: Yawman et al. 2006.

circumcision rates: Alanis and Lucidi 2004.

effect of circumcision on male newborn behavior: Richards et al. 1976. A subsequent report (Brackbill and Schroder 1980) reviewed thirty-eight studies and concluded that circumcision has little effect on newborn behavior. Interestingly, this review also found little evidence for any behavioral differences between newborn boys and girls.

sex difference in newborn olfaction: Balogh and Porter 1986.

61   *sex difference in response to breast-milk odor:* Makin and Porter 1989.

olfactory sex difference at all ages: Doty et al. 1984.

no sex difference in frontal-lobe activity: Bartocci et al. 2000.

auditory EEG sex difference: McFadden 1998; in an e-mail correspondence, Dr. McFadden replied to my question about whether the sex difference in hearing significantly alters boys' and girls' experience in the classroom, writing, "This verges on the absurd. The 3 dB difference at threshold almost surely disappears at suprathreshold levels. That is, there is no obvious benefit from having teachers speak louder to boys than girls . . . [and] arguing for separate schools because of the small sex diff. in hearing sensitivity is just dopey."

faster wave V response: Ribeiro and Carvallo 2008; Stuart and Yang 2001; Eldredge and Salamy 1996. According to Ribeiro and Carvallo, the sex difference in newborns' wave V latency is mostly attributable to girls' smaller cochleas. Stuart and Yang summarize the existing evidence for sex differences in infant auditory brain-stem responses as "equivocal."

d value of 0.26: This value is from Eldredge and Salamy 1996.

more sensitive in boys: Sininger et al. 1998. This study is also cited by Leonard Sax (2005, p. 17), who erroneously extrapolates their finding into an "80 percent greater" auditory brain-stem response in girls than boys. Sax's faulty reasoning is thoroughly dissected by Liberman (2006a).

62   *sex difference in OAEs:* Berninger 2007.

OAEs in girls who have twin brothers: McFadden 1993.

shorter female cochlea: Berninger 2007.

d value of about 0.15: This effect size was calculated by Liberman (2008) based on data in Berninger (2007).

girls hear shouting: Sax 2005, pp. 17–18, 87–88. Sax's extrapolation from auditory thresholds to classroom behavior is also discussed on his website http://www.singlesexschools.org/research-learning.htm. For a more thorough deconstruction of Sax's exaggerated claims about auditory sex differences, see Liberman (2006a, 2006b, 2008). At the 2008 posting, Liberman writes: "In my opinion, there's no doubt that there are group differences between the sexes in hearing. But there are two points in contention. The first question is whether these differences are so large and so consistent—especially in relation to variation among boys and among girls—as to create a significant problem in mixed-sex classrooms. My somewhat-informed opinion is that this is very unlikely to be true. In fact, I think that the word 'preposterous' might be appropriate. The second question is whether the presentation of scientific evidence in Dr. Leonard Sax's book *Why Gender Matters* can be trusted. On the basis of a careful examination of an admittedly small sample of cases, my opinion is that the answer is 'no.'"

"The bulk of the evidence": Maccoby and Jacklin 1974, pp. 25–26. Among studies of infants between three and a half and thirteen months old, two experiments showed that girls responded more than boys to auditory stimuli; one experiment showed that boys responded more than girls; and in six other experiments, no sex difference was found.

63 *no sex difference in newborn vision:* Ibid., pp. 29–32.

girls' visual systems develop a few weeks earlier: Malcolm et al. 2002; Peterzell et al. 1995.

girls' stereovision begins earlier: Gwiazda et al. 1989.

testosterone and infants' 3-D vision: Held et al. 1988.

no evidence for sex differences in infants' learning: I corresponded with Carolyn Rovee-Collier about sex differences in infant learning. Her lab has published dozens of papers on operant conditioning in infants using a variety of stimuli. She replied: "We have analyzed for sex differences repeatedly with infants of all ages and never find any."

64 *girls are sometimes described as ahead in fine motor control:* Nagy et al. 2007. A lack of sex difference in the pincer grasp was reported by Butterworth and Morissette 1996.

studies finding a gross motor advantage in boys: Reinisch and Sanders 1992; Capute et al. 1985.

grip strength: Jacklin et al. 1984.

65 *reflex asymmetry:* Grattan et al. 1992.

boys' higher activity levels: Eaton and Enns 1986.

66 *slope-crawling experiment:* Mondschein et al. 2000.

67 *sex difference in rhesus monkey infant vocalizations:* Wallen 2005.

68   *girls begin talking one month earlier than boys:* Halpern 2000, p. 96.
      different pattern of brain waves: Molfese and Molfese 1979.
      one early study said that baby girls' brains are more lateralized: Shucard et al. 1981.
      frequently cited study has been disputed: Molfese and Radtke 1982.
      stronger left-hemisphere activation in boys: Friederici et al. 2008.
      more lateralized OAEs in boys: Berninger 2007; Newmark et al. 1997.
69   *more lateralized hearing sensitivity in boys:* Sininger et al. 1998.
      researchers at San Diego State University: Fenson et al. 1994.
70   *girls point earlier:* Butterworth 1998.
      5 percent more gestures than boys: Berglund et al. 2005.
      sex-typed infant gestures: Fenson et al. 1994, pp. 74–79.
      being male is no excuse for talking late: Ibid., p. 117. In another study, this one involving Chinese children, two-year-old girls were found to lead boys slightly in vocabulary and grammatical development, but gender accounted for less than 1 percent of the total variance in language development (Zhang et al. 2008).
      Language in = Language out: Eliot 2000, pp. 382–90.
72   *one 1979 study found that girls held eye contact longer:* Hittleman and Dickes 1979.
      larger 2004 study found no sex difference in the newborn period: Leeb 2004. Another study, although not conducted blind, also found no sex difference in eye contact at either one or three months of age (Moss and Robson 1968).
73   *sex difference in newborn face preference:* Connellan et al. 2000. The authors did not report effect sizes, but I calculated the d value for girls' greater attention to Connellan's face as -0.17 (a small difference), while the d value for boys' greater attention to the mobile was +0.47 (a moderate difference).
      seemingly every commentator on sex differences: Sax 2005, p. 19; Brizendine 2006, p. 15; Pinker 2005b; Pinker 2008, p. 35.
      researcher was not always blind to the sex of the babies: When I e-mailed Dr. Baron-Cohen about this issue, he replied that Connellan was aware of the baby's gender "in just a few cases" and emphasized that they followed the standard practice of making sure that the coders who measured the babies' eye movements from videotapes were indeed blind to their sex.
      stands in opposition to many similar studies: Spelke 2005. I corresponded with Mark Johnson, a developmental psychologist at the University of London who has published many papers on face perception in infants. He replied: "We have not observed any clear sex differences in our newborn studies on face preferences."
      Canadian researchers found that both boys and girls gazed longer at a toy: Cossette et al. 1996. The infants in this report were studied at both two and a half and five months old. At both ages, both sexes spent considerably more

time looking at the mobile than at their mothers; however, at two and a half months (but not at five months), the girls spent more time smiling at their mothers than the boys did. This was the only significant sex difference out of thirty-two comparisons of emotional expression in this study, but the researchers who measured the babies' videotaped expressions were not blind to their sex.

74   *boys, not girls, paid greater attention to faces:* Moss and Robson 1968.

"there is no evidence that girls are more interested in social": Maccoby and Jacklin 1974, pp. 36–37. The authors identified thirteen experiments (published in ten papers) up to that date in which a sex difference had been found in infants' attention to either social or nonsocial stimuli. Among the nine experiments with nonsocial stimuli, five showed girls fixating longer than boys, and the other four showed boys fixating longer than girls. Among the four experiments that used social stimuli, in three, boys fixated longer, and in one, girls did. Maccoby and Jacklin summed up with "the conservative conclusion that the sexes do not differ in their interest in visual social stimuli."

"testosterone . . . shrinks the centers for communication": Brizendine 2006, p. 15.

testosterone increases amygdala size: Cooke et al. 1999.

boys startle, cry, and grimace more: Haviland and Malatesta 1981.

stable sleeping pattern: Nagy et al. 2001.

newborn irritability or consolability: Four studies have looked for sex differences using the Brazelton Neonatal Behavioral Assessment. The largest, by Boatella-Costa et al. (2007), reported greater irritability in boys. Lundqvist and Sabel (2000) found no significant difference in irritability, but girls scored higher on the ability to console themselves. However, two other studies reported no sex difference on either measure: Davis and Emory 1995; Canals et al. 2003.

greater cortisol surge in boys: Davis and Emory 1995. However, this study did not find any significant sex differences in actual newborn behavior.

75   *negative message to boys about their expressiveness:* Haviland and Malatesta 1981.

Italian study of face-to-face infant-mother communication: Lavelli and Fogel 2002.

gaze aversion: Haviland and Malatesta 1981.

76   *girls express greater interest:* Malatesta and Haviland 1982. As the authors point out, the difference in interest expressions may actually be due to subtle variations in facial structure: girls' eyes are a bit wider and their brows a bit higher than boys', which may give the viewer the impression of greater interest than is seen in boys' narrower eyes and lower brow structure, perhaps making girls seem more receptive to social advances.

mothers of girls tend to ignore anger: Manstead 1992.

parents help shape sex difference in eye contact: Leeb 2004; Lavelli and Fogel 2002.

imitative crying in newborns: Simner 1971; two other studies with comparable findings are reviewed in Eisenberg and Lennon 1983.

77   *"hard-wired for empathy":* Baron-Cohen 2003, p. 1.

two dozen studies of babies' abilities to recognize expression: McClure 2000.

"Anyone who has raised boys and girls": Brizendine 2006, p. 16.

78   *d value of 0.26:* McClure 2000.

area TE develops more slowly in male monkeys: Bachevalier et al. 1989; Hagger and Bachevalier 1991.

no sex differences in attachment: Maccoby 1998, p. 16.

79   *boys receive less positive caregiving:* NICHD Early Child Care Research Network 2005, p. 125.

boys may suffer from infant daycare: Belsky and Rovine 1988.

girls may benefit from infant daycare: Baydar and Brooks-Gunn 1991.

prevalence of autism: NIMH fact sheet 2004/2007.

signs of autism in the first year: Mandell et al. 2005.

80   *autism possibly underdiagnosed in girls* (footnote): Flora 2006.

more objective measures to measure empathy: Eisenberg and Lennon 1983.

eye region of the face: Baron-Cohen's group failed to find a statistically significant difference in empathy between men and women in this study (Baron-Cohen et al. 2001).

results often did not hold up: Here are synopses of the several studies from Baron-Cohen's lab addressing the relationship of fetal testosterone to various behavioral measures. In most cases, a significant relationship was found when boys and girls were analyzed together, but the effects were not significant within either sex. Several times, the authors stated that their study was statistically "underpowered" to detect such differences, but the solution—sampling a larger population—was not attempted. Knickmeyer et al. 2006b: Although the authors found a significant correlation between fetal testosterone and one particular measure of empathy (the number of intentional terms a child used to describe a cartoon character's actions), they did not find this correlation for other measures of empathy, and the correlation with intentional terms used did not hold when boys and girls were analyzed separately. Knickmeyer et al. 2005a: In this study, the authors found a significant relationship between prenatal testosterone and restricted interests in boys, but not in girls; they also found a significant relationship between fetal testosterone and social relationships, but it did not hold within either sex alone. Lutchmaya et al. 2002a: This study found a relationship between fetal testosterone and gazing at the mother's face in boys, but not in girls. Lutchmaya et al. 2002b: There was no relationship between amniotic testosterone level and vocabulary size within the separate populations of boys and girls, although there was a significant relationship when all children were analyzed together.

81   *fetal testosterone and "autistic traits"* (footnote): Aueyeung et al. 2008.

    girls with CAH are not any more likely to be diagnosed with autism: Knickmeyer et al. 2006a.

    postnatal brain overgrowth in autism: Courchesne et al. 2003.

82   *girls with CAH do not have larger brains:* Merke et al. 2003.

    genetics of autism: Muhle et al. 2004; Ronald et al. 2006.

    autism not clearly linked to the X chromosome: Pickles et al. 2000.

    fragile X syndrome and autism (footnote): Muhle et al. 2004.

    autism and TV exposure: DeNoon 2006.

    reason for starting autistic treatment as young as possible: Landa 2007.

    "recovery" from autism with early intervention: Helt et al. 2008.

83   *classic study of parents' judgments of newborns:* Rubin et al. 1974.

84   *one study of girls mislabeled as boys and boys mislabeled as girls:* Haviland and Malatesta 1981.

    dozens of gender-disguise studies: Stern and Karraker 1989.

    different birth congratulations: Bridges 1993.

85   *comment more on the physical activity of infants:* This observation was likely accurate, given what we've seen about their different activity levels (Pomerleau et al. 1997).

    gender stereotyping by parents is declining: Karraker 1995.

    prefer the higher-pitched female voice: DeCasper and Prescott 1984.

    preference for female faces except among babies with stay-at-home fathers: Quinn et al. 2002.

86   *male/female as second dichotomy:* Coates and Wolfe 1995.

87   *estrogen and newborn girls' breast tissue:* Schmidt et al. 2002.

    possible behavioral effects of neonatal hormone surges: Quigley 2002.

88   *graph of testosterone and estrogen levels:* based on data in Winter et al. 1976.

    neonatal testosterone and rat behavior: Meaney and Stewart 1981.

89   *little effect of neonatal testosterone on rhesus behavior:* Brown and Dixson 1999.

    neonatal testosterone and object discrimination in monkeys: Hagger and Bachevalier 1991.

    neonatal testosterone and effects on puberty: Mann and Fraser 1996.

90   *cryptorchidism and neonatal testosterone:* Raivio et al. 2003.

    cryptorchidism and later fertility problems: Mann and Fraser 1996.

    visual development and neonatal testosterone: Held et al. 1988.

    language processing and neonatal testosterone: Friederici et al. 2008. Both this study and Held et al. were very small and only correlational. An equally likely scenario is that vision and language processing (or brain maturation

332 *Notes to pp. 91–94*

generally) is slowed by *prenatal* testosterone, and that variations among different boys in postnatal testosterone reflect similar variations prenatally.

91   *lower sperm count:* Swan et al. 2003.

cryptorchidism and hypospadia: Fernandez et al. 2007.

evidence is weaker: Rogan and Ragan 2003; Schoeters et al. 2008.

92   *evidence of harm in humans is sparse:* The strongest evidence linking phthalates to altered testosterone action in boys comes from one study by Shanna Swan and colleagues at the University of Rochester. They measured the distance between the anus and penis in baby boys (a parameter that's been shown in rodents to be sensitive to androgen-disrupting chemicals) and found that the distance correlated with the mothers' exposure to phthalates during pregnancy. While provocative, it's important to point out that this study found no evidence of actual harm to infants, and it has yet to be replicated (Swan et al. 2005).

concerns about DEHP in baby boys and pregnant women: Bucher 2008. One study has followed children who were exposed to high levels DEHP during medical treatment as neonates; no deleterious effects were found on their growth, endocrine function, or pubertal development (Rais-Bahrami et al. 2004).

soy isoflavones are less potent: Rogan and Ragan 2003.

ten times higher than the dose known to disrupt women's menstrual cycles: Setchell et al. 1997.

93   *adverse effects of exposure to soy phytoestrogens in rats:* Chen and Rogan 2004.

masculinization of the female brain and behavior: Kouki et al. 2003.

infant monkeys' testosterone levels suppressed by soy: Sharpe et al. 2002.

little evidence of harm in adults reared on soy formula: Strom et al. 2001.

advised against the feeding of soy formula: British Dietetic Association 2003.

94   *effectiveness of "parentese":* This higher-pitched, slower, and more singsong style of speaking is used by parents throughout the world, and with good reason. Research by Patricia Kuhl and her colleagues has shown that such baby talk, or infant-directed speech, actually helps babies learn language better than the normal pitch and cadence adults use when addressing other adults. In this study, mothers from three countries (the United States, Sweden, and Russia) were found to exaggerate different vowel sounds when speaking in baby talk (that is, while actually speaking to their babies) as compared to their normal voices (speaking to other adults). This exaggeration stretches the acoustic differences between sounds like *ah* and *oo*, providing babies a better model for learning speech (Kuhl et al. 1997). Considering their generally slower language acquisition, boys may benefit even more than girls from hearing their parents and other caregivers speak to them in this more interesting, soothing, and informative style.

95   *17 percentile drop in vocabulary development:* Zimmerman et al. 2007.
       live, interactive storyteller: Kuhl et al. 2003.
       babies whose parents respond to their babbling: Kent 1984.
96   *dialogic reading:* Debaryshe 1993; Whitehurst et al. 1988.
       boys are more susceptible to ear infections: Paradise et al. 1997.
97   *antibiotics for ear infections:* Rosenfeld 1995.
       tympanostomy tubes: Johnston et al. 2004.
98   *benefits of infant massage:* Field 1995.
101   *breastfeeding rates in United States:* Ryan et al. 2002.
       breastfeeding promotes eye contact: Lavelli and Poli 1998.
       cognitive advantages of breast milk: Horwood and Fergusson 1998.
       safer plastics: University of California Pediatric Environmental Health Specialty Unit, 2008.

3. LEARNING THROUGH PLAY IN THE PRESCHOOL YEARS

106   *picking presents for children of the opposite sex:* Halpern 2000, p. 252.
       Swedish study of toy gender preference: Servin et al. 1999.
107   *parents' reports of toy preferences in six-month-olds:* Furby and Wilke 1982.
       six-month-old boys show no toy-picture preference: Alexander et al. 2008.
       Canadian study of toy preference: Serbin et al. 2001. However, a similar British study (Campbell et al. 2000) found a marginally significant preference as early as nine months of age, but only in boys. Girls this age did not prefer to look at girl toys, nor was there any difference in visual toy preference at three months.
       Dutch study of toy preference: van de Beek et al. 2007.
       Swedish study of toy preference at different ages: Servin et al. 1999.
       toy preference in girls with CAH: Servin et al. 2003; Berenbaum and Snyder 1995.
108   *British study of prenatal testosterone and play interests:* Hines et al. 2002. However, two other studies failed to find a relationship between amniotic testosterone and the later play interests of either girls or boys (van de Beek et al. 2007; Knickmeyer et al. 2005b). Also, a study of girls with male twins failed to reveal any influence of prenatal testosterone on toy choice (Henderson and Berenbaum 1997).
       toy preference in vervet monkeys: Alexander and Hines 2002.
       toy preference in rhesus monkeys: Hassett et al. 2008. In this study, the masculine toys were various wheeled vehicles and the feminine toys were a variety of stuffed animals or soft dolls. Male rhesus monkeys showed a strong preference for the vehicles over the plush toys, but females played about equally with both types of toys. Note, however, that the vervet monkeys in

Alexander and Hines's study did not show a sex difference in preference for a stuffed animal, so such toys seem to be genuinely less gender specific than vehicles or balls.

sex difference in rhesus monkeys' interest in infants: Herman et al. 2003. This paper shows a robust difference between male and female monkeys, with an effect size, or d value, of 1.94.

109   *behavioral genetic study of gender-typical play:* Iervolino et al. 2005. Interestingly, this study determined that sex-specific behavior is *more* strongly determined by genes in girls than in boys: 57 percent of the variance in girls' play was accounted for by genes, versus 34 percent of boys' variance. The difference may reflect the fact that boys are more constrained by parents and society (that is, nurture) to avoid girl toys. By contrast, girls are less discouraged to avoid male playthings and so are freer to express their full genetic potential in play choice.

110   *boys grow increasingly adamant about avoiding girl toys:* Katz and Boswell 1986.

111   *parents discourage boys from girl toys:* Lytton and Romney 1991.
boys themselves voice the awareness: Raag and Rackliff 1998.
gender labels dampen children's motivation: Martin and Dinella 2002.

113   *Dutch study of gender-nonconforming children:* van Beijsterveldt et al. 2006.
American survey of gender-atypical behavior: Sandberg et al. 1993.

114   *U.S. twin study of gender nonconformity:* Coolidge et al. 2002. Interestingly, one large study of preschoolers found much lower heritability, especially for gender-atypical boys, and a correspondingly larger role of environment as compared to the findings with older children (Knafo et al. 2005). The discrepancy probably indicates that parents counteract children's gender-atypical tendencies when they are very young, especially in boys. Gender-atypical behavior may also appear less heritable in young children simply because it's harder to define: most younger boys and girls do explore each other's domains—whether it's dress-up clothes, dolls, sports, or toy cars—more than older children do, making it hard to identify truly gender-atypical children at this age.

homosexual or bisexual orientation as high as 80 percent: The reported number of "sissy boys" who became homosexual or bisexual in Richard Green's study has ranged from 68 to 80 percent in different reports of this research. The higher figure is from LeVay (1996, p. 98); Green's original article (1985) puts it at 68 percent, but six of the boys were still under sixteen years old and may not have reached their mature orientation.

most tomboys end up heterosexual: Peplau et al. 1999.

115   *cruel training of Kyle:* LeVay 1996, pp. 100–01.
gender identification is a strikingly universal milestone: Best and Williams 1993.

gender awareness in traditional versus less traditional families: Fagot et al. 1992.

116 *two-year-old boys who could not yet pass a test of gender labeling:* Fagot and Leinbach 1993.

gender self-labeling and toy preference in the third year: Weinraub et al. 1984.

gender constancy: Maccoby 1998, p. 165.

117 *gender choices seen as right and wrong:* Lobel and Menashri 1993.

118 *"Everyone has a penis; only girls wear barrettes":* Bem 1998, p. 109.

two-year-old boys engage in more conflict and negative assertiveness: Campbell et al. 2002.

119 *9 percent of their play time with opposite sex:* Maccoby 1998, p. 22.

universal nature of gender-segregated play in children and other young primates: Maccoby 1998, pp. 98, 288.

120 *kindness and intelligence as top qualities:* Best and Williams 1993. This analysis is based on the widely cited data of David Buss, the University of Texas psychologist whose study of men and women in thirty-seven contemporary cultures is more typically used to demonstrate that women seek a "strong, rich, provider" while men seek a "beautiful, compassionate, nurturing" woman.

121 *study of peer influence on toy choice in three- and four-year-olds:* Serbin et al. 1979.

122 *British study of siblings and gender development:* Rust et al. 2000. The effect size of older siblings' sex on second-born children's masculinity or femininity was considerable in this study, with a d value of 0.57 for girls 0.66 for boys. For another study that reveals the sizable impact of sibling gender on preschoolers' play behavior, see Iervolino et al. 2005.

sibling influence is largely one-way: Henderson and Berenbaum 1997.

Monica Kunkel: American Automobile Association 2008.

123 *sibling differentiation:* Berk 2003, p. 536.

boys' higher activity level begins in infancy: Campbell and Eaton 1999.

sex differences in motor activity level throughout life: Eaton and Enns 1986.

124 *modest correlation between maternal testosterone and male-type play:* Hines et al. 2002.

Baron-Cohen's study of amniotic testosterone and later play interests: Knickmeyer et al. 2005b.

benefits of rough-and-tumble play: Pellis and Pellis 2007.

125 *social variables that influence the frequency of rough play:* Wallen 1996.

126 *violent play among Yanomami and Sioux boys:* Geary 1998, pp. 226–27.

128 *teachers are more likely than other adults to see superhero play as aggressive:* Connor 1989.

129 *violence is fun, cool, glamorous:* Carlsson-Paige and Levin 1999.

girls with CAH have less interest in babies: Leveroni and Berenbaum 1998.

130   *unable to alter interest in babies by manipulating prenatal testosterone:* Herman et al. 2003.

not until age two or three that girls are more strongly attracted to real babies: Blakemore 1992.

no difference in the amount of praise mothers bestow: Blakemore 1990.

role modeling to shape nurturing behavior: Blakemore 1998.

"private speech": Berk 1994.

131   *preschool-age girls are better able to identify expressions:* McClure 2000.

girls just three and a half years old performing as well as boys of five: Boyatzis et al. 1993.

girls start talking a month or two earlier: Fenson et al. 1994.

girls speak more clearly: Halpern 2000, pp. 95–97; Joseph 2000.

girls with CAH are no less verbal: Hines 2000; Malouf et al. 2006.

study of twelve very different cultures: Whiting and Edwards 1992.

133   *boys begin outdistancing girls in physical strength, speed:* Zaichowsky et al. 1980, pp. 44–45.

University of Chicago study: Levine et al. 1999.

135   *girls' fine motor advantage:* By age three, girls have been found to be either faster or more accurate than boys at two fine motor tasks: touching each finger in succession against the thumb and copying an adult's hand postures. By contrast, boys this age are more accurate at targeting skills, such as throwing a ball to a precise location (Kimura 2000, pp. 35–38).

cautioning daughters more than sons: Morrongiello and Dawber 2000.

136   *female gender identity in girls with CAH:* Zucker et al. 1996.

activity level varies more with type of play than gender of playmate: O'Brien and Huston 1985.

138   *benefits of high-quality preschool:* Espinosa 2002; Duncan et al. 2007.

society earns back some eight dollars for each dollar spent: Rolnick and Grunewald 2003.

140   *benefits of vestibular stimulation:* Eliot 2000, pp. 154–56.

142   *math and musical keyboard training:* Rauscher et al. 1997.

4. STARTING SCHOOL

144   *9 percent of kindergartners are redshirted:* West et al. 2000.

proportion of six-year-olds entering kindergarten: Deming and Dynarski 2008. According to this analysis, about two-thirds of the increase in number of six-year-olds beginning kindergarten is due to redshirting, while one-third of the change is due to earlier school cutoff deadlines.

older children in a class may have a modest advantage in kindergarten: NICHD Early Child Care Research Network 2007.

fades by later elementary school: West et al. 2000; Stipek 2002.

problem behavior in children who were held back: Byrd et al. 1997.

145 *children who attend more preschool score higher on IQ tests:* Stipek 2002; Eliot 2000, pp. 456–59.

an extra year of school raises a child's IQ: Cahan and Cohen 1989.

young first-graders made far greater progress: Morrison et al. 1997.

later average starting age of boys and college completion: Deming and Dynarski 2008.

146 *verbal ability equivalent to a mere two IQ points:* More precisely, the difference in verbal skills amounts to 0.13 of a standard deviation (or d value) in children under six, according to a meta-analysis by Hyde and Linn (1988); this is equivalent to about two points on a standardized IQ test. The difference shrinks to a d value of 0.06 between six and ten years of age.

no sex difference in vocabulary: Halpern 2000, pp. 97–98.

boys perform slightly better on visuospatial tasks: Levine et al. 1999; Voyer et al. 1995.

study of mathematically gifted four- and five-year-olds: Robinson et al. 1996.

147 *women prove about 6 percent faster than men on fine motor skills:* Peters et al. 1990.

sex difference in pegboard performance begins at age five: Smith et al. 2000.

imitating adult hand gestures: Kimura 2000, pp. 35–38.

the average woman scores higher than about two-thirds of men: Based on the effect size of 0.33 for the sex difference in speech production reported in Hyde and Linn 1988.

verbal fluency sex difference in children: Brocki and Bohlin 2004.

boys are more likely to suffer from stuttering: Halpern 1989.

149 *ADHD is two to nine times more prevalent in boys:* American Psychiatric Association 2000.

inhibiting their urge to uncover a hidden toy: Diamond 1985.

boys are more mischievous: Maccoby 1998, p. 116.

study of self-control, impulsivity, and rule following in toddlers: Kochanska et al. 1996.

sex differences in inhibitory control throughout childhood: Bendersky et al. 2003.

girls' advantage in inhibitory control: Else-Quest et al. 2006. The d value for girls' advantage in inhibitory control was 0.41, and as high as 1.01 for effortful control, which requires children's conscious, willful attention toward overcoming impulses.

150 *girls reach their peak level of frontal gray matter earlier:* Giedd et al. 1999a.

boys' decision-making abilities suggest earlier orbitofrontal maturation: Overman et al. 1996.

EEGs show earlier maturation in boys: Barry et al. 2004; Clarke et al. 2001.

EEG study suggests girls process mistakes more efficiently: Liotti et al. 2007.
stronger verbal skills and better inhibitory control: Berk 2003, p. 504.

151   *Tools of the Mind:* Diamond et al. 2007.
self-verbalization technique to improve attention: Reid and Harris 1993.
more physical breaks in Japanese schools: Pellegrini and Smith 1998.
exercise-induced brain changes: Cotman and Berchtold 2002.

152   *different play styles in young boys and girls:* Fabes et al. 2003.

153   *stable dominance hierarchies in first- and third-grade boys:* Pettit et al. 1990.

154   *gender intensification:* Martin and Fabes 2001.
kaleidoscope versus Fisher-Price movie viewer: Frey and Ruble 1992.
gender segregation and progressive schools: Maccoby 1998, pp. 287–88.
boys better behaved and better liked the more time they spent playing with girls:
Fabes et al. 1997.

155   *enhancement of spatial skills in girls through play:* Robert and Héroux 2004.

156   *Building Blocks curriculum:* This program was developed by Douglas Clements and Julie Sarama at SUNY Buffalo and is described at http://www
.gse.buffalo.edu/org/buildingblocks.

157   *"Expectations . . . can engender the brain":* Hines 2004, p. 227.
study of parents' expectations in three nations: Lummis and Stevenson 1990.
parents are likelier to credit a son's success to innate talent: Räty et al. 2002.
"Pygmalion effect": Schugurensky 2002.

158   *teachers favored boys in the 1970s and '80s:* Sadker and Sadker 1995.
longer wait time: Gore and Roumagoux 1983.
teachers misjudged girls as outtalking boys: Lafrance 1991.

159   *teacher training and gender equity:* Sanders 1997.
large British study and laddism: Plewis 1997.
research on teacher bias "does not show consistent favoritism toward boys or girls": Kleinfeld 1998.

161   *young boys benefit from playing with girls:* Fabes et al. 1997.
alternate seating for boys and girls: BBC News 1999.

162   *Taiwan and Japan incorporate twice as many recesses:* Bjorklund and Brown 1998.

163   *Tools of the Mind curriculum:* Diamond et al. 2007.

164   *merits of Montessori classroom:* For an excellent overview, see Lillard 2005.

167   *low percentage of male early childhood teachers:* National Education Association 2003; Center for Early Childhood Leadership, National-Louis University, 2004.

169   *high proportion of male teachers in Greece:* Hopf and Hatzichristou 1999.
teachers paid more today: National Education Association 2003. According to this report, male teachers in 2001 earned an average salary of $46,326 and female teachers $42,440. However, more recent data from the American Federation of Teachers (2005) show that teacher salaries leveled out and even declined slightly (relative to inflation) between 2003 and 2005.

men make up 13 percent of current nursing students: Icon 2005.

"Just Say NO, to No Touch": Piburn 2005.

170   many benefits of touch: Eliot 2000, pp. 138–44.

5. THE WONDER OF WORDS

173   Gurian warns parents: Gurian and Stevens 2005, p. 52.

174   "a woman uses about 20,000 words": Quote is from the front jacket of Brizendine 2006.

urban legend was debunked: Liberman 2006c.

16,215 for women and 15,669 for men: Mehl et al. 2007.

fifty-six studies of mixed-sex conversations: Gleason and Ely 2002.

affiliative versus assertive speech: Hyde 2005.

women's advantage is one-tenth of a standard deviation: Hyde and Linn 1988.

175   NAEP reading scores in grades four and eight: Lee et al. 2007. This 2007 Nation's Report Card is based on tests of 191,000 fourth-graders and 160,700 eighth-graders from across the United States.

twelfth-grade reading scores: Grigg et al. 2007, p. 8; Corbett et al. 2008, p. 74.

global percentage of illiterate women: Verner 2005; Doyle 1997.

PISA reading assessment: Organization for Economic Cooperation and Development (OECD) 2005, p. 150; OECD 2004, p. 285.

176   PISA results and single-sex schooling: Archer 2004b.

leisure reading predicts PISA scores: Kirsch et al. 2002.

177   GPA gap: Halpern 2002; Halpern 2000, pp. 82, 259; Corbett et al. 2008, p. 52.

three-to-one sex ratio of foreign-language majors: Morris 2003.

SAT gap: College Board 2008.

178   verbal analogies: Halpern, pp. 127–28.

demographics of girls taking the SAT: College Board 2005.

factored-in differences in family income: Burton et al. 1988.

179   gender gap does look discriminatory: In 1999, women won a lawsuit against the National Merit Scholarship Corporation, which bases its $28 million in annual awards on the Preliminary SAT. In the years leading up to the lawsuit, girls were winning less than 40 percent of National Merit Scholarships. To settle the suit, the test makers agreed to add a writing-skills section to the exam, which dramatically narrowed the gender gap (FairTest Examiner 1999). The ETS made a similar change to the SAT exam in 2005.

180   cross-cultural research and parental expectations: Lummis and Stevenson 1990.

children understand an impressive thirteen thousand words: Pinker 1995, p. 151.

181   phonological skills needed to master reading: Whitehurst and Lonigan 2001.

182   *girls score higher on tests of phonological awareness:* Kovas et al. 2005.

pleasure reading in girls and boys: Neuman 1986.

gifted boys and reading: Anderson et al. 1985.

184   *number of girls who perform at or above the proficient level:* NAEP Data Explorer, http://nces.ed.gov/nationsreportcard/nde.

"alarming" sex difference in writing skill: Hedges and Nowell 1995.

"Wordiness and confusing grammar are crippling": Peterson 2004.

initial 1995 study: Shaywitz et al. 1995.

186   *the Shaywitzes' study still highlighted in popular works:* For example, Simon Baron-Cohen hailed it as a "landmark study" in his book *The Essential Difference* (2003, p. 58).

the idea that females process language more bilaterally than males: Even Sally Shaywitz is not afraid to admit that the work has been overinterpreted by the popular press: "It is important to note that not all women in our study were bilateral and that despite the differences in the ways they performed the tasks, men and women performed equally well. . . . Sometimes I am concerned we will be caught in the morass of making too much of little differences when we don't really know what they mean" (Poppick 2005).

no significant sex difference in hemisphere activation: Sommer et al. 2008.

187   *stroke and dichotic listening studies:* Sommer et al. 2008; Sommer et al. 2004; Frost et al. 1999; Hiscock et al. 1994.

11 percent greater density of neurons: Witelson et al. 1995.

another study finding sex differences in the same brain area: Jacobs et al. 1993.

188   *I wish I could figure out where this one comes from:* The source is probably a small study (twenty-one subjects) that found a proportionally larger planum temporale in women as compared to men (Harasty et al. 1997). However, subsequent research, based on much larger numbers of men and women, has not confirmed this finding, nor has a difference in planum temporale size been demonstrated between boys and girls. These results are summarized in Knaus et al. 2004.

no difference in size or symmetry of planum temporale: Witelson and Pallie 1973; Wada et al. 1975; Vadlamudi et al. 2006; Eckert et al. 2001. However, one study of children between three and fourteen years old did report an "unexpected gender difference" of greater planum temporale asymmetry in girls (Preis et al. 1999). In adults, one review of seven studies reported that a woman's planum temporale is 2.5 percent more symmetrical than a man's (Shapleske et al. 1999), whereas Sommer et al. (2008) found slightly greater symmetry in men.

conflicted evidence for sex difference in newborns' language processing: Molfese and Molfese 1979; Shucard et al. 1981; Molfese and Radtke 1982; Friederici et al. 2008.

no lateralization difference in large fMRI study of older children: Plante et al. 2006.

189   *slightly more mature brain activation in girls:* Burman et al. 2008 studied children ranging from nine to fifteen years old and found a more mature pattern of brain activation in girls, which correlated with their higher scores on spelling and rhyming tests.

language experience best predicts neural response to speech: Mills et al. 2005.

EEG and testosterone levels: Friederici et al. 2008.

girls with twin brothers: Cohen-Bendahan et al. 2004.

little evidence for effect of prenatal testosterone: For example, Lutchmaya et al. (2002b) did not find a correlation between prenatal testosterone level and vocabulary size within separate populations of toddler boys or girls.

normal verbal skills in girls with CAH and women exposed to DES: Hines 2004, pp. 168–69.

190   *5 to 10 percent faster at reading aloud near peak monthly estrogen:* Hampson 1990.

replication of Hampson's findings: Maki et al. 2002; Symonds et al. 2004.

changes in activation of linguistic circuits: Two recent studies have found differences in brain activation during verbal tasks at different phases of the menstrual cycle (Konrad et al. 2008; Craig et al. 2008). However, it is important to note that in neither of these studies did women's actual verbal performance change between the low- and high-hormonal phases of the menstrual cycle, so it's not clear that the changes in blood flow have anything to do with verbal processing.

no change in verbal abilities over menstrual cycle: Gordon and Lee 1993; Mordecai et al. 2008; Konrad et al. 2008.

no relationship between verbal fluency and levels of six different sex hormones: Halari et al. 2005.

191   *estrogen improves memory in "menopausal" rats:* Fitch and Bimonte 2002.

estrogen improves memory in elderly female monkeys: Lacreuse 2006.

WHIMS trial revealed increased risk of dementia: Rapp et al. 2003; Shumaker et al. 2004. It is thought that earlier studies of HRT, which were not randomized or placebo controlled, suffered from a healthy-user bias: women who *chose* to take hormones showed better mental function than those who did not, but they were also healthier and better educated to begin with, both factors that are associated with better cognitive aging.

HRT may be beneficial if restricted to shortly after menopause: Maki 2006.

192   *girls ahead in vocabulary growth during toddler years:* Fenson et al. 1994.

20 percent larger expressive vocabulary in two-year-old girls: Galsworthy et al. 2000.

girls are some 15 percent more verbally fluent: Kovas et al. 2005.

no sex difference in receptive vocabulary: Maccoby and Jacklin 1974, pp. 79–85.

parents talk just as much to their sons as their daughters: Huttenlocher et al. 1991; Hart and Risley 1992.

mothers talk more to their daughters: Leaper et al. 1998.

girls talk more with one another: Maccoby 1998, p. 107.

social mimicry and gender differences in vocal pitch: Gleason and Ely 2002.

193   *gender and environment contributions to variance in verbal skills:* Kovas et al. 2005.

MCDI scores versus twin gender in Plomin study: Galsworthy et al. 2000.

195   *quantity and quality of language immersion:* Walker et al. 1994.

SES affects verbal skills much more than gender does: Eckert et al. 2001.

environment's contributions to verbal skills: Harlaar et al. 2005.

Tom Cruise on dyslexia: Cruise 2003.

196   *"myth" of sex differences in dyslexia (footnote):* Shaywitz et al. 1990. This debate continues to rage; see, e.g., Liederman et al. 2005 versus Siegel and Smyth 2005.

two-to-one sex ratio in dyslexia: Rutter et al. 2004; Liederman et al. 2005.

197   *sex does not influence heritability of dyslexia:* Schumacher et al. 2007; Fisher and Francks 2006.

dyslexia due to temporal-parietal dysfunction: Temple 2002.

198   *brain imaging of dyslexia remediation:* Simos et al. 2002; Temple et al. 2003; Eden et al. 2004.

199   *earlier remediation is better for both reading skills and emotional health:* Eden and Moats 2002; Alexander et al 2004.

dyslexia diagnoses could be reduced with early intervention: Vellutino and Scanlon 2001.

200   *study from North Vancouver:* Siegel and Smyth 2005.

critical period for language learning: Eliot 2000, pp. 358–64.

word learning shapes the left hemisphere: Mills et al. 2005.

201   *child SES predicts left-hemisphere specialization for language:* Raizada et al. 2008.

202   *children need phonological skills:* Whitehurst and Lonigan 2001.

formal literacy training by parents facilitates reading: Levy et al. 2006.

203   *reading material for boys:* This and other tips for increasing boys' literacy can be found in the excellent online resource published by the Ontario Ministry of Education (2004).

mentors for boys: Ontario Ministry of Education (2004).

204   *"gun in his holster and a book in his hand":* Tyre 2008, p. 153.

6. SEX, MATH, AND SCIENCE

207   *female physicians' earnings:* Much of this difference has to do with women's choices of the lower-earning medical specialties, such as pediatrics and family practice, along with fewer hours worked per week (forty-nine average for women versus fifty-seven for men, according to Kane and Loeblich 2003). Nonetheless, women more than ten years out of medical school earn only

85 percent of what men earn, even after correcting for such differences as type of practice and working hours (Baker 1996).

only one-fifth of accounting-firm partners are women: Work/life & Women's Initiative Executive Committee 2004.

208   *females initially outperform males in computation:* Halpern 2000, p. 113.

NAEP math and science data: Grigg et al. 2006. Math scores can also be viewed at http://nces.ed.gov/nationsreportcard/ltt/results2004/sub-math-gender.asp. Recent analysis of math standardized achievement tests in ten U.S. states found no overall difference between boys and girls, but these tests may be too easy to pick up subtle differences (Hyde et al. 2008).

percentage of girls taking physics and calculus (footnote): Freeman 2004, p. 8.

girls earn higher grades: Kleinfeld 1998.

grade-test disparity: Halpern et al. 2007.

PISA results in math and science: OECD 2004, pp. 97, 297.

209   *"critical filter":* The term was coined by sociologist Lucy Sells and is cited in Halpern 2000, p. 113.

math and science are where the money is: Hedges and Nowell 1995.

AP exam gaps: Freeman 2004, p. 63.

math SAT gap: College Board 2008.

210   *61 percent in accounting:* Freeman 2004, p. 13.

74 percent of vet students: Maines 2007.

statistics on women in engineering and computer science: U.S. Department of Labor, 2002; Freeman 2004, p. 79.

211   *engineering or computer science graduates paid more:* Bae et al. 2000, p. 84.

proportion of women faculty members in biology and math: These numbers, reported in Handelsman et al. (2005), are based on the top 50 U.S. universities. The proportion of women faculty members in science and math is higher at lower-tier universities and 2-year colleges.

female attrition in all careers: The glass ceiling also exists in more verbal fields, such as law, where only 17 percent of partners are women, as compared to 48 percent of law students and 44 percent of associates in law firms, according to 2005 data from the National Association for Law Placement, cited in Katz 2006.

212   *"More geniuses, more idiots":* Pinker's quote appears in Johnson 2005.

sex variability and childbearing (footnote): Pinker 2005a.

four-to-one ratio for scores above six hundred: Benbow and Stanley 1983.

SMPY coverage in Newsweek and Time: Kessel 2006.

213   *female Chinese math Olympians:* Halpern 2000, p. 115.

214   *ratio now at 2.8 boys per girl:* Association for Women in Mathematics, 2006.

only a fraction of the news coverage: Monastersky 2005.

biased SMPY recruiting materials: Ruskai 1991.

students have to apply to SMPY program: Hyde and McKinley 1997, p. 39.

mothers who recalled media reports about the SMPY: Eccles and Jacobs 1986.
Steve Olson quote: This comes from an interview about his book *Count Down* and can be found at http://www.houghtonmifflinbooks.com/features/countdown/.

215   *most research finds no evidence of a sex difference in physical reasoning:* Spelke 2005.

no evidence that boys are "born for math": Ibid.

216   *Francis Crick quote:* Wade 2006.

Rosalind Franklin: Rosalind Franklin produced the x-ray photograph of crystallized DNA that Watson and Crick used, without her knowledge, to confirm their theory. Many historians believe that she would have come up with the double helix herself within a few weeks of Watson and Crick. In any case, her subsequent research on the structure of virus particles proves that spatial imagination was not a problem for her, in spite of being female (Maddox 2002).

217   *d statistic for mental rotation ranges:* Voyer et al. 1995; Masters and Sanders 1993.

brain-imaging studies of mental rotation: Spatial awareness is well known to depend on the right hemisphere of the brain, so one theory is that males are better at spatial skills because their right hemispheres are more active during such tasks than females'. Several studies do indeed support this idea (e.g., Frings et al. 2006). However, a few studies have found the opposite result—stronger *left* hemisphere activation in men (Roberts and Bell 2002)—while a separate review of seven studies found only one supporting the theory that males use the right hemisphere more than females for spatial processing (Jager and Postma 2003).

four-and-a-half-year-old boys outperform girls at mental rotation: Levine et al. 1999.

studies that found no sex difference in infant spatial ability: Hespos and Rochat 1997; Örnkloo and von Hofsten 2007.

infant mental rotation experiments: Moore and Johnson 2008; Quinn and Liben 2008.

218   *sex difference in mental rotation shows up early and grows throughout childhood:* Voyer et al. 1995. However, researchers have found few differences in *brain activation* during spatial tasks in children under eight years old (Roberts and Bell, 2000, 2002).

females misestimate the horizontal: Hecht and Proffitt 1995.

sex differences in the water-level task have steadily declined: Voyer et al. 1995, p. 261.

219   *sex difference in infant stereovision:* Gwiazda et al. 1989.

women's visual acuity declines earlier: Baker 1987, p. 11.

EEG of infants' visual processing: Malcolm et al. 2002.

boys' motion detection is actually poorer than girls': Schrauf et al. 1999.

220 *testosterone delays male stereovision*: Held 1989.

221 *male advantage in virtual maze navigation*: Moffatt et al. 1998.

hippocampus versus frontal and parietal lobes in route finding: Weiss et al. 2003; Grön et al. 2000; Thomsen et al. 2000.

men and women use same brain areas when they engage in similar route-finding strategies: Blanch et al. 2004.

222 *meta-analysis of object-location memory studies*: Voyer et al. 2007.

object-location memory depends on the type of objects: Cherney and Ryalls 1999.

no sex difference in object location memory before puberty: Voyer et al. 2007.

223 *Fred and Wilma scenario*: Francis 2004, pp. 150–74.

no fossil tools hinting at changes in spatial skills: Wynn et al. 1996.

antlers versus nipples (footnote): Francis 2004, p. 161.

224 *sex difference in meadow voles' hippocampi*: Jacobs et al. 1990.

mating style and sex difference in human spatial skills: Jones et al. (2003) reviewed the existing evidence for theories of sex differences in spatial skills and concluded there is weak support for the range theory but no evidence to support the hunter-gatherer hypothesis.

human hippocampus is no larger in men: Yurgelun-Todd et al. 2003; Gur et al. 2002; Pruessner et al. 2001.

studies finding larger hippocampi in women: Goldstein et al. 2001; Giedd et al. 1997; Caviness et al. 1996.

225 *study of London taxi drivers*: Maguire et al. 2000.

no archaeological evidence to support these theories: Wynn et al. 1996.

no evidence that sex-linked genes underlie different spatial abilities: Jardine and Martin 1984.

effect of neonatal testosterone on rats' navigation: Williams and Meck 1991.

226 *prenatal testosterone and later spatial skills*: These studies actually found an *inverse* relationship between fetal testosterone and later spatial or math ability—that is, stronger ability in children exposed to *lower* levels of prenatal testosterone: Jacklin et al. 1988; Finegan et al. 1992. A third study (Grimshaw et al. 1995) did find a positive correlation between amniotic testosterone and the *speed* of mental rotation in girls; however, it did not find a relationship between testosterone and the *accuracy* of mental rotation, which is the parameter that usually differs between the sexes.

no difference in math or visuospatial skills in girls with male twins: Luciano et al. 2004; Hines 2004, pp. 174–75. One preliminary study did report enhanced spatial abilities in girls with male twins (Cole-Harding et al. 1988). However, this finding was never published in a peer-reviewed journal.

digit ratio does not correlate with spatial skills: Puts et al. 2008.

better spatial skills in girls with CAH: Ibid.

recent study of virtual navigation in women with CAH: Mueller et al. 2008.

improved spatial skills with testosterone treatment: These include studies of elderly men (Janowsky et al. 1994) and young adult women (Aleman et al. 2004). Conversely, a study of men with prostate cancer (Cherrier et al. 2003) found their mental rotation skills declined (and their verbal memory improved) when they took drugs to block the action of androgens.

227   *1995 Dutch study:* Van Goozen et al. 1995.

"small cognitive effects despite the very strong hormonal manipulations": Slabbekoorn et al. 1999.

testosterone measurements similarly equivocal: Hines 2004, p. 178.

no relationship between testosterone levels and various cognitive skills: Halari et al. 2005.

no changes in boys' performance with testosterone injections: Liben et al. 2002.

menstrual cycle and spatial skills: Studies that have found a link between high-estrogen phases of the menstrual cycle and low spatial skills are Hausmann et al. 2004; Maki et al. 2002; Halpern and Tan 2001; Moody 1997. Studies that have failed to find a relationship are Rosenberg and Park 2002; Epting and Overman 1998; Gordon and Lee 1993.

228   *estrogen improves spatial skills in female rats:* Sandstrom and Williams 2004.

estrogen increases hippocampal synapses in female rats: Yankova et al. 2001.

Turner's syndrome and visuospatial skills: Ross et al. 2002.

male-to-female transsexuals and spatial ability: Hines 2004, p. 179.

visuospatial skills unaltered in women taking hormone replacement: Sherwin 2003.

"probably would not be detected outside the laboratory": Halpern et al. 2005, p. 58.

testosterone not the reason why males consistently outperform females: As Melissa Hines (2004, p. 180), who has studied this issue for decades, concludes, "Taken together, the results of the many investigations of activational influences of androgen and estrogen on human cognitive performance do not provide convincing evidence that such influences exist."

229   *boys' advantage in geography:* Cohn 2000.

230   *visuospatial skills are important in many areas of math:* Casey et al. 1995.

women mathematicians gravitate toward algebra and statistics : Halpern 2000, p. 118.

"if a student can visualize what a pizza slice would look like": Beth Casey, as quoted in Olson 2005, p. 73.

female fashion-design majors and mental rotation: Esgate and Flynn 2005.

231   *"We believe that strategies . . .":* Casey et al. 2004.

232   *identifying girls who could benefit from intensive spatial training:* Casey et al. 2001.

spatial skills in Eskimo: Berry 1966, 1971.

Huttenlocher study of SES and spatial skills: Levine et al. 2005.

233   *building toys and improved spatial skills:* Brosnan 1998; Connor and Serbin 1977.

student athletes exhibit superior visuospatial skills: Ozel et al. 2004.

no sex difference in reaction time among serious athletes: Lum et al. 2002.

Nature *study of spatial awareness and video games:* Green and Bavelier 2003.

234   *video games improve spatial skills:* Cherney 2008; De Lisi and Wolford 2002; Newcombe et al. 2002; Subrahmanyam et al. 2000; Okagaki and Frensch 1994.

sex difference eliminated by video-game experience: Feng et al. 2007.

mental rotation and sign language: Newcombe et al. 2002.

Piaget's water-level task was "unlearnable": Thomas et al. 1973.

exercises for mastering water-level task: Vasta et al. 1996.

235   *correlation between water-level task and math SAT scores:* Li et al. 1999.

Chinese girls surpass American boys in math skills: Li 2000.

236   *Barbie Liberation Organization:* Zimmer 2006.

equal number of boys and girls liked science and math: Bae et al. 2000, p. 62.

parents' beliefs that six-year-old boys are better at math: Lummis and Stevenson 1990.

gender and expectations regarding math performance in young children: Halpern 2000, p. 28; Entwisle and Baker 1983.

"I am just not good in mathematics": Ginsburg et al. 2005, p. 20.

237   *confidence versus spatial skills in girls' math performance:* Casey et al. 2001.

talent versus hard work: Halpern 2000, p. 292.

attribution gap in science performance: Howe 1996.

gender-stereotype-threat experiment (Figure 6.7 data): Spencer et al. 1999.

238   *stereotype threat in white versus Asian males' math performance:* Aronson et al. 1999.

stereotype threat impairs white males in athleticism and social sensitivity: Stone et al. 1999; Koenig and Eagly 2005.

239   *effect of ditzy ads on women's career interests:* Davies et al. 2002. In addition to affecting women's stated career interests, the stereotypic commercials also depressed women's performance on a challenging math test, confirming other demonstrations of stereotype threat.

240   *Bob Fischer quote:* Olson 2004, pp. 30–31.

"I heard numerous stories of how being a girl does make a difference": McDonnell 2005.

241   *as one microbiologist described:* Wallon 2005.

Ben Barres quote: Dean 2006.

242   *25 percent is the tipping point:* Babcock and Laschever 2007, p. 101.

243   *U.S. ranks twenty-second in student science proficiency:* OECD 2004, p. 294.

girls can catch up to boys when taught spatial skills: Connor et al. 1978.

244   *88 percent of LEGO sets purchased for boys:* Brosnan 1998.

male-to-female ratio of chess experts: Chabris and Glickman 2006.

245 *Angela Smith quote:* Storm and Jenkins 2002, p. 58.

246 *the digital divide is very real:* Terlecki and Newcombe 2005.

AAUW suggestions for improving girls' computer skills: AAUW Educational Foundation, 2000.

247 *value of hands-on experience for high-school science:* Burkam et al. 1997.

248 *"Math needs a Stephen Jay Gould":* Vanderkam 2005.

249 *stereotyping by fathers and lower interest in math among daughters:* Jacobs et al. 2005.

mentoring students that intelligence is not fixed: Good et al. 2003.

250 *smaller gender gap in math among Asians and the role of mothers:* Huntsinger and Jose 1995.

middle-school girls perform better in science classes taught by women: Dee 2005.

7. LOVE AND WAR

253 *little evidence of significant sex difference in emotional expression:* Malatesta and Haviland 1982; Maccoby and Jacklin 1974, pp. 177–82.

newborn males are more irritable: Haviland and Malatesta 1981.

254 *boys cry a bit less than girls by age four or five:* Quas et al. 2000.

men undergo greater increases in heart rate, blood pressure, and sweating: Manstead 1992.

anonymous survey of emotional experiences in men and women: Simon and Nath 2004. While men and women reported similar amounts of emotional experience overall, there were differences in the types of experiences: men reported more positive emotions, such as excitement, pride, and calmness, while women were more prone to anxiety and sadness, differences that Simon and Nath found were largely attributable to the lower social and economic status of the females in this large sample.

frequency of crying is not affected by onset of menstruation: Van Tilburg et al. 2002.

255 *unforgiving boy code:* Pollack 1998; a similar perspective on boys' emotional suppression is described in Kindlon and Thompson 1999.

256 *Martha Stewart quote:* Rosenbloom 2005.

257 *male-female differences in neurochemistry, genes, or brain anatomy that might correlate with depression:* Campbell and MacQueen 2006; Kendler et al. 2006.

serotonin metabolized differently in men and in women: Jans et al. 2006.

estrogen and testosterone levels and depression in adolescent girls: Angold et al. 1999.

testosterone buffers males from mood and anxiety disorders: Holden 2005.

258   *no effect of hormone therapy on adolescents' mood:* Susman et al. 1998.

low self-esteem the biggest risk factor for depression in early adolescence: MacPhee and Andrews 2006.

sex difference in self-esteem across the life span: Kling et al. 1999.

masculine traits predict self-esteem: Lippa 2002, p. 50.

body image is rapidly declining among young Asian women: Lee 2000.

259   *sex difference in body image is growing in recent decades:* Feingold and Mazzella 1998.

260   *social experiences play a larger role in depression than hormones:* Piccinelli and Wilkinson 1999.

males take an active and assertive approach to maintaining their self-esteem: Wichstrøm 1999.

coping mechanisms that prevent depression: Piccinelli and Wilkinson 1999; Kuehner 2003.

sex difference in physician empathy: Hojat et al. 2002.

261   *self-report versus objective measures of empathy:* Eisenberg and Lennon 1983.

women's brains process emotions more symmetrically than men's: Wager et al. 2003.

d value of 0.40 for sex difference in recognizing others' emotion: Hall 1978.

women detect emotions on men's faces better than women's: This is based on two studies (Erwin et al. 1992; Rahman et al. 2004). Another study, however, reported that men and women were equally able to detect expressions on men's faces, but women were considerably better at detecting other women's expressions (Lewin and Herlitz 2002).

men detect anger in other men better than women do: Geary 2002.

brains of parents versus nonparents: Seifritz et al. 2003; Proverbio et al. 2006.

262   *sex difference in smiling emerges at puberty:* A review of twenty studies found that in half of them, girls smiled more than boys, and in the other half, boys smiled more than girls. By contrast, in adults, women smiled more in eighteen out of nineteen studies (Hall and Halberstadt 1986). In yet another study, the sex difference in smiling was found to be largest in adolescents and then declined progressively through adulthood, but no data for children were reported (LaFrance et al. 2003).

empathy sex difference is smaller in children than adults: McClure 2000.

263   *negative impact on the brain when deprived of normal nurturing:* Cushing and Kramer 2005.

fathers who are the primary caregivers become more sensitive: Barnett and Rivers 2004, p. 213.

men experience many of the same changes in nurturing hormones: Cushing and Kramer 2005.

children raised in emotionally intelligent homes score higher on empathy: Dunn et al. 1991.

Emory University study to improve emotion recognition: Grinspan et al. 2003.

264   *males are more likely to strike, shoot, and commit homicide:* Campbell 1999.
       d statistic ranging from 0.42 to 0.63: Archer 2004a; Hyde 1984.
265   *sex difference in aggression emerges by age two:* Campbell et al. 1998.
       rough-and-tumble play may promote interpersonal skills: Pellis and Pellis 2007.
       sex difference in aggressiveness declines after age thirty: Archer 2004a.
       males battle to gain mating rights and to guard females: Geary 1998, pp. 59–62.
266   *sexual selection for hunting ability and male visuospatial skills* (footnote): Jones et al. 2003.
       boys jostle for status as early as age three: Thorne 1993, p. 58.
267   *digit ratio correlates with aggression:* Hampson et al. 2008; Bailey and Hurd 2005; Benderlioglu and Nelson 2004.
268   *no jump in aggressiveness when testosterone levels rise at puberty:* Archer 2006.
       challenge hypothesis of testosterone and aggression: Wingfield 2005.
       testosterone rises with challenge in both men and women: Archer 2006; Edwards et al. 2006.
269   *low serotonin is linked to both aggression and depression:* Carver et al. 2008.
       prolactin level rises in fathers and correlates with sensitivity to their infants: Fleming et al. 2002.
       oxytocin improves men's ability to discern emotional expressions: Domes et al. 2007.
       vasopressin important in bonding and paternal care: Wang et al 1998.
       testosterone declines when men become fathers: Gray et al. 2007.
270   *girls resort to hidden tactics:* Letendre 2007.
       relational aggression begins in preschool: Crick et al. 1997.
       relational aggression is universal and peaks in teen years: Mealey 1999.
       harassment by high-ranking female monkeys can suppress ovulation in lower-ranking females: Campbell 1999.
       no sex difference in feelings of anger; assertiveness disparaged in females: Potegal and Archer 2004.
271   *relational aggression in girls and how to combat it:* Simmons 2002; Wiseman 2002.
       both victims and relational aggressors at greater risk for anxiety and depression: Crick et al. 1997.
272   *aggression and violence vary dramatically by culture:* Bussey and Bandura 1999.
       most crime is committed by a small proportion of delinquents; most men are not natural-born killers: Maxson 1999.
       violence between heterosexual partners varies: Archer 2000.
       girls' increasing physical aggression: Crick 1997.

women's increasing physical aggression: Lagerspetz 1999. For example, data from Finland show that the number of assaults by women has increased five-fold, from about 2 percent of total assaults in the late 1960s to nearly 10 percent in 1994.

decline in violence over history: Pinker 2007.

273   *Melanie Wood on competition:* Wiegand 1998.

274   *competitive encounters among British university students:* Cashdan 1998.

275   *men are more comfortable than women in asserting their desire to be the best in sports:* Kivlighan et al. 2005.

facial symmetry is universally regarded as attractive: Rhodes 2006.

female competition and male selection explains the desire for certain physical traits: Geary 1998, p. 151.

female weight and standards of beauty (footnote): LaFraniere 2007.

276   *eating disorders arise from female competition:* Mealey 1999.

patriarchy as basis for covert female competition: Campbell 1999.

277   *no sex difference in competitiveness before age five:* Maccoby and Jacklin 1974, p. 250.

sex difference in children's competitiveness varies across cultures: Strube 1981.

278   *study of gender and competitiveness in fourth-grade running races:* Gneezy and Rustichini 2004.

279   *importance of childhood competition for development of women leaders:* Rimm 1999, pp. 170–74, 185–90.

280   *sex differences in various types of risk taking:* Byrnes et al. 1999.

no sex difference in cocaine abuse: Quiñones-Jenab 2006.

risk-taking difference emerges between two and three years of age: Morrongiello et al. 2000.

d values for risk taking: Byrnes et al. 1999.

boys are 73 percent more likely to die in accidents: UNICEF 2001.

emergency-room visits for boys versus girls: Linakis et al. 2006.

boys are likelier to die in automobile accidents: Data from the National Center for Health Statistics, http://www.childtrendsdatabank.org/indicators/77Vehicle Deaths.cfm.

boys are likelier to die in ATV crashes: Brown et al. 2002.

boys cheat more on exams than girls do: Whitley et al. 1999.

boys likelier than girls to be expelled from school and arrested for selling drugs: McCord et al. 2001, pp. 56, 85.

men arrested four times more often than women: Greenfeld and Snell 2000.

281   *women are more financially risk averse:* Embrey and Fox 1997.

females more likely to leave a question blank: Ben-Shakhar and Sinai 1991; von Schrader and Ansley 2006. The latter study examined a very large population of children taking the Iowa Test of Basic Skills. While girls were significantly more likely to omit questions than boys were, particularly in younger grades, the number of omissions was so small that it had little significant impact on their overall scores.

females' but not males' test scores improved when time constraint was lifted: American Physical Society 1996.

282   *genes involved in sensation seeking:* Stoel et al. 2006.

the more active form of COMT is associated with greater sensation seeking in women: Lang et al. 2007.

COMT activity higher in frontal lobes of men than women: Chen et al. 2004.

relationship between testosterone and sensation seeking: Hampson et al. 2008; Aluja and Torrubia 2004; Kerschbaum et al. 2006. However, testosterone did *not* correlate with sensation seeking in Rosenblitt et al. 2001.

evolutionary rationale for greater fearfulness in women: Campbell 1999.

283   *fear of success:* Fried-Buchalter 1997.

amygdala is larger in males: Brierley et al. 2002; Goldstein et al. 2001; Durston et al. 2001. However, several other recent studies report no sex difference in amygdala size, and yet another study reports that girls with CAH, who are exposed to elevated levels of androgens before birth and are more inclined toward rough-and-tumble play, do not have enlarged amygdalae (Merke et al. 2003).

women show stronger activation in the left amygdala, men in the right: Cahill et al. 2004; Canli et al. 2002.

prenatal testosterone and amygdala activation: A relationship is suggested by Ernst et al. 2007, who found a male-typical pattern of amygdala activation in women with CAH. However, another recent study by Ciumas et al. (2008) found typical female limbic function in women with CAH, so it remains unclear whether prenatal testosterone contributes to sex differences in adults' neural processing of emotion.

284   *testosterone enhances object-reversal performance in young female monkeys:* Clark and Goldman-Rakic 1989.

sex differences in orbitofrontal activity and cognition that persist into adulthood: Overman 2004; Bolla et al. 2004.

minuscule difference in timidity between infant boys and girls: Jacklin et al. 1983.

girls more inhibited by fourteen months: Kagan et al. 1987.

girls more likely to express fear: Maccoby and Jacklin 1974, pp. 182–90. However, while girls are likelier than boys to *admit* feelings of fear or anxiety, observational and physiological studies do not reveal an obvious sex difference in the presence of these feelings.

sex difference in inhibition in Chinese children: Chen et al. 1998.

infant-crawling experiment: Mondschein et al. 2000. See chapter 2.

mothers of boys are slower to intervene on the playground: Morrongiello and Dawber 2000.

parents are more tolerant of fear in daughters: Potegal and Archer 2004; Morrongiello and Dawber 2000.

285 *Chinese infants appear more inhibited than Canadian infants:* Chen et al. 1998.

286 *love potion in the chocolate cauldrons:* Rowling 2005, pp. 389–96.

d values of 3.5 and 3.99 for sexual attraction: Lippa 2002, pp. 21–22.

287 *"magic age of ten":* Herdt and McClintock 2000.

adrenarche and sexual attraction: McClintock and Herdt 1996.

288 *androgen levels rise four-fold in girls and peak between fourteen and sixteen:* Ankarberg and Norjavaara 1999.

androgens are not neuropoisons: Richter 2006.

289 *rising testosterone and emergence of sexual behavior:* Halpern et al. 1997, 1998.

no significant change in aggressiveness in twelve-to-sixteen-year-old boys: Halpern et al. 1993.

pubertal testosterone does not significantly increase male aggression: Archer 2006.

hormone treatments increased teens' interest in sex: Finkelstein et al. 1998.

hormone treatments promoted feeling more attractive and competent: Schwab et al. 2001.

estrogen or testosterone increased self-reported aggressive behaviors in adolescents: Finkelstein et al. 1997.

hormones had little effect on verbal aggression, mood, or overall behavior: Susman et al. 1998.

the hormones did not alter spatial ability: Liben et al. 2002.

intelligence grows with chronological age rather than pubertal development: Graber and Petersen 1991.

MRI studies of hundreds of children: Giedd et al. 1996. In this article, Giedd concludes about the brain: "The hypothesis of increased maturational changes around the time of puberty is not supported."

neither sex undergoes any dramatic neural shift at the time of puberty: Lenroot et al. 2007.

290 *sex difference in the SDN-POA does not emerge until about ten:* Swaab and Hofman 1988.

39 percent larger BNST in males emerging around the time of puberty: Chung et al. 2002.

291 *sex difference in the amygdala emerges during later childhood or puberty:* Giedd et al. 1996.

testosterone increases amygdala size in adult male rats: Cooke et al. 1999.

amygdala an important center for sexual arousal in men but not women: Hamann 2005.

sex-hormone receptors sparser in the cerebral cortex than limbic brain areas: Abdelgadir et al. 1999.

292   *girls who go through precocious puberty wait until their teen years:* Finkelstein et al. 1998.

not the hormonal changes but their impact on peer relations that sculpts the adolescent psyche: Graber and Sontag 2006.

294   *sex accounted for only 16 percent of the variance in moral reasoning:* Jaffee and Hyde 2000.

295   *value of coed sports:* Pollack 1998, p. 299.

298   *heavy TV viewing associated with stronger gender stereotypes:* Lippa 2002, p. 145.

violence on TV and influence on children's real-life aggression: Huesmann et al. 2003.

growing sexualization of girls: American Psychological Association, Task Force on the Sexualization of Girls (2007).

8. TRUCE TIME

301   *women in their twenties are actually outearning men:* based on research by Andrew Beveridge at Queen's College in New York and cited in Roberts 2007.

dropout rate by gender (graph): accessed at http://nces.ed.gov/programs/digest/do6/tables/dto6_104.asp.

302   *"It's the daycare, stupid":* quoted in Rimer 2005.

when women made up a mere 20 percent of college students: This is based on data from the Pell Institute and National Center for Education Statistics and published by the *Washington Post* at http://www.washingtonpost.com/wp-srv/education/daily/graphics/gender_062502.html.

304   *females of all ages can detect quieter sounds than males:* Sininger et al. 1998.

deconstruction of Sax's claim (footnote): Liberman 2006b and 2008. The quote is from Sax 2005, p. 18.

not relevant to what goes on in a classroom: Dennis McFadden of the University of Texas, who has conducted extensive research on sex differences in the auditory system, wrote to me by e-mail that "arguing for separate schools because of the small sex difference in hearing sensitivity is just dopey."

305   *differences in boys' and girls' color vision not dramatic:* Leonard Sax (2005, pp. 19–22) states that boys and girls prefer different colors because of a fundamental sex difference in the retina, but as David Hilbert (2006) points out, this claim is based on data from rats, which are nearly colorblind. According to Hilbert, the only data from humans show no sex difference in input from the two retinal cell types (known as M and P), refuting Sax's claim that "Boys prefer colors such as black, gray, silver, and blue because that's the way the M cells are wired."

Newsweek description of Foust Elementary School: Tyre 2005.

no significant difference in testosterone between prepubescent boys and girls: Elmlinger et al. 2005.

no difference in serotonin level between boys and girls: Flachaire et al. 1990.

no difference in oxytocin level between boys and girls: Fries et al. 2005.

cognitive differences do not justify single-sex education: Halpern 1997; Hyde and Linn 2006.

306 Canadian report on single-sex schooling: Thompson and Ungerleider 2004.

New Zealand report on single-sex schooling: Harker 2000.

Australian report on single-sex schooling: Marsh and Rowe 1996.

"not sufficiently weighty to yield definitive conclusions": Salomone 2003, p. 239.

benefits of single-sex education are "equivocal": U.S. Department of Education, Office of Planning, Evaluation and Policy Development 2005.

paradox of single-sex education: Smithers and Robinson 2006.

307 SAT scores at Smith College declined during the 1970s: Kaminer 1998.

only 3 percent of girls who take the SAT express interest in all-women colleges: Schemo 2006.

309 pro-academic choice is sufficient to boost achievement and test scores: Salomone, pp. 218–219; Smithers and Robinson 2006.

results of California pilot program: Datnow et al. 2001.

310 "High-performing young women . . . special care": Schemo 2006.

disadvantaged and at-risk children may benefit most from single-sex schooling: Salomone 2003, p. 235.

311 weak evidence that single-sex education promotes nontraditional interests: Spielhofer et al. (2004) found an effect of single-sex education on girls' academic interests, but not boys'. The study by the U.S. Department of Education (Office of Planning, Evaluation and Policy Development, 2005) found very modest support for such an effect in girls but no evidence that it works in boys.

any time the ratio gets above three to one, the minority gender tends to disengage: Babcock and Laschever (2007, p. 101) cite a study in which MBAs rated female job applicants lower and also judged them as more stereotypically feminine when women made up 25 percent or less of the applicant pool. Similarly, research on stereotype threat has shown that students' actual performance declines when they are aware of being in a minority that is not expected to do as well.

evidence from Belgium and Nigeria for benefit of same-sex teachers: Haag 2000.

benefit of same-sex teachers in middle-school science, English, and social studies classes: Dee 2005.

312 coed classes can deflate stereotypes: In their analysis of Australian schools, Marsh and Rowe (1996, p. 153) found "that boys and girls changed their at-

titudes more in the direction of gender equality when they were actually in mixed-sex classes where they were forced to confront their preconceptions than when they were in single-sex classes."

cultural neuroscience: Han and Northoff 2008.

314   *plasticity as a double-edged sword:* According to one line of research, sex differences in personality traits appear larger the more developed and freer a culture is (Schmitt et al. 2008).

ILLUSTRATION CREDITS

Figure 1.1 (p. 21): Image courtesy of Clinical Tools, Inc. **Figure 1.3 (p. 44):** Illustration adapted from image courtesy of Emma Nelson, University of Liverpool, and reproduced by permission of Western Academic & Specialist Press Ltd. First published in *Before Farming: The Archaeology and Anthropology of Hunter Gatherers* [online version] 2006/1 article 6. **Figure 2.2 (p. 66):** Adapted from *Journal of Experimental Child Psychology,* Vol. 77, E. R. Mondschein, K. E. Adolph, and C. S. Tamis-LeMonda, "Gender bias in mothers' expectations about infant crawling," 304–16, copyright © 2000, with permission from Elsevier. **Figure 3.1 (p. 134):** Copyright © 1999 by the American Psychological Association. Adapted with permission from S. C. Levine, J. Huttenlocher, A. Taylor, and A. Langrock. (1999). "Early sex differences in spatial skill," *Developmental Psychology* 35, 940–49. **Figure 5.4 (p. 198):** Modified and reprinted from *Current Opinion in Neurobiology,* Vol. 12, E. Temple, "Brain mechanisms in normal and dyslexic readers," 178–83, copyright © 2002, with permission from Elsevier. **Figure 6.3 (p. 217):** Images are courtesy of Dr. Peter Meijer. **Figure 6.5 (p. 221):** Reprinted from *Evolution and Human Behavior,* Vol. 19, S. D. Moffatt, E. Hampson, and M. Hatzipantelis, "Navigation in a 'virtual' maze: sex differences and correlation with psychometric measures of spatial ability in humans," 73–87, copyright © 1998, with permission from Elsevier.

Bibliography

AAUW Educational Foundation. 2000. Tech-savvy: Educating girls in the new computer age. Washington DC: American Association of University Women, http://www.aauw.org/research/tech_savvy.cfm.

Abdelgadir, S. E., C. E. Roselli, J. V. Choate, and J. A. Resko. 1999. Androgen receptor messenger ribonucleic acid in brains and pituitaries of male rhesus monkeys: studies on distribution, hormonal control, and relationship to luteinizing hormone secretion. *Biology of Reproduction,* 60:1251–56.

Abramovich, D. R., I. A. Davidson, A. Longstaff, and C. K. Pearson. 1987. Sexual differentiation of the human midtrimester brain. *European Journal of Obstetrics, Gynecology, and Reproductive Biology,* 25:7–14.

Alanis, M. C., and R. S. Lucidi. 2004. Neonatal circumcision: a review of the world's oldest and most controversial operation. *Obstetrical and Gynecological Survey,* 59:379–95.

Aleman, A., E. Bronk, R. P. Kessels, et al. 2004. A single administration of testosterone improves visuospatial ability in young women. *Psychoneuroendocrinology,* 29:612–17.

Alexander, A. W., and A. M. Slinger-Constant. 2004. Current status of treatments for dyslexia: critical review. *Journal of Child Neurology,* 19:7404–58.

Alexander, G. M., and M. Hines. 2002. Sex differences in response to children's toys in nonhuman primates (cercopithecus aethiops sabaeus). *Evolution and Human Behavior,* 23:467–79.

Alexander, G. M., T. Wilcox, and R. Woods. 2008. Sex difference in infants' visual interest in toys. *Archives of Sexual Behavior,* 38:427–33.

Allen, L. S., M. Hines, J. E. Shryne, and R. A. Gorski. 1989. Two sexually dimorphic cell groups in the human brain. *Journal of Neuroscience,* 9:497–506.

Almond, D., and L. Edlund. 2006. Trivers-Willard at birth and one year: evidence from U.S. natality data 1983–2001. *Proceedings of the Royal Society B,* 274:2491–96.

———. 2008. Son-biased sex ratios in the 2000 United States Census. *Proceedings of the National Academy of Sciences USA,* 105:5681–82.

Aluja, A., and R. Torrubia. 2004. Hostility-aggressiveness, sensation seeking, and sex hormones in men: re-exploring their relationship. *Neuropsychobiology,* 50:102–7.

American Academy of Pediatrics. Task Force on Circumcision. 1999. Circumcision policy statement. *Pediatrics,* 103:686–93.

American Automobile Association. 2008. Auto ace. *AAA Living (Illinois/N. Indiana,* July–August).

American Federation of Teachers. 2005. Survey and analysis of teacher salary trends, http://www.aft.org/salary/2005/download/AFT2005SalarySurvey.pdf.

American Physical Society. 1996. Fighting the gender gap: Standardized tests are poor indicators of ability in physics. *APS News* (July), http://www.aps.org/publications/apsnews/199607/upload/jul96.pdf.

American Psychiatric Association. 2000. *Diagnostic and Statistical Manual of Mental Disorders,* 4th Ed. (DSM-IV), Washington DC.

American Psychological Association, Task Force on the Sexualization of Girls. 2007. *Report of the APA Task Force on the Sexualization of Girls.* Washington DC: American Psychological Association, http://www.apa.org/pi/wpo/sexualization.html.

Anderson, M. A., N. A. Tollefson, and E. C. Gilbert. 1985. Giftedness and reading: a cross-sectional view of differences in reading attitudes and behaviors. *Gifted Child Quarterly,* 2:186–89.

Angold, A., E. J. Costello, A. Erkanli, and C. M. Worthman. 1999. Pubertal changes in hormone levels and depression in girls. *Psychological Medicine,* 29:1043–53.

Ankarberg, C., and E. Norjavaara. 1999. Diurnal rhythm of testosterone secretion before and throughout puberty in healthy girls: Correlation with 17ß-estradiol and dehydroepiandrosterone sulfate. *Journal of Clinical Endocrinology and Metabolism,* 84:975–84.

Archer, J. 2000. Sex differences in aggression between heterosexual partners: a meta-analytic review. *Psychological Bulletin,* 126:651–80.

———. 2004a. Sex differences in aggression in real-world settings: A meta-analytic review. *Review of General Psychology,* 8:291–322.

———. 2006. Testosterone and human aggression: an evaluation of the challenge hypothesis. *Neuroscience and Biobehavioral Reviews,* 30:319–45.

Archer, L. 2004b. Mixed sex or single sex? In *Gender in Education, 3–19: A Fresh Approach,* ed. H. Claire, 50–56. Association of Teachers and Lecturers, http://www.atl.org.uk/publications-and-resources/research-publications/gender-in-education.asp.

Arnold, A. P. 2003. The gender of the voice within: The neural origin of sex differences in the brain. *Current Opinion in Neurobiology,* 13:759–64.

Aronson, J., M. J. Lustina, C. Good, et al. 1999. When white men can't do math: Necessary and sufficient factors in stereotype threat. *Journal of Experimental Social Psychology*, 35:29–46.

Askling, J., G. Erlandsson, M. Kaijser, et al. 1999. Sickness in pregnancy and sex of child. *Lancet*, 354:2053.

Association for Women in Mathematics. 2006. Background for the AWM petition concerning the inclusion of Dr. Camilla Benbow on the National Mathematics Advisory Panel, http://www.awm-math.org/benbow_petition/background.html.

Aueyeung, B., S. Baron-Cohen, E. Ashwin, et al. 2008. Fetal testosterone and autistic traits. *British Journal of Psychology*, 100:1–22.

Babcock, L., and S. Laschever. 2007. *Women Don't Ask: The High Cost of Avoiding Negotiation—and Positive Strategies for Change.* New York: Bantam.

Bachevalier, J., C. Hagger, and B. Bercu. 1989. Gender differences in visual habit formation in 3-month-old rhesus monkeys. *Developmental Psychobiology*, 22:585–99.

Bae, Y., S. Choy, C. Geddes, et al. 2000. *Trends in Educational Equity of Girls & Women.* U.S. Department of Education, National Center for Education Statistics (NCES 2000–030), http://nces.ed.gov/pubs2000/2000030.pdf.

Bailey, A. A., and P. L. Hurd. 2005. Finger length ratio (2D:4D) correlates with physical aggression in men but not in women. *Biological Psychology*, 68:215–22.

Baker, L. C. 1996. Differences in earnings between male and female physicians. *New England Journal of Medicine*, 334:960–64.

Baker, M. A. 1987. Sensory functioning. In *Sex Differences in Human Performance*, ed. M. A. Baker, 5–36. New York: John Wiley.

Bakker, J., C. De Mees, Q. Douhard, et al. 2006. Alpha-fetoprotein protects the developing female mouse brain from masculinization and defeminization by estrogens. *Nature Neuroscience*, 9:220–26.

Balogh, R. D., and R. H. Porter. 1986. Olfactory preferences resulting from mere exposure in human neonates. *Infant Behavior and Development*, 9:395–401.

Barnett, R., and C. Rivers. 2004. *Same Difference: How Gender Myths Are Hurting Our Relationships, Our Children, and Our Jobs.* New York: Basic Books.

Baron-Cohen, S. 2003. *The Essential Difference: The Truth About the Male & Female Brain,* New York: Basic Books.

Baron-Cohen, S., S. Wheelwright, J. Hill, et al. 2001. The "Reading the Mind in the Eyes" Test revised version: a study with normal adults, and adults with Asperger syndrome or high-functioning autism. *Journal of Child Psychology & Psychiatry*, 42:241–51.

Barry, R. J., A. R. Clarke, R. McCarthy, et al. 2004. Age and gender effects in EEG coherence: I. Developmental trends in normal children. *Clinical Neurophysiology*, 115:2252–58.

Bartocci, M., L. L. Bergqvist, H. Lagercrantz, and K. J. Anand. 2006. Pain activates cortical areas in the preterm newborn brain. *Pain*, 122:109–17.

Bartocci, M., J. Winberg, C. Ruggiero, et al. 2000. Activation of olfactory cortex in newborn infants after odor stimulation: a functional near-infrared spectroscopy study. *Pediatric Research,* 48:18–23.

Baydar, N., and J. Brooks-Gunn. 1991. Effects of maternal employment and childcare arrangements on preschoolers' cognitive and behavioral outcomes: Evidence from the children of the National Longitudinal Survey of Youth. *Developmental Psychology,* 27:932–45.

BBC News. 1999. Closing the school gender gap (May 14), http://news.bbc.co.uk/1/hi/education/343348.stm.

Bearman, P. S., and H. Brückner. 2002. Opposite-sex twins and adolescent same-sex attraction. *American Journal of Sociology,* 107:1179–205.

Beatty, W. W. 1992. Gonadal hormones and sex differences in nonreproductive behaviors. In *Handbook of Behavioral Neurobiology,* Vol. 2: *Sexual Differentiation,* eds. A. A. Gerall, H. Moltz, and I. L. Ward, 85–128. New York: Plenum.

Begley, S. 1995. Gray matters: New technologies that catch the mind in the very act of thinking show how men and women use their brains differently. *Newsweek* (March 27), http://www.newsweek.com/id/110064.

Bekedam, D. J., S. Engelsbel, B.W.J. Mol, et al. 2002. Male predominance in fetal distress during labor. *American Journal of Obstetrics and Gynecology,* 187:1605–7.

Bell, A. D., and S. Variend. 1985. Failure to demonstrate sexual dimorphism of the corpus callosum in childhood. *Journal of Anatomy,* 143:143–47.

Belsky, J., and M. J. Rovine. 1988. Nonmaternal care in the first year of life and the security of infant-parent attachment. *Child Development,* 59:157–67.

Bem, S. 1998. *An Unconventional Family.* New Haven, CT: Yale University Press.

Benbow, C. P., and J. C. Stanley. 1983. Sex differences in mathematical reasoning ability: more facts. *Science,* 222:1029–31.

Benderlioglu, Z., and R. J. Nelson. 2004. Digit length ratios predict reactive aggression in women, but not in men. *Hormones and Behavior,* 46:558–64.

Bendersky, M., G. Gambini, A. Lastella, et al. 2003. Inhibitory motor control at five years as a function of prenatal cocaine exposure. *Journal of Developmental and Behavioral Pediatrics,* 24:345–51.

Benninger, C., P. Matthis, and D. Scheffner. 1984. EEG development of healthy boys and girls. Results of a longitudinal study. *Electroencephalography and Clinical Neurophysiology,* 57:1–12.

Ben-Shakhar, G., and Y. Sinai. 1991. Gender differences in multiple-choice tests: the role of differential guessing tendencies. *Journal of Educational Measurement,* 28:23–35.

Berenbaum, S. A. 1999. Effects of early androgens on sex-typed activities and interests in adolescents with congenital adrenal hyperplasia. *Hormones and Behavior,* 35:102–10.

Berenbaum, S. A., and E. Snyder. 1995. Early hormonal influences on childhood sex-typed activity and playmate preferences: Implications for the development of sexual orientation. *Developmental Psychology,* 31:31–42.

Berglund, E., M. Eriksson, and M. Westerlund. 2005. Communicative skills in relation to gender, birth order, childcare and socioeconomic status in 18-month-old children. *Scandinavian Journal of Psychology*, 46:485–91.

Berk, L. E. 1994. Why children talk to themselves. *Scientific American* (November), 78–83.

———. 2003. *Child Development*, 6th ed. Boston: Allyn and Bacon.

Berninger, E. 2007. Characteristics of normal newborn transient-evoked otoacoustic emissions: ear asymmetries and sex effects. *International Journal of Audiology*, 46:661–69.

Berry, J. 1971. Ecological and cultural factors in spatial development. *Canadian Journal of Behavioral Science*, 3:324–36.

Berry, J. W. 1966. Temne and Eskimo perceptual skills. *International Journal of Psychology*, 1:207–29.

Best, D. L., and J. E. Williams. 1993. A cross-cultural viewpoint. In *The Psychology of Gender*, eds. A. E. Beall and R. J. Sternberg, 215–48. New York: Guilford.

Bhaumik, U., I. Aitken, I. Kawachi, et al. 2004. Narrowing of sex differences in infant mortality in Massachusetts. *Journal of Perinatology*, 24:94–99.

Bishop, K. M., and D. Whalsten. 1997. Sex differences in the human corpus callosum: Myth or reality? *Neuroscience and Biobehavioral Reviews*, 21:581–601.

Bjorklund, D. F., and R. D. Brown. 1998. Physical play and cognitive development: Integrating activity, cognition, and education. *Child Development*, 69:604–6.

Blakemore, J.E.O. 1990. Children's nurturant interactions with their infant siblings: An exploration of gender differences and maternal socialization. *Sex Roles*, 22:43–57.

———. 1992. The influence of age, gender, and having a younger sibling on children's knowledge about babies. *Journal of Genetic Psychology*, 153:139–54.

———. 1998. The influence of gender and parental attitudes on preschool children's interest in babies: Observations in natural settings. *Sex Roles*, 38:73–94.

Blanch, R. J., D. Brennan, B. Condon, et al. 2004. Are there gender-specific neural substrates of route learning from different perspectives? *Cerebral Cortex*, 14:1207–13.

Boatella-Costa, E., C. Costas-Moragas, F. Botet-Mussons, et al. 2007. Behavioral gender differences in the neonatal period according to the Brazelton scale. *Early Human Development*, 83:91–97.

Bolla, K. I., D. A. Eldreth, J. A. Matochik, and J. L. Cadet. 2004. Sex-related differences in a gambling task and its neurological correlates. *Cerebral Cortex*, 14:1226–32.

Boyatzis, C. J., E. Chazan, and C. Z. Ting. 1993. Preschool children's decoding of facial emotions. *Journal of Genetic Psychology*, 154:375–82.

Brackbill, Y., and K. Schroder. 1980. Circumcision, gender differences, and neonatal behavior: An update. *Developmental Psychobiology*, 13:607–14.

Bradley, S. J., G. D. Oliver, A. B. Chernick, and K. J. Zucker. 1998. Experiment of nature: Ablatio penis at 2 months, sex reassignment at 7 months, and a psychosexual follow-up in young adulthood. *Pediatrics*, 102:e9.

Bridges, J. S. 1993. Pink or blue? Gender-stereotypic perceptions of infants as conveyed by birth congratulations cards. *Psychology of Women Quarterly*, 17:193–205.

Brierley, B., P. Shaw, and A. S. David. 2002. The human amygdala: a systematic review and meta-analysis of volumetric magnetic resonance imaging. *Brain Research: Brain Research Reviews*, 39:84–105.

British Dietetic Association. 2003. Paediatric group position statement on the use of soya protein for infants. *Journal of Family Health Care*, 13:93.

Brizendine, L. 2006. *The Female Brain*. New York: Morgan Road Books.

Brocki, K. C., and G. Bohlin. 2004. Executive functions in children aged 6 to 13: a dimensional and developmental study. *Developmental Neuropsychology*, 26:571–93.

Brosnan, M. J. 1998. Spatial ability in children's play with Lego blocks. *Perceptual and Motor Skills*, 87:19–28.

Brown, G. R., and A. F. Dixson. 1999. Investigation of the role of postnatal testosterone in the expression of sex differences in behavior of infant rhesus macaques (*Macaca mulatta*). *Hormones and Behavior*, 35:186–94.

Brown, R. L., M. E. Koepplinger, C. T. Mehlman, et al. 2002. All-terrain vehicle and bicycle crashes in children: epidemiology and comparison of injury severity. *Journal of Pediatric Surgery*, 37:375–80.

Brown, W. M., M. Hines, B. A. Fane, and S. M. Breedlove. 2002. Masculinized finger length patterns in human males and females with congenital adrenal hyperplasia. *Hormones and Behavior*, 42:380–86.

Bucher, J. 2008. Testimony on "Effects of bisphenol A and phthalates" before the U.S. House of Representatives, Committee on Energy and Commerce, Subcommittee on Commerce, Trade, and Consumer Protection (June 10), http://www.hhs.gov/asl/testify/2008/06/t20080610a.html.

Buck, J. J., R. M. Williams, I. A. Hughes, and C. L. Acerini. 2003. In-utero androgen exposure and 2nd to 4th digit length ratio—comparisons between healthy controls and females with classical congenital adrenal hyperplasia. *Human Reproduction*, 18:976–79.

Burkam, D. T., V. E. Lee, and B. A. Smerdon. 1997. Gender and science learning early in high school: Subject matter and laboratory experiences. *American Educational Research Journal*, 34:297–331.

Burman, D. D., T. Bitan, and J. R. Booth. 2008. Sex differences in neural processing of language among children. *Neuropsychologia*, 46:1349–62.

Burton, N. W., C. Lewis, and N. Robertson. 1988. *College Board Report No. 88–9: Sex Differences in SAT Scores*. New York: College Entrance Examination Board.

Bussey, K., and A. Bandura. 1999. Social cognitive theory of gender development and differentiation. *Psychological Review*, 106:676–713.

Butterworth, G. 1998. What is special about pointing in babies? In *The Development of Sensory, Motor and Cognitive Capacities in Early Infancy: From Perception to Cognition*, eds. F. Simion and G. Butterworth, 171–90. East Sussex, England: Psychology Press.

Butterworth, G., and P. Morissette. 1996. Onset of pointing and the acquisition of language in infancy. *Journal of Reproductive & Infant Psychology*, 14:219–31.

Byne, W., S. Tobet, L. A. Mattiace, et al. 2001. The interstitial nuclei of the human anterior hypothalamus: an investigation of variation with sex, sexual orientation, and HIV status. *Hormones and Behavior*, 40:86–92.

Byrd, R. S., M. Weitzman, and P. Auinger. 1997. Increased behavior problems associated with delayed school entry and delayed school progress. *Pediatrics*, 100:654–61.

Byrnes, J. P., D. C. Miller, and W. D. Schafer. 1999. Gender differences in risk taking: A meta-analysis. *Psychological Bulletin*, 125:367–83.

Cahan, S., and N. Cohen. 1989. Age versus schooling effects on intelligence development. *Child Development*, 60:1239–49.

Cahill L. 2006. Why sex matters for neuroscience. *Nature Reviews Neuroscience*, 7:477–84.

Cahill, L., M. Uncapher, L. Kilpatrick, et al. 2004. Sex-related hemispheric lateralization of amygdala function in emotionally influenced memory: an FMRI investigation. *Learning and Memory*, 11:261–66.

Cameron, D. 2007. *The Myth of Mars and Venus. Do Men and Women Really Speak Different Languages?* Oxford University Press.

Campbell, A. 1999. Staying alive: evolution, culture, and women's intrasexual aggression. *Behavioral and Brain Sciences*, 22:203–52.

Campbell, A., S. Muncer, and J. Odber. 1998. Primacy of organizing effects of testosterone. *Behavioral and Brain Sciences*, 21:365.

Campbell, A., L. Shirley, and L. Caygill. 2002. Sex-typed toy preferences in three domains: Do two-year-olds need cognitive variables? *British Journal of Psychology*, 93:203–17.

Campbell, A., L. Shirley, C. Heywood, and C. Crook. 2000. Infants' visual preference for sex-congruent babies, children, toys and activities: a longitudinal study. *British Journal of Developmental Psychology*, 18:479–98.

Campbell, D. W., and W. O. Eaton. 1999. Sex differences in the activity level of infants. *Infant and Child Development*, 8:1–17.

Campbell, S., and G. MacQueen. 2006. An update on regional brain volume differences associated with mood disorders. *Current Opinion in Psychiatry*, 19:25–33.

Canals, J., J. Fernández-Ballart, and G. Esparó. 2003. Evolution of neonatal behavior assessment scale scores in the first month of life. *Infant Behavior and Development*, 26:227–37.

Canli, T., J. E. Desmond, Z. Zhao, and J. D. Gabrieli. 2002. Sex differences in the neural basis of emotional memories. *Proceedings of the National Academy of Sciences USA*, 99:10789–94.

Capute, A. J., B. K. Shapiro, F. B. Palmer, et al. 1985. Normal gross motor development: the influences of race, sex and socio-economic status. *Developmental Medicine and Child Neurology*, 27:635–43.

Carlsson-Paige, N., and D. E. Levin. 1987. *The War Play Dilemma: Balancing Needs and Values in the Early Childhood Classroom*. New York: Teachers College Press.

———. 1999. The war-toy connection. *Christian Science Monitor*, vol. 91, Issue 217 (October 5).

Carver, C. S., S. L. Johnson, and J. Joormann. 2008. Serotonergic function, two-mode models of self-regulation, and vulnerability to depression: what depression has in common with impulsive aggression. *Psychological Bulletin*, 134:912–43.

Casey, B., B. Pezaris, K. Anderson, and J. Bassi. 2004. Gender differences in spatial skills: Implications for the mathematics education of young children. In *Challenging Young Children Mathematically*, eds. C. Greenes and J. Tsankova, 28–39. Golden, CO: National Council of Supervisors of Mathematics.

Casey, M. B., R. L. Nuttall, and E. Pezaris. 2001. Spatial-mechanical reasoning skills versus mathematics self-confidence as mediators of gender differences on mathematics subtests using cross-national gender-based items. *Journal for Research in Mathematics Education*, 32:28–57.

Casey, M. B., R. Nuttall, E. Pezaris, and C. P. Benbow. 1995. The influence of spatial ability on gender differences in mathematics college entrance test scores across diverse samples. *Developmental Psychology*, 31:697–705.

Cashdan, E. 1998. Are men more competitive than women? *British Journal of Social Psychology*, 37:213–29.

Caviness, V. S., D. N. Kennedy, C. Richelme, et al. 1996. The human brain age 7–11 years: A volumetric analysis based on magnetic resonance images. *Cerebral Cortex*, 6:726–36.

Center for Disease Control. 1997. Rates of homicide, suicide, and firearm-related death among children—26 industrialized countries. *MMWR Weekly*, 46:101–5 (February 7), http://www.cdc.gov/mmwr/preview/mmwrhtml/00046149 .htm.

———. 2002. Youth risk behavior surveillance—United States, 2001. *MMWR Surveillance Summaries*, 51(SS04):1–64 (June 28), http://www.cdc.gov/mmwr/preview/mmwrhtml/ss5104a1.htm.

Center for Early Childhood Leadership, National-Louis University. 2004. Directors' perceptions about male involvement in early childhood programs. *Research Notes* (Summer), http://cecl.nl.edu/research/issues/rnsu04.pdf.

Chabris, C. F., and M. E. Glickman. 2006. Sex differences in intellectual performance: Analysis of a large cohort of competitive chess players. *Psychological Science*, 17:1040–46.

Chen, A., and W. J. Rogan. 2004. Isoflavones in soy infant formula: a review of evidence for endocrine and other activity in infants. *Annual Review of Nutrition*, 24:33–54.

Chen, J., B. K. Lipska, N. Halim, et al. 2004. Functional analysis of genetic variation in catechol-O-methyltransferase (COMT): effects on mRNA, protein, and enzyme activity in postmortem human brain. *American Journal of Human Genetics*, 75:807–21.

Chen, X., P. D. Hastings, K. H. Rubin, et al. 1998. Child-rearing attitudes and behavioral inhibition in Chinese and Canadian toddlers: A cross-cultural study. *Developmental Psychology*, 34:677–86.

Cherney, I. D. 2008. Mom, let me play more computer games: They improve my mental rotation skills. *Sex Roles*, 59:776–86.

Cherney, I. D., and B. O. Ryalls. 1999. Gender-linked differences in the incidental memory of children and adults. *Journal of Experimental Child Psychology*, 72:305–28.

Cherrier, M. M., A. L. Rose, and C. Higano. 2003. The effects of combined androgen blockade on cognitive function during the first cycle of intermittent androgen suppression in patients with prostate cancer. *Journal of Urology*, 170:1808–11.

Choate, J.V.A., O. D. Slayden, and J. A. Resko. 1998. Immunocytochemical localization of androgen receptors in brains of developing and adult male rhesus monkeys. *Endocrine*, 8:51–60.

Chung, W.C.J., G. J. De Vries, and D. F. Swaab. 2002. Sexual differentiation of the bed nucleus of the stria terminalis in humans may extend into adulthood. *Journal of Neuroscience*, 22:1027–33.

Ciumas, C., A. L. Hirschberg, and I. Savic. 2009. High fetal testosterone and sexually dimorphic cerebral networks in females. *Cerebral Cortex*, 19:1167–74.

Clark, A. S., and P. S. Goldman-Rakic. 1989. Gonadal hormones influence the emergence of cortical function in nonhuman primates. *Behavioral Neuroscience*, 103:1287–95.

Clarke, A. R., R. J. Barry, R. McCarthy, and M. Selikowitz. 2001. Age and sex effects in the EEG: development of the normal child. *Clinical Neurophysiology*, 112:806–14.

Clarke, S., R. Kraftsik, H. Van der Loos, and G. M. Innocenti. 1989. Forms and measures of adult and developing human corpus callosum: is there sexual dimorphism? *Journal of Comparative Neurology*, 280:213–30.

Coates, S. W., and S. M. Wolfe. 1995. Gender identity disorder in boys: The interface of constitution and early experience. *Psychoanalytic Inquiry*, 15:6–38.

Cohen-Bendahan, C. C., J. K. Buitelaar, S. H. van Goozen, and P. T. Cohen-Kettenis. 2004. Prenatal exposure to testosterone and functional cerebral lateralization: a study in same-sex and opposite-sex twin girls. *Psychoneuroendocrinology*, 29:911–16.

Cohen-Bendahan, C. C., J. K. Buitelaar, S. H. van Goozen, et al. 2005a. Is there an effect of prenatal testosterone on aggression and other behavioral traits? A study comparing same-sex and opposite-sex twin girls. *Hormones and Behavior*, 47:230–37.

Cohen-Bendahan, C. C., S. H. van Goozen, J. K. Buitelaar, and P. T. Cohen-Kettenis. 2005b. Maternal serum steroid levels are unrelated to fetal sex: a study in twin pregnancies. *Twin Research and Human Genetics*, 8:173–77.

Cohn, E. 2000. Are men's fingers faster? *The American Prospect*, Vol. 11, Issue 11 (April 24).

Colapinto, J. 2001. *As Nature Made Him: The Boy Who Was Raised as a Girl.* New York: Harper Perennial.

Cole-Harding, S., A. L. Morstad, and J. R. Wilson. 1988. Spatial ability in members of opposite-sex twin pairs [Abstract]. *Behavior Genetics,* 18:710.

Collaer, M. L., S. Reimers, and J. T. Manning. 2007. Visuospatial performance on an Internet line judgment task and potential hormonal markers: Sex, sexual orientation, and 2D:4D. *Archives of Sexual Behavior,* 36:177–92.

College Board. 2005. *College-Bound Seniors: Total Group Profile Report,* http://www .collegeboard.com.

———. 2008. *College-Bound Seniors: Total Group Profile Report,* http://professionals .collegeboard.com/profdownload/Total_Group_Report.pdf.

Connellan, J., S. Baron-Cohen, S. Wheelwright, et al. 2000. Sex differences in human neonatal social perception. *Infant Behavior and Development,* 23:113–18.

Connor, J. M., M. Schackman, and L. A. Serbin. 1978. Sex-related differences in response to practice on a visual-spatial test and generalization to a related test. *Child Development,* 49:24–29.

Connor, J. M., and L. A. Serbin. 1977. Behaviorally based masculine- and feminine-activity-preference scales for preschoolers: Correlates with other classroom behaviors and cognitive tests. *Child Development,* 48:1411–16.

Connor, K. 1989. Aggression: Is it in the eye of the beholder? *Play and Culture,* 2:213–17.

Cooke, B. M., G. Tabibnia, and S. M. Breedlove. 1999. A brain sexual dimorphism controlled by adult circulating androgens. *Proceedings of the National Academy of Sciences USA,* 96:7538–40.

Coolidge, F. L., L. L. Thede, and S. E. Young. 2002. The heritability of gender identity disorder in a child and adolescent twin sample. *Behavioral Genetics,* 32:251–57.

Cooperstock, M., and J. Campbell. 1996. Excess males in preterm birth: interactions with gestational age, race, and multiple birth. *Obstetrics and Gynecology,* 88:189–93.

Corbett, C., C. Hill, and A. St. Rose. 2008. *Where the Girls Are: The Facts About Gender Equity in Education.* Washington DC: AAUW Educational Foundation, http:// www.aauw.org/research/upload/whereGirlsAre.pdf.

Cossette, L., A. Pomerleau, G. Malcuit, and J. Kaczorowski. 1996. Emotional expressions of female and male infants in a social and a nonsocial context. *Sex Roles,* 35:693–709.

Cotman, C. W., and N. C. Berchtold. 2002. Exercise: a behavioral intervention to enhance brain health and plasticity. *Trends in Neuroscience,* 25:295–301.

Courchesne, E., R. Carper, and N. Akshoomoff. 2003. Evidence of brain overgrowth in the first year of life in autism. *JAMA,* 290:337–44.

Craig, M. C., P. C. Fletcher, E. M. Daly, et al. 2008. Physiological variation in estradiol and brain function: a functional magnetic resonance imaging study of verbal memory across the follicular phase of the menstrual cycle. *Hormones and Behavior*, 53:503–8.

Crick, N. R. 1997. Engagement in gender normative versus nonnormative forms of aggression: Links to social-psychological adjustment. *Developmental Psychology*, 33:610–17.

Crick, N. R., J. F. Casas, and M. Mosher. 1997. Relational and overt aggression in preschool. *Developmental Psychology*, 33:579–88.

Cronk, L. 2007. Boy or girl: gender preferences from a Darwinian point of view. *Reproductive BioMedicine Online*, 15 Suppl. 2:23–32.

Cruise, T. 2003. My struggle to read. *People* magazine (July 21).

Cushing, B. S., and K. M. Kramer. 2005. Mechanisms underlying epigenetic effects of early social experience: The role of neuropeptides and steroids. *Neuroscience and Biobehavioral Reviews*, 29:1089–105.

Datnow, A., L. Hubbard, and E. Woody. 2001. Is single gender schooling viable in the public sector? Lessons from California's pilot program. Final Report, http://www.oise.utoronto.ca/depts/tps/adatnow/final.pdf.

Davies, P. G., S. J. Spencer, D. M. Quinn, and R. Gerhardstein. 2002. Consuming images: How television commercials that elicit stereotype threat can restrain women academically and professionally. *Personality and Social Psychology Bulletin*, 28:1615–28.

Davis, M., and E. Emory. 1995. Sex differences in neonatal stress reactivity. *Child Development*, 66:14–27.

De Lisi, R., and J. L. Wolford. 2002. Improving children's mental rotation accuracy with computer game playing. *Journal of Genetic Psychology*, 163:272–82.

Dean, C. 2006. Dismissing "sexist opinions" about women's place in science. (Interview with Ben A. Barres, Stanford professor of neurobiology), *New York Times* (July 18).

Debaryshe, B. D. 1993. Joint picture-book reading correlates of early oral language skill. *Journal of Child Language*, 20:455–61.

DeCasper, A. J., and P. A. Prescott. 1984. Human newborns' perception of male voices: preference, discrimination, and reinforcing value. *Developmental Psychobiology*, 17:481–91.

Dee, T. S. 2005. Teachers and the gender gaps in student achievement. National Bureau of Economic Research, Working Paper 11660 (www.nber.org/papers/w11660).

del Mar Melero-Montes, M., and H. Jick. 2000. Hyperemesis gravidarum and the sex of the offspring. *Epidemiology*, 12:123–24.

Deming, D., and S. Dynarski. 2008. The lengthening of childhood. National Bureau of Economic Research, Working Paper 14124 (www.nber.org/papers/w14124).

Dempsey, P. J., G. C. Townsend, and L. C. Richards. 1999. Increased tooth crown size in females with twin brothers: Evidence for hormonal diffusion between human twins in utero. *American Journal of Human Biology*, 11:577–86.

DeNoon, D. 2006. TV implicated in autism rise. *WebMD Health* (October 18), http://www.webmd.com/brain/autism/news/20061019/tv-implicated-in-autism-rise.

Diamond, A. 1985. Development of the ability to use recall to guide action, as indicated by infants' performance on "A not B." *Child Development,* 56:868–83.

Diamond, A., W. S. Barnett, J. Thomas, and S. Munro. 2007. Preschool program improves cognitive control. *Science,* 318:1387–88.

Domes, G., M. Heinrichs, A. Michel, et al. 2007. Oxytocin improves "mind-reading" in humans. *Biological Psychiatry,* 61:731–33.

Doty, R. L., P. Shaman, S. L. Applebaum, et al. 1984. Smell identification ability: changes with age. *Science,* 226:1441–43.

Doyle, R. 1997. Female illiteracy worldwide. *Scientific American* (May), p. 20.

Duncan, G. J., C. J. Dowsett, A. Claessens, et al. 2007. School readiness and later achievement. *Developmental Psychology,* 43:1428–46.

Dunn, J., J. Brown, C. Slomkowski, et al. 1991. Young children's understanding of other people's feelings and beliefs: individual differences and their antecedents. *Child Development,* 62:1352–66.

Durston, S., H. E. Hulshoff Pol, B. J. Casey, et al. 2001. Anatomical MRI of the developing human brain: what have we learned? *Journal of the American Academy of Child and Adolescent Psychiatry,* 40:1012–20.

Eaton, W. O., and L. R. Enns. 1986. Sex differences in human motor activity level. *Psychological Bulletin,* 100:19–28.

Eccles, J., and J. Jacobs. 1986. Social forces shape math participation. *Signs: Journal of Women in Culture and Society,* 11:367–80.

Eckert, M. A., L. J. Lombardino, and C. M. Leonard. 2001. Planar asymmetry tips the phonological playground and environment raises the bar. *Child Development,* 72:988–1002.

Eden, G. F., K. M. Jones, K. Cappell, et al. 2004. Neural changes following remediation in adult developmental dyslexia. *Neuron,* 44:411–22.

Eden, G. F., and L. Moats. 2002. The role of neuroscience in the remediation of students with dyslexia. *Nature Neuroscience,* Suppl. 5:1080–84.

Edwards, D. A., K. Wetzel, and D. R. Wyner. 2006. Intercollegiate soccer: saliva cortisol and testosterone are elevated during competition, and testosterone is related to status and social connectedness with team mates. *Physiology and Behavior,* 87:135–43.

Efrat, Z., O. O. Akinfenwa, and K. H. Nicolaides. 1999. First-trimester determination of fetal gender by ultrasound. *Ultrasound in Obstetrics and Gynecology,* 13:305–7.

Eisenberg, N., and R. Lennon. 1983. Sex differences in empathy and related capacities. *Psychological Bulletin,* 94:100–31.

Eldredge, L., and A. Salamy. 1996. Functional auditory development in preterm and full term infants. *Early Human Development,* 45:215–28.

Eliot, L. 2000. *What's Going on in There? How the Brain and Mind Develop in the First Five Years of Life.* New York: Bantam.

Ellis, L., and S. Bonin. 2002. Social status and the secondary sex ratio: New evidence on a lingering controversy. *Social Biology*, 49:35–43.

Elmlinger, M. W., K. Werner, W. Henning, and D. P. Claus. 2005. Reference intervals for testosterone, androstenedione and SHBG levels in healthy females and males from birth until old age. *Clinical Laboratory*, 51:625–32.

Else-Quest, N. M., J. S. Hyde, H. H. Goldsmith, and C. Van Hulle. 2006. Gender differences in temperament: A meta-analysis. *Psychological Bulletin*, 132:33–72.

Embrey, L. C., and J. J. Fox. 1997. Gender differences in the investment decision-making process. *Association for Financial Counseling and Planning Education*, 8:33–39.

Entwisle, D. R., and D. P. Baker. 1983. Gender and young children's expectations for performance in arithmetic. *Developmental Psychology*, 19:200–9.

Eogan, M. A., M. P. Geary, M. P. O'Connell, and D. P. Keane. 2003. Effect of fetal sex on labour and delivery: retrospective review. *British Medical Journal*, 326:137.

Epting, L. K., and W. H. Overman. 1998. Sex-sensitive tasks in men and women: A search for performance fluctuations across the menstrual cycle. *Behavioral Neuroscience*, 112:1304–17.

Ernst, M., F. S. Maheu, E. Schroth, et al. 2007. Amygdala function in adolescents with congenital adrenal hyperplasia: a model for the study of early steroid abnormalities. *Neuropsychologia*, 45:2104–13.

Erwin, R. J., R. C. Gur, R. E. Gur, et al. 1992. Facial emotion discrimination: I. Task construction and behavioral findings in normal subjects. *Psychiatry Research*, 42:231–40.

Esgate, A., and M. Flynn. 2005. The brain-sex theory of occupational choice: a counterexample. *Perceptual and Motor Skills*, 100:25–37.

Espinosa, L. M. 2002. High-quality preschool: Why we need it and what it looks like. *National Institute for Early Education Research*, Issue 1 (November), http://nieer.org/resources/policybriefs/1.pdf.

Fabes, R. A., C. L. Martin, and L. D. Hanish. 2003. Young children's play qualities in same-, other-, and mixed-sex peer groups. *Child Development*, 74:921–32.

Fabes, R. A., S. A. Shepard, I. K. Guthrie, and C. L. Martin. 1997. Roles of temperamental arousal and gender-segregated play in young children's social adjustment. *Developmental Psychology*, 33:693–702.

Fagot, B. I., and M. D. Leinbach. 1993. Gender-role development in young children: From discrimination to labeling. *Developmental Review*, 13:205–24.

Fagot, B. I., M. D. Leinbach, and C. O'Boyle. 1992. Gender labeling, gender stereotyping, and parenting behaviors. *Developmental Psychology*, 28:225–30.

FairTest Examiner. 1999. Gender bias victory wins millions for females but National Merit Test remains biased (Spring), http://www.fairtest.org/gender-bias-victory-wins-millions-females-national.

Feingold, A., and R. Mazzella. 1998. Gender differences in body image are increasing. *Psychological Science*, 9:190–95.

Feng, J., I. Spence, and J. Pratt. 2007. Playing an action video game reduces gender differences in spatial cognition. *Psychological Science*, 18:850–55.

Fenson, L., P. S. Dale, J. S. Reznick, et al. 1994. Variability in early communicative development. *Monographs of the Society for Research in Child Development,* Serial No. 242, Vol. 59(5).

Fernandez, M. F., B. Olmos, A. Granada, et al. 2007. Human exposure to endocrine-disrupting chemicals and prenatal risk factors for cryptorchidism and hypospadias: A nested case-control study. *Environmental Health Perspectives,* 115 Suppl. 1:8–14.

Field, T. 1995. Massage therapy for infants and children. *Journal of Developmental and Behavioral Pediatrics:* 16:105–11.

Finegan, J-A. K., G. A. Niccols, and G. Sitarenios. 1992. Relations between prenatal testosterone levels and cognitive abilities at 4 years. *Developmental Psychology,* 28:1075–89.

Finkelstein, J. W., E. J. Susman, V. M. Chinchilli, et al. 1997. Estrogen or testosterone increases self-reported aggressive behaviors in hypogonadal adolescents. *Journal of Clinical Endocrinology and Metabolism,* 82:2433–38.

———. 1998. Effects of estrogen or testosterone on self-reported sexual responses and behaviors in hypogonadal adolescents. *Journal of Clinical Endocrinology and Metabolism,* 83:2281–85.

Fisher, S. E., and C. Francks. 2006. Genes, cognition and dyslexia: learning to read the genome. *Trends in Cognitive Sciences,* 10:250–57.

Fitch, R. H., and H. A. Bimonte. 2002. Hormones, brain, and behavior: Putative biological contributions to cognitive sex differences. In *Biology, Society and Behavior: The Development of Sex Differences in Cognition,* eds. A. McGillicuddy-De Lisi and R. De Lisi, 55–91. Westport CT: Ablex.

Fitch, R. H., and V. H. Denenberg. 1998. A role for ovarian hormones in sexual differentiation of the brain. *Behavioral and Brain Sciences,* 21:311–52.

Flachaire, E., C. Beney, A. Berthier, et al. 1990. Determination of reference values for serotonin concentration in platelets of healthy newborns, children, adults, and elderly subjects by HPLC with electrochemical detection. *Clinical Chemistry,* 36:2117–20.

Fleming, A. S., C. Corter, J. Stallings, M. Steiner. 2002. Testosterone and prolactin are associated with emotional responses to infant cries in new fathers. *Hormones and Behavior,* 42:399–413.

Flora, C. 2006. An Aspie in the City. *Psychology Today* (November/December), http://www.psychologytoday.com/articles/pto-20061103-000002.html.

Flory, C. D. 1935. Sex differences in skeletal development. *Child Development,* 6:205–12.

Francis, R. C. 2004. *Why Men Won't Ask for Directions: The Seductions of Sociobiology.* Princeton University Press.

Freeman, C. E. 2004. *Trends in Educational Equity of Girls & Women.* U.S. Department of Education, National Center for Education Statistics (NCES 2005–016), http://nces.ed.gov/pubs2005/2005016.pdf.

Freese, J., and B. Powell. 1999. Sociobiology, status, and parental investment in sons and daughters: Testing the Trivers-Willard hypothesis. *American Journal of Sociology*, 106:1704–43.

Frey, K. S., and D. N. Ruble. 1992. Gender constancy and the "cost" of sex-typed behavior: A test of the conflict hypothesis. *Child Development*, 28:714–21.

Fried-Buchalter, S. 1997. Fear of success, fear of failure, and the imposter phenomenon among male and female marketing managers. *Sex Roles*, 37:847–59.

Friederici, A. D., A. Pannekamp, C-J. Partsch, et al. 2008. Sex hormone testosterone affects language organization in the infant brain. *NeuroReport*, 19:283–86.

Fries, A. B. W., T. E. Ziegler, J. R. Kurian, et al. 2005. Early experience in humans is associated with changes in neuropeptides critical for regulating social behavior. *Proceedings of the National Academy of Sciences USA*, 102:17237–40.

Frings, L., K. Wagner, J. Unterrainer, et al. 2006. Gender-related differences in lateralization of hippocampal activation and cognitive strategy. *NeuroReport*, 17:417–21.

Frost, J. A., J. R. Binder, J. A. Springer, et al. 1999. Language processing is strongly left lateralized in both sexes. *Brain*, 122:199–208.

Fugger, E. F., S. H. Black, K. Keyvanfar, and J. D. Schulman. 1998. Births of normal daughters after MicroSort sperm separation and intrauterine insemination, in-vitro fertilization, or intracytoplasmic sperm injection. *Human Reproduction*, 13:2367–70.

Fujimoto, V. Y., and M. R. Soules. 1998. Normal and abnormal sexual development. In *Textbook of Reproductive Medicine*, 2nd Ed., eds. B. R. Carr and R. E. Blackwell, 113–35. Stamford CT: Appleton & Lange.

Furby, L., and M. Wilke. 1982. Some characteristics of infants' preferred toys. *Journal of Genetic Psychology*, 140:207–19.

Galsworthy, M. J., G. Dionne, P. S. Dale, and R. Plomin. 2000. Sex differences in early verbal and non-verbal cognitive development. *Developmental Science*, 3:206–15.

Garcia-Falgueras, A., and D. F. Swaab. 2008. A sex difference in the hypothalamic uncinate nucleus: relationship to gender identity. *Brain*, 131:3132–46.

Gaulin, S.J.C., and C. J. Robbins. 1991. Trivers-Willard effect in contemporary North American society. *American Journal of Physical Anthropology*, 85:61–69.

Geary, D. C. 1998. *Male, Female: The Evolution of Human Sex Differences*. Washington DC: American Psychological Association.

———. 2002. Sexual selection and sex differences in social cognition. In *Biology, Society, and Behavior: The Development in Sex Differences in Cognition*, eds. A. McGillicuddy-De Lisi and R. De Lisi, 23–53. Westport CT: Ablex.

Giedd, J. N., J. Blumenthal, N. O. Jeffries, F. X. Castellanos, et al. 1999a. Brain development during childhood and adolescence: a longitudinal MRI study. *Nature Neuroscience*, 2:861–63.

———. 1999b. Development of the human corpus callosum during childhood and adolescence: A longitudinal MRI study. *Progress in Neuro-Psychopharmacology and Biological Psychiatry*, 23:571–88.

Giedd, J. N., F. X. Castellanos, J. C. Rajapakse, et al. 1997. Sexual dimorphism of the developing human brain. *Progress in Neuro-Psychopharmacology and Biological Psychiatry*, 21:1185–1201.

Giedd, J. N., A. C. Vaituzis, S. D. Hamburger, et al. 1996. Quantitative MRI of the temporal lobe, amygdala, and hippocampus in normal human development: Ages 4–18 years. *Journal of Comparative Neurology*, 366:223–30.

Gilmore, J. H., W. Lin, M. W. Prastawa, et al. 2007. Regional gray matter growth, sexual dimorphism, and cerebral asymmetry in the neonatal brain. *Journal of Neuroscience*, 27:1255–60.

Ginsburg, A., G. Cooke, S. Leinwand, et al. 2005. *Reassessing U.S. International Mathematics Performance: New Findings from the 2003 TIMSS and PISA*. Washington DC: American Institutes for Research, http://www.air.org/news/documents/TIMSS_PISA%20math%20study.pdf.

Gleason, J. B., and R. Ely. 2002. Gender differences in language development. In *Biology, Society and Behavior: The Development of Sex Differences in Cognition*, eds. A. McGillicuddy-De Lisi and R. De Lisi, 127–54. Westport CT: Ablex.

Gneezy, U., and A. Rustichini. 2004. Gender and competition at a young age. *American Economic Review* (Papers and Proceedings of the 116th Annual Meeting), 94:377–81.

Goldstein, J. M., L. J. Seidman, N. J. Horton, et al. 2001. Normal sexual dimorphism of the adult human brain assessed by *in vivo* magnetic resonance imaging. *Cerebral Cortex*, 11:490–97.

Good, C., J. Aronson, and M. Inzlicht. 2003. Improving adolescents' standardized test performance: An intervention to reduce the effects of stereotype threat. *Applied Developmental Psychology*, 24:645–62.

Gooren, L. 2006. The biology of human psychosexual differentiation. *Hormones and Behavior*, 50:589–601.

Gordon, H. W., and P. A. Lee. 1993. No difference in cognitive performance between phases of the menstrual cycle. *Psychoneuroendocrinology*, 18:521–31.

Gore, D., and D. Roumagoux. 1983. Wait-time as a variable in sex-related differences during fourth-grade mathematics instruction. *Journal of Educational Research*, 76:273–75.

Graber, J. A., and A. C. Petersen. 1991. Cognitive changes at adolescence: Biological perspective. In *Brain Maturation and Cognitive Development: Comparative and Cross-Cultural Perspectives*, eds. K. R. Gibson and A. C. Petersen, 253–79. Hawthorne NY: Aldine de Gruyter.

Graber, J. A., and L. M. Sontag. 2006. Puberty and girls' sexuality: Why hormones are not the complete answer. *New Directions for Child and Adolescent Development*, 112:23–38.

Grattan, M. P., E. De Vos, J. Levy, and M. K. McClintock. 1992. Asymmetric action in the human newborn: sex differences in patterns of organization, *Child Development*, 63:273–89.

Gray, P. B., J. C. Parkin, and M. E. Samms-Vaughan. 2007. Hormonal correlates of human paternal interactions: A hospital-based investigation in urban Jamaica. *Hormones and Behavior*, 52:499–507.

Green, C. S., and D. Bavelier. 2003. Action video game modifies visual selective attention. *Nature*, 423:534–37.

Green, R. 1985. Gender identity in childhood and later sexual orientation: Follow-up of 78 males. *American Journal of Psychiatry*, 142:339–41.

Greenfeld, L. A., and T. L. Snell. 2000. Bureau of Justice Statistics. U.S. Department of Justice, http://www.ojp.usdoj.gov/bjs/pub/ascii/wo.txt.

Grigg, W., P. Donahue, and G. Dion. 2007. *The Nation's Report Card: 12th-Grade Reading and Mathematics 2005*. U.S. Department of Education, National Center for Education Statistics (NCES 2007–468), http://nces.ed.gov/nationsreportcard/pdf/main2005/2007468.pdf.

Grigg, W. S., M. A. Lauko, and D. M. Brockway. 2006. *The Nation's Report Card: Science 2005*. U.S. Department of Education, National Center for Education Statistics (NCES 2006–466), http://nces.ed.gov/nationsreportcard/pdf/main2005/2006466.pdf.

Grimshaw, G. M., G. Sitarenios, and J. A. Finegan. 1995. Mental rotation at 7 years: relations with prenatal testosterone levels and spatial play experiences. *Brain and Cognition*, 29:85–100.

Grinspan, D., A. Hemphill, and S. Nowicki. 2003. Improving the ability of elementary school-age children to identify emotion in facial expression. *Journal of Genetic Psychology*, 164:88–100.

Grön, G., A. P. Wunderlich, M. Spitzer, et al. 2000. Brain activation during human navigation: gender-different neural networks as substrate of performance. *Nature Neuroscience*, 3:404–8.

Gualtieri, T., and R. E. Hicks. 1985. An immunoreactive theory of selective male affliction. *Behavioral and Brain Sciences*, 8:427–41.

Guinsburg, R., C. A. Peres, M.F.B. de Almedia, et al. 2000. Differences in pain expression between male and female newborn infants. *Pain*, 85:127–33.

Gur, R. C., F. Gunning-Dixon, W. B. Bilker, and R. E. Gur. 2002. Sex differences in temporo-limbic and frontal brain volumes of healthy adults. *Cerebral Cortex*, 12:998–1003.

Gurian, M. 2001. *Boys and Girls Learn Differently!* San Francisco: Jossey-Bass.

Gurian, M., and K. Stevens. 2005. *The Minds of Boys, Saving Our Sons from Falling Behind in School and Life*. San Francisco: Jossey-Bass.

Gwiazda, J., J. Bauer, and R. Held. 1989. Binocular function in human infants: correlation of stereoptic and fusion-rivalry discriminations. *Journal of Pediatric Ophthalmology and Strabismus*, 26:128–32.

Haag, P. 2000. K–12 single-sex education: What does the research say? *ERIC Digest*, ED444758.

Hagger, C., and J. Bachevalier. 1991. Visual habit formation in 3-month-old monkeys (*Macaca mulatta*): Reversal of sex difference following neonatal manipulations of androgens. *Behavioral Brain Research*, 45:57–63.

Halari, R., M. Hines, V. Kumari, et al. 2005. Sex differences and individual differences in cognitive performance and their relationship to endogenous gonadal hormones and gonadotropins. *Behavioral Neuroscience* 119:104–17.

Hall, J. A. 1978. Gender effects in decoding nonverbal cues. *Psychological Bulletin,* 85:845–57.

Hall, J. A., and J. Halberstadt. 1986. Smiling and gazing. In *The Psychology of Gender: Advances through Meta-analysis,* eds. J. S. Hyde, M. C. Linn, 136–58. Baltimore: Johns Hopkins University Press.

Halpern, C. T., J. R. Udry, B. Campbell, and C. Suchindran. 1993. Relationships between aggression and pubertal increases in testosterone: a panel analysis of adolescent males. *Social Biology,* 40:8–24.

Halpern, C. T., J. R. Udry, and C. Suchindran. 1997. Testosterone predicts initiation of coitus in adolescent females. *Psychosomatic Medicine,* 59:161–71.

———. 1998. Monthly measures of salivary testosterone predict sexual activity of adolescent males. *Archives of Sexual Behavior,* 27:445–65.

Halpern, D. F. 1989. The disappearance of cognitive gender differences: What you see depends on where you look. *American Psychologist,* 44:1156–58.

———. 1997. Sex differences in intelligence. Implications for education. *American Psychologist,* 52:1091–102.

———. 2000. *Sex Differences in Cognitive Abilities,* 3rd Ed., Mahwah NJ: Lawrence Erlbaum.

———. 2002. Sex differences in achievement scores: Can we design assessments that are fair, meaningful, and valid for girls and boys? *Issues in Education,* 8:2–21.

Halpern, D. F., C. P. Benbow, D. C. Geary, et al. 2007. The science of sex differences in science and mathematics. *Psychological Science in the Public Interest,* 8:1–51.

Halpern, D. F., and U. Tan. 2001. Stereotypes and steroids: Using a psychobiosocial model to understand cognitive sex differences. *Brain and Cognition,* 45:392–414.

Halpern, D. F., J. Wai, and A. Saw. 2005. A psychobiosocial model: Why females are sometimes greater than and sometimes less than males in math achievement. In *Gender Differences in Mathematics: An Integrative Psychological Approach,* eds. A. M. Gallagher and J. C. Kaufman, 3–72. Cambridge University Press.

Hamann, S. 2005. Sex differences in the responses of the human amygdala. *Neuroscientist,* 11:288–93.

Hampson, E. 1990. Variations in sex-related cognitive abilities across the menstrual cycle. *Brain and Cognition,* 14:26–43.

Hampson, E., C. L. Ellis, and C. M. Tenk. 2008. On the relation between 2D:4D and sex-dimorphic personality traits. *Archives of Sexual Behavior,* 37:133–44.

Han, S., and G. Northoff. 2008. Culture-sensitive neural substrates of human cognition: a transcultural neuroimaging approach. *Nature Reviews Neuroscience,* 9:646–54.

Han, T. L., S. P. Flaherty, J. H. Ford, and C. D. Matthews. 1993. Detection of X- and Y-bearing human spermatozoa after motile sperm isolation swim-up. *Fertility and Sterility,* 60:1046–51.

Handelsman, J., N. Cantor, M. Carnes, et al. 2005. More women in science. *Science,* 309:1190–91.

Harasty, J., K. L. Double, G. M. Halliday, et al. 1997. Language-associated cortical regions are proportionally larger in the female brain. *Archives of Neurology,* 54:171–76.

Harker, R. 2000. Achievement, gender and the single-sex/coed debate. *British Journal of Sociology of Education,* 21:203–18.

Harlaar, N., F. M. Spinath, P. S. Dale, and R. Plomin. 2005. Genetic influences on early word recognition abilities and disabilities: a study of 7-year-old twins. *Journal of Child Psychology and Psychiatry,* 46:373–84.

Harlow, B. L., F. D. Frigoletto, D. W. Cramer, et al. 1995. Epidemiologic predictors of cesarean section in nulliparous patients at low risk. *American Journal of Obstetrics and Gynecology,* 172:156–62.

Hart, B., and T. R. Risley. 1992. American parenting of language-learning children: Persisting differences in family-child interactions observed in natural home environments. *Developmental Psychology,* 28:1096–105.

———. 1995. *Meaningful Differences in the Everyday Experience of Young American Children.* Baltimore: Paul H. Brooks.

Hassett, J. M., E. R. Siebert, and K. Wallen. 2008. Sex differences in rhesus monkey toy preferences parallel those of children. *Hormones and Behavior,* 54:359–64.

Hausmann, M., D. Slabbekoorn, S.H.M. Van Goozen, et al. 2004. Sex hormones affect spatial abilities during the menstrual cycle. *Behavioral Neuroscience,* 114:1245–50.

Haviland, J. J., and C. Z. Malatesta. 1981. The development of sex differences in nonverbal signals: fallacies, fact, and fantasies. In *Gender and Nonverbal Behavior,* eds. C. Mayo and N. M. Henley, 183–208. New York: Springer-Verlag.

Hecht, H., and D. R. Proffitt. 1995. The price of expertise: Effects of experience on the water-level task. *Psychological Science,* 6:90–95.

Hedges, L. V., and A. Nowell. 1995. Sex differences in mental test scores, variability, and numbers of high-scoring individuals. *Science,* 269:41–45.

Held, R. 1989. Development of cortically mediated visual process in human infants. In *Neurobiology of Early Infant Behavior,* eds. C. von Euler, H. Forssberg, and H. Lagercrantz, 155–64. New York: Stockton.

Held, R., J. Bauer, and J. Gwiazda. 1988. Age of onset of binocularity correlates with level of plasma testosterone in infant males. *Investigative Ophthalmology and Visual Science,* Suppl. 29:60.

Helt, M., E. Kelley, M. Kinsbourne, et al. 2008. Can children with autism recover? If so, how? *Neuropsychology Review,* 18:339–66.

Henderson, B. A., and S. A. Berenbaum. 1997. Sex-typed play in opposite-sex twins. *Developmental Psychobiology,* 31:115–23.

Hepper, P. G., E. A. Shannon, and J. C. Dornan. 1997. Sex differences in fetal mouth movements. *Lancet,* 350:1820.

Herdt, G., and M. McClintock. 2000. The magical age of 10. *Archives of Sexual Behavior,* 29:587–606.

Herman, R. A., M. A. Measday, and K. Wallen. 2003. Sex differences in interest in infants in juvenile rhesus monkeys: relationship to prenatal androgen. *Hormones and Behavior*, 43:573–83.

Herman, R. A., and K. Wallen. 2007. Cognitive performance in rhesus monkeys varies by sex and prenatal androgen exposure. *Hormones and Behavior*, 51: 496–507.

Hespos, S. J., and P. Rochat. 1997. Dynamic mental representation in infancy. *Cognition*, 64:153–88.

Hilbert, D. 2006. More on rats and men and women. *Language Log* (August 22), http://itre.cis.upenn.edu/~myl/languagelog/archives/003488.html.

Hines, M. 2000. Gonadal hormones and sexual differentiation of human behavior: Effects on psychosocial and cognitive development. In *Sexual Differentiation of the Brain*, ed. A. Matsumoto, 257–78. Boca Raton FL: CRC Press.

———. 2004. *Brain Gender*. Oxford University Press.

Hines, M., S. F. Ahmed, and I. A. Hughes. 2003. Psychological outcomes and gender-related development in complete androgen insensitivity syndrome. *Archives of Sexual Behavior*, 32:93–101.

Hines, M., S. Golombok, J. Rust, et al. 2002. Testosterone during pregnancy and gender role behavior of preschool children: a longitudinal, population study. *Child Development*, 73:1678–87.

Hiscock, M., R. Inch, C. Jacek, et al. 1994. Is there a sex difference in human laterality? I. An exhaustive survey of auditory laterality studies from six neuropsychology journals. *Journal of Clinical and Experimental Neuropsychology*, 16:423–35.

Hittleman, J. H., and R. Dickes. 1979. Sex differences in neonatal eye contact time. *Merrill-Palmer Quarterly*, 25:171–84.

Hojat, M., J. S. Gonnella, T. J. Nasca, et al. 2002. Physician empathy: definition, components, measurement, and relationship to gender and specialty. *American Journal of Psychiatry*, 159:1563–69.

Holden, C. 2005. Sex and the suffering brain. *Science*, 308:1574–77.

Hopf, D., and C. Hatzichristou. 1999. Teacher gender-related influence in Greek schools. *British Journal of Educational Psychology*, 69:1–18.

Horwood, L. J., and D. M. Fergusson. 1998. Breastfeeding and later cognitive and academic outcomes. *Pediatrics*, 101:e9-e15.

Howe, A. C. 1996. Adolescents' motivation, behavior and achievement in science. *Research Matters—to the Science Teacher*, No. 9603 (December 6), http://www.narst.org/publications/research/Adolescent.cfm.

Hrdy, S. B. 1999. *Mother Nature: A History of Mothers, Infants, and Natural Selection*. New York: Pantheon.

Huesmann, L. R., J. Moise-Titus, C-L. Podolski, and L. D. Eron. 2003. Longitudinal relations between children's exposure to TV violence and their aggressive and violent behavior in young adulthood: 1977–1992. *Developmental Psychology*, 39:201–21.

Huntsinger, C. S., and P. E. Jose. 1995. Chinese American and Caucasian American family interaction patterns in spatial rotation puzzle solutions. *Merrill-Palmer Quarterly*, 41:471–96.

Huttenlocher, J., W. Haight, A. Bryk, et al. 1991. Early vocabulary growth: relation to language input and gender. *Developmental Psychology*, 27:236–48.

Hyde, J. S. 1984. How large are gender differences in aggression? A developmental meta-analysis. *Developmental Psychology*, 20:722–36.

———. 2005. The gender similarities hypothesis. *American Psychologist*, 60:581–92.

Hyde, J. S., S. M. Lindberg, M. C. Linn, et al. 2008. Gender similarities characterize math performance. *Science*, 321:494–95.

Hyde, J. S., and M. C. Linn. 1988. Gender differences in verbal ability: A meta-analysis. *Psychological Bulletin*, 104:53–69.

———. 2006. Diversity: Gender similarities in mathematics and science. *Science*, 314:599–600.

Hyde, J. S., and N. M. McKinley. 1997. Gender differences in cognition: Results from meta-analysis. In *Gender Differences in Human Cognition*, eds. P. J. Caplan, M. Crawford, J. S. Hyde, and J.T.E. Richardson, 30–51. New York: Oxford University Press.

Icon, E. 2005. Men in nursing today. *Working Nurse* (May 6).

Iervolino, A. C., M. Hines, S. E. Golombok, et al. 2005. Genetic and environmental influences on sex-typed behavior during the preschool years. *Child Development*, 76:826–40.

Ingemarsson, I. 2003. Gender aspects of preterm birth. *BJOG: An International Journal of Obstetrics and Gynaecology*, 110 Suppl. 20:34–38.

Jacklin, C. N., E. E. Maccoby, and C. H. Doering. 1983. Neonatal sex-steroid hormones and timidity in 6–18-month-old boys and girls. *Developmental Psychobiology*, 16:163–68.

Jacklin, C. N., E. E. Maccoby, C. H. Doering, and D. R. King. 1984. Neonatal sex-steroid hormones and muscular strength of boys and girls in the first three years. *Developmental Psychobiology*, 17:301–10.

Jacklin, C. N., K. T. Wilcox, and E. E. Maccoby. 1988. Neonatal sex-steroid hormones and cognitive abilities at six years. *Developmental Psychobiology*, 21:567–74.

Jacobs, B., M. Schall, and A. B. Scheibel. 1993. A quantitative dendritic analysis of Wernicke's area in humans. II. Gender, hemispheric, and environmental factors. *Journal of Comparative Neurology*, 327:97–111.

Jacobs, J. E., P. Davis-Kean, M. Bleeker, et al. 2005. "I can, but I don't want to." The impact of parents, interests, and activities on gender differences in math. In *Gender Differences in Mathematics: An Integrative Psychological Approach*, eds. A. M. Gallagher and J. C. Kaufman, 246–63. Cambridge University Press.

Jacobs, L. F., S. J. Gaulin, D. F. Sherry, and G. E. Hoffman. 1990. Evolution of spatial cognition: sex-specific patterns of spatial behavior predict hippocampal size. *Proceedings of the National Academy of Sciences USA*, 87:6349–52.

Jaffee, S., and J. S. Hyde. 2000. Gender differences in moral orientation: A meta-analysis. *Psychological Bulletin,* 126:703–26.

Jager, G., and A. Postma. 2003. On the hemispheric specialization for categorical and coordinate spatial relations: a review of the current evidence. *Neuropsychologia,* 41:504–15.

Janowsky, J. S., S. K. Oviatt, and E. S. Orwoll. 1994. Testosterone influences spatial cognition in older men. *Behavioral Neuroscience,* 108:325–32.

Jans, L. A., W. J. Riedel, C. R. Markus, and A. Blokland. 2006. Serotonergic vulnerability and depression: assumptions, experimental evidence and implications. *Molecular Psychiatry,* 12:522–43.

Jardine, R., and N. G. Martin. 1984. No evidence for sex-linked or sex-limited gene expression influencing spatial orientation. *Behavioral Genetics,* 14:345–53.

Jing, H., R. T. Pivik, J. M. Gilchrist, and T. M. Badger. 2008. No difference indicated in electroencephalographic power spectral analysis in 3- and 6-month-old infants fed soy- or milk-based formula. *Maternal and Child Nutrition,* 4:136–45.

Johnson, C. Y. 2005. Do genes play a role in science and math gender gap? *Boston Globe* (January 25).

Johnston, L. C., H. M. Feldman, J. L. Paradise, et al. 2004. Tympanic membrane abnormalities and hearing levels at the ages of 5 and 6 years in relation to persistent otitis media and tympanostomy tube insertion in the first 3 years of life: A prospective study incorporating a randomized clinical trial. *Pediatrics,* 114: e58-e67.

Joint LWPES/ESPE CAH Working Group. 2002. Consensus statement on 21-hydroxylase deficiency from the Lawson Wilkins Pediatric Endocrine Society and the European Society for Paediatric Endocrinology. *Journal of Clinical Endocrinology & Metabolism,* 87:4048–53.

Jones, C. M., V. A. Braithwaite, and S. D. Healy. 2003. The evolution of sex differences in spatial ability. *Behavioral Neuroscience,* 117:403–11.

Joseph, R. 2000. The evolution of sex differences in language, sexuality, and visual-spatial skills. *Archives of Sexual Behavior,* 29:35–66.

Kagan, J., J. S. Reznick, and N. Snidman. 1987. Temperamental variation in response to the unfamiliar. In *Perinatal Development: A Psychobiological Perspective,* eds. N. A. Krasnegor, E. M. Blass, M. A. Hofer, and W. P. Smotherman, 421–40. Orlando: Academic Press.

Kaiser, J. 2005. An earlier look at baby's genes. *Science,* 309:1476–78.

Kaminer, W. 1998. The trouble with single-sex schools. *Atlantic Monthly* (April), http://www.theatlantic.com/doc/prem/199804/single-sex.

Kane, C. K., and H. Loeblich. 2003. Physician income: The decade in review. In *Physician Socioeconomic Statistics,* eds. J. D. Wassenaar and S. L. Thran, 5–11. American Medical Association, https://catalog.ama-assn.org/MEDIA/ProductCatalog/m350028_PSStats_2003.pdf.

Kaplan, K. 2008. Accuracy of gender test kits in question. *Los Angeles Times* (February 28).

Karraker, K. H. 1995. Parents' gender-stereotyped perceptions of newborns: the eye of the beholder, revisited. *Sex Roles*, 33:687–701.

Katz, E. 2006. Gender influences law firm hiring, promotion, sociology professor says. *Virginia Law* (February 17), http://www.law.virginia.edu/home2002/html/news/2006_spr/gorman.htm.

Katz, P. A., and S. Boswell. 1986. Flexibility and traditionality in children's gender roles. *Genetic, Social and General Psychology Monographs*, 112:103–47.

Kelly, S. J., N. L. Ostrowski, and M. A. Wilson. 1999. Gender differences in brain and behavior: hormonal and neural bases. *Pharmacology, Biochemistry and Behavior*, 64:655–64.

Kendler, K. S., M. Gatz, C. O. Gardner, and N. L. Pedersen. 2006. A Swedish national twin study of lifetime major depression. *American Journal of Psychiatry*, 163:109–14.

Kent, R. D. 1984. Psychobiology of speech development: coemergence of language and a movement system. *American Journal of Physiology*, 246:R888–94.

Kerschbaum, H. H., M. Ruemer, S. Weisshuhn, and W. Klimesch. 2006. Gender-dependent differences in sensation seeking and social interaction are correlated with saliva testosterone titre in adolescents. *Neuroendocrinology Letters*, 27:315–20.

Kessel, C. 2006. Statement from Cathy Kessel, president-elect of the Association for Women in Mathematics, to be read into the public record on June 29, 2006 at the second meeting of the National Mathematics Advisory Panel, http://www.awm-math.org/benbow_petition/statement_UNC.html.

Kimura, D. 2000. *Sex and Cognition*. Cambridge MA: MIT Press.

Kindlon, D., and M. Thompson. 1999. *Raising Cain: Protecting the Emotional Lives of Boys*. New York: Ballantine.

Kirsch, I., J. de Jong, D. Lafontaine, et al. 2002. *Reading for Change: Performance and Engagement across Countries. Results from PISA 2000*. Organization of Economic Cooperation and Development, http://www.oecd.org/dataoecd/43/54/33690904.pdf.

Kivlighan, K. T., D. A. Granger, and A. Booth. 2005. Gender differences in testosterone and cortisol response to competition. *Psychoneuroendocrinology*, 30:58–71.

Kleinfeld, J. 1998. The myth that schools shortchange girls: Social science in the service of deception. Washington DC: Women's Freedom Network (ERIC Document Reproduction Service No. ED423210).

Kling, K. C., J. S. Hyde, C. J. Showers, and B. N. Buswell. 1999. Gender differences in self-esteem: a meta-analysis. *Psychological Bulletin*, 125:470–500.

Klinga, K., E. Bek, and B. Runnebaum. 1978. Maternal peripheral testosterone levels during the first half of pregnancy. *American Journal of Obstetrics and Gynecology*, 131:60–62.

Knafo, A., A. C. Iervolino, and R. Plomin. 2005. Masculine girls and feminine boys: genetic and environmental contributions to atypical gender development in early childhood. *Journal of Personality and Social Psychology*, 88:400–12.

Knaus, T. A., A. M. Bollich, D. M. Corey, et al. 2004. Sex-linked differences in the anatomy of the perisylvian language cortex: a volumetric MRI study of gray matter volumes. *Neuropsychology,* 18:738–47.

Knickmeyer, R., S. Baron-Cohen, B. A. Fane, et al. 2006a. Androgens and autistic traits: A study of individuals with congenital adrenal hyperplasia. *Hormones and Behavior,* 50:148–53.

Knickmeyer, R., S. Baron-Cohen, P. Raggatt, et al. 2006b. Fetal testosterone and empathy. *Hormones and Behavior,* 49:282–92.

Knickmeyer, R., S. Baron-Cohen, P. Raggatt, and K. Taylor. 2005a. Foetal testosterone, social relationships and restricted interests in children. *Journal of Child Psychology and Psychiatry,* 46:198–210.

Knickmeyer, R. C., S. Wheelwright, K. Taylor, et al. 2005b. Gender-typed play and amniotic testosterone. *Developmental Psychology,* 41:517–28.

Kochanska, G., K. Murray, T. Y. Jacques, et al. 1996. Inhibitory control in young children and its role in emerging internalization. *Child Development,* 67:490–507.

Koenig, A. M., and A. H. Eagly. 2005. Stereotype threat in men on a test of social sensitivity. *Sex Roles,* 52:489–96.

Konrad, C., A. Engelien, and S. Schöning. 2008. The functional anatomy of semantic retrieval is influenced by gender, menstrual cycle, and sex hormones. *Journal of Neural Transmission,* 115:1327–37.

Koshi, R., T. Koshi, L. Jeyaseelan, and S. Vettivel. 1997. Morphology of the corpus callosum in human fetuses. *Clinical Anatomy,* 10:22–26.

Kouki, T., M. Kishitake, M. Okamoto, et al. 2003. Effects of neonatal treatment with phytoestrogens, genistein and daidzein, on sex difference in female rat brain function: estrous cycle and lordosis. *Hormones and Behavior,* 44:140–45.

Kovas, Y., C. M. Haworth, S. A. Petrill, and R. Plomin. 2007. Mathematical ability of 10-year-old boys and girls: genetic and environmental etiology of typical and low performance. *Journal of Learning Disabilities,* 40:554–67.

Kovas, Y., M. E. Hayiou-Thomas, B. Oliver, et al. 2005. Genetic influences in different aspects of language development: The etiology of language skills in 4.5-year-old twins. *Child Development,* 76:632–51.

Koziel, S., and S. J. Ulijaszek. 1991. Waiting for Trivers and Willard: Do the rich really favor sons? *American Journal of Physical Anthropology,* 115:71–79.

Kranz, G. 2001. Moms and dads thank heaven, and Microsort, for little girls. *Small Times* (August 21), http://www.smalltimes.com.

Kuehner, C. 2003. Gender differences in unipolar depression: an update of epidemiological findings and possible explanations. *Acta Psychiatrica Scandinavica,* 108:163–74.

Kuhl, P. K., J. E. Andruski, I. A. Chistovich, et al. 1997. Cross-language analysis of phonetic units in language addressed to infants. *Science,* 277:684–86.

Kuhl, P. K., F. M. Tsao, and H. M. Liu. 2003. Foreign-language experience in infancy: effects of short-term exposure and social interaction on phonetic learning. *Proceedings of the National Academy of Sciences USA,* 100:9096–101.

Lacreuse, A. 2006. Effects of ovarian hormones on cognitive function in nonhuman primates. *Neuroscience*, 138:859–67.

LaFrance, M. 1991. School for scandal: different educational experiences for females and males. *Gender and Education*, 3:3–13.

LaFrance, M., M. A. Hecht, and E. L. Paluck. 2003. The contingent smile: A meta-analysis of sex differences in smiling. *Psychological Bulletin*, 129:305–34.

LaFraniere, S. 2007. In Mauritania, seeking to end an overfed ideal. *New York Times* (July 4), http://www.nytimes.com/2007/07/04/world/africa/04mauritania.html.

Lagerspetz, K.M.J. 1999. Theories of male and female aggression. *Behavioral and Brain Sciences*, 22:229–30.

Landa, R. 2007. Early communication development and intervention for children with autism. *Mental Retardation and Developmental Disabilities Research Reviews*, 13:16–25.

Lang, U. E., M. Bajbouj, T. Sander, and J. Gallinat. 2007. Gender-dependent association of the functional catechol-O-methyltransferase Val158Met genotype with sensation seeking personality trait. *Neuropsychopharmacology*, 32:1950–55.

Lavelli, M., and A. Fogel. 2002. Developmental changes in mother-infant face-to-face communication: birth to 3 months. *Developmental Psychology*, 38:288–305.

Lavelli, M., and M. Poli. 1998. Early mother-infant interaction during breast- and bottle-feeding. *Infant Behavior and Development*, 21:667–84.

Lazarus, G. M. 2001. Gender-specific medicine in pediatrics. *Journal of Gender-Specific Medicine*, 4:50–53.

Leader, L. R., P. Baillie, B. Martin, and E. Vermeulen. 1982. The assessment and significance of habituation to a repeated stimulus by the human fetus. *Early Human Development*, 7:211–19.

Leaper, C., K. J. Anderson, and P. Sanders. 1998. Moderators of gender effects on parents' talk to their children: A meta-analysis. *Developmental Psychology*, 34:3–27.

Lee, J., W. Grigg, and P. Donahue. 2007. *The Nation's Report Card: Reading 2007*. U.S. Department of Education, National Center for Education Statistics, Institute of Education Sciences (NCES 2007–496), http://nces.ed.gov/nationsreportcard/pdf/main2007/2007496.pdf.

Lee, S. 2000. Eating disorders are becoming more common in the East too. *British Medical Journal*, 321:1023.

Leeb, R. T. 2004. Here's looking at you, kid! A longitudinal study of perceived gender differences in mutual gaze behavior in young infants. *Sex Roles*, 50:1–14.

Lenroot, R. K., N. Gogtay, D. K. Greenstein, et al. 2007. Sexual dimorphism of brain developmental trajectories during childhood and adolescence. *NeuroImage*, 36:1065–73.

Letendre, J. 2007. "Sugar and spice but not always nice": Gender socialization and its impact on development and maintenance of aggression in adolescent girls. *Child and Adolescent Social Work Journal*, 24:353–68.

LeVay, S. 1991. A difference in hypothalamic structure between heterosexual and homosexual men. *Science*, 253:1034–37.

———. 1996. *Queer Science: The Use and Abuse of Research into Homosexuality.* Cambridge MA: MIT Press.

LeVay, S., and S. Valente. 2002. *Human Sexuality.* Sunderland MA: Sinauer.

Leveroni, C., and S. A. Berenbaum. 1998. Early androgen effects on interest in infants: evidence from children with congenital adrenal hyperplasia. *Developmental Neuropsychology,* 14:321–40.

Levine, S. C., J. Huttenlocher, A. Taylor, and A. Langrock. 1999. Early sex differences in spatial skill. *Developmental Psychology,* 35:940–49.

Levine, S. C., M. Vasilyeva, S. F. Lourenco, et al. 2005. Socioeconomic status modifies the sex difference in spatial skills. *Psychological Science,* 16:841–45.

Levy, B. A., Z. Gong, S. Hessels, et al. 2006. Understanding print: Early reading development and the contributions of home literacy experiences. *Journal of Experimental Child Psychology,* 93:63–93.

Lewin, C., and A. Herlitz. 2002. Sex differences in face recognition—women's faces make the difference. *Brain and Cognition,* 50:121–28.

Li, C. 2000. Instruction effect and developmental levels: A study on water-level task with Chinese children ages 9–17. *Contemporary Education Psychology,* 25:488–98.

Li, C., R. L. Nuttall, and S. Zhao. 1999. The effect of writing Chinese characters on success on the water-level task. *Journal of Cross-Cultural Psychology,* 30:91–105.

Liben, L. S., E. J. Susman, J. W. Finkelstein, et al. 2002. The effects of sex steroids on spatial performance: A review and an experimental clinical investigation. *Developmental Psychology,* 38:236–53.

Liberman, M. 2006a. Leonard Sax on hearing. *Language Log* (August 22), http://itre.cis.upenn.edu/~myl/languagelog/archives/003487.html.

———. 2006b. Girls and boys and classroom noise. *Language Log* (September 9), http://158.130.17.5/~myl/languagelog/archives/003561.html.

———. 2006c. Sex on the brain. *Boston Globe* (September 24), http://www.boston.com/news/globe/ideas/articles/2006/09/24/sex_on_the_brain/.

———. 2008. Liberman on Sax on Liberman on Sax on hearing (May 19), http://languagelog.ldc.upenn.edu/nll/?p=171.

Lieberman, E., J. M. Lang, A. P. Cohen, et al. 1997. The association of fetal sex with the rate of cesarean section. *American Journal of Obstetrics and Gynecology,* 176:667–71.

Liederman, J., L. Kantrowitz, and K. Flannery. 2005. Male vulnerability to reading disability is not likely to be a myth: A call for new data. *Journal of Learning Disabilities,* 38:109–29.

Lillard, A. S. 2005. *Montessori: The Science Behind the Genius.* New York: Oxford University Press.

Linakis, J. G., S. Amanullah, and M. J. Mello. 2006. Emergency department visits for injury in school-aged children in the United States: A comparison of nonfatal injuries occurring within and outside of the school environment. *Academic Emergency Medicine,* 13:567–70.

Liotti, M., S. R. Pliszka, R. Perez, et al. 2007. Electrophysiological correlates of response inhibition in children and adolescents with ADHD: influence of gender, age, and previous treatment history. *Psychophysiology*, 44:936–48.

Lippa, R. 1998. Gender-related individual differences and National Merit Test Performance: Girls who are "masculine" and boys who are "feminine" tend to do better. In *Males, Females, and Behavior: Toward a Biological Understanding*, ed. L. Ellis and L. Ebertz 177–93. Westport CT: Praeger.

Lippa, R. A. 2002. *Gender, Nature and Nurture*. Mahwah NJ: Lawrence Erlbaum.

———. 2003. Are 2D:4D finger-length ratios related to sexual orientation? Yes for men, no for women. *Journal of Personality and Social Psychology*, 85:179–88.

Lobel, T. E., and J. Menashri. 1993. Relations of conceptions of gender-role transgressions and gender constancy to gender-typed toy preferences. *Developmental Psychology*, 29:150–55.

Loehlin, J. C., and N. G. Martin. 2000. Dimensions of psychological masculinity-femininity in adult twins from opposite-sex and same-sex pairs. *Behavior Genetics*, 30:19–28.

Luciano, M., M. J. Wright, G. M. Geffen, et al. 2004. Multivariate genetic analysis of cognitive abilities in an adolescent twin sample. *Australian Journal of Psychology*, 56:79–88.

Lueptow, L. B. 2005. Increasing differentiation of women and men: gender trait analysis 1974–1997. *Psychological Reports* 97:277–87.

Lum, J., J. Enns, and J. Pratt. 2002. Visual orienting in college athletes: Explorations of athlete type and gender. *Research Quarterly for Exercise and Sport*, 73:156–67.

Lummaa, V., J. E. Pettay, and A. F. Russell. 2007. Male twins reduce fitness of female co-twins in humans. *Proceedings of the National Academy of Sciences USA*, 104:10915–20.

Lummis, M., and H. W. Stevenson. 1990. Gender differences in beliefs and achievement: A cross-cultural study. *Developmental Psychology*, 26:254–63.

Lundqvist, C., and M. Hafström. 1999. Non-nutritive sucking in full-term and pre-term infants studied at term conceptional age. *Acta Paediatrica*, 88:1287–89.

Lundqvist, C., and K-G. Sabel. 2000. Brief report: The Brazelton Neonatal Behavioral Assessment Scale detects differences among newborn infants of optimal health. *Journal of Pediatric Psychology*, 25:577–82.

Lutchmaya, S., S. Baron-Cohen, and P. Raggatt. 2002a. Foetal testosterone and eye contact in 12-month-old human infants. *Infant Behavior and Development*, 25:327–35.

———. 2002b. Foetal testosterone and vocabulary size in 18- and 24-month-old infants. *Infant Behavior and Development*, 24:418–24.

Lutchmaya, S., S. Baron-Cohen, P. Raggatt, et al. 2004. 2nd to 4th digit ratios, fetal testosterone and estradiol. *Early Human Development*, 77:23–28.

Lytton, H., and D. M. Romney. 1991. Parents' differential socialization of boys and girls: A meta-analysis. *Psychological Bulletin*, 109:267–96.

Maccoby, E. E. 1998. *The Two Sexes: Growing Up Apart, Coming Together*. Cambridge MA: Harvard University Press.

Maccoby, E. E., and C. N. Jacklin. 1974. *The Psychology of Sex Differences*. Stanford University Press.

MacPhee, A. R., and J.J.W. Andrews. 2006. Risk factors for depression in early adolescence. *Adolescence,* 41:435–66.

Maddox, B. 2002. *Rosalind Franklin: The Dark Lady of DNA*. New York: Harper-Collins.

Mage, D. T., and E. M. Donner. 2004. The fifty percent male excess of infant respiratory mortality. *Acta Paediatrica,* 93:1210–15.

Maguire, E. A., D. G. Gadian, I. S. Johnsrude, et al. 2000. Navigation-related structural change in the hippocampi of taxi drivers. *Proceedings of the National Academy of Sciences USA,* 97:4398–403.

Maines, R. 2007. Why women become veterinarians but not engineers. *Chronicle of Higher Education* (May 25), 53:57.

Maki, P. M. 2006. Hormone therapy and cognitive function: Is there a critical period for benefit? *Neuroscience,* 138:1027–30.

Maki, P. M., J. B. Rich, and R. S. Rosenbaum. 2002. Implicit memory varies across the menstrual cycle: estrogen effects in young women. *Neuropsychologia,* 40:518–29.

Makin, J. W., and R. H. Porter. 1989. Attractiveness of lactating females' breast odors to neonates. *Child Development,* 60:803–10.

Malas, M. A., S. Dogan, E. H. Evcil, and K. Desdicioglu. 2006. Fetal development of the hand, digits and digit ratio (2D:4D). *Early Human Development,* 82:469–75.

Malatesta, C. Z., and J. M. Haviland. 1982. Learning display rules: the socialization of emotion expression in infancy. *Child Development,* 53:991–1003.

Malcolm, C. A., D. L. McCulloch, and A. J. Shepherd. 2002. Pattern-reversal visual evoked potentials in infants: gender differences during early visual maturation. *Developmental Medicine & Child Neurology,* 44:345–51.

Maligaya, M. L., C. A. Chan, J. D. Jacobson, et al. 2006. A follow-up expanded study of the correlation of sperm velocity in seminal plasma and offspring gender. *Archives of Andrology,* 52:39–44.

Malouf, M. A., C. J. Migeon, K. A. Carson, et al. 2006. Cognitive outcome in adult women affected by congenital adrenal hyperplasia due to 21-hydroxylase deficiency. *Hormone Research,* 65:142–50.

Mandell, D. S., M. M. Novak, and C. D. Zubritsky. 2005. Factors associated with age of diagnosis among children with autism spectrum disorders. *Pediatrics,* 116:1480–86.

Mann, D. R., and H. M. Fraser. 1996. The neonatal period: a critical interval in male primate development. *Journal of Endocrinology,* 149:191–97.

Manning, J. T., A.J.G. Churchill, and M. Peters. 2007. The effects of sex, ethnicity, and sexual orientation on self-measured digit ratio (2D:4D). *Archives of Sexual Behavior,* 36:223–33.

Manning, J. T., D. Scutt, J. Wilson, and D. I. Lewis-Jones. 1998. The ratio of 2nd to

4th digit length: a predictor of sperm numbers and concentrations of testosterone, luteinizing hormone and oestrogen. *Human Reproduction*, 13:3000–3004.

Manstead, A.S.R. 1992. Gender differences in emotion. In *Handbook of Individual Differences: Biological Perspectives*, eds. A. Gale and M. W. Eysenck, 355–87. Chichester, England: John Wiley.

Marsh, H. W., and K. J. Rowe. 1996. The effects of single-sex and mixed-sex mathematics classes within a coeducational school: A reanalysis and comment. *Australian Journal of Education*, 40:147–62.

Martin, C. L., and L. M. Dinella. 2002. Children's gender cognitions, the social environment, and sex differences in cognitive domains. In *Biology, Society and Behavior: The Development of Sex Differences in Cognition*, eds. A. McGillicuddy-De Lisi, R. De Lisi, 207–39. Westport CT: Ablex.

Martin, C. L., and R. A. Fabes. 2001. The stability and consequences of young children's same-sex peer interactions. *Child Development*, 37:431–46.

Masters, M. S., and B. Sanders. 1993. Is the gender difference in mental rotation disappearing? *Behavior Genetics*, 23:337–41.

Matken, R., D. Karabinus, G. L. Harton, et al. 2003. MicroSort separation of X- and Y-chromosome bearing sperm: ongoing clinical trial results [abstract]. *American College of Obstetrics and Gynecology*.

Matthis, P., D. Scheffner, C. Benninger, et al. 1980. Changes in the background activity of the electroencephalogram according to age. *Electroencephalography and Clinical Neurophysiology*, 49:626–35.

Maxson, S. C. 1999. Some reflections on sex differences in aggression and violence. *Behavioral and Brain Sciences*, 22:232–33.

McClintock, M., and G. Herdt. 1996. Rethinking puberty: the development of sexual attraction. *Current Directions in Psychological Science*, 5:178–83.

McClure, E. B. 2000. A meta-analytic review of sex differences in facial expression processing and their development in infants, children and adolescents. *Psychological Bulletin*, 126:424–53.

McCord, J., C. S. Widom, and N. A. Crowell. 2001. *Juvenile Crime, Juvenile Justice*. Panel on Juvenile Crime: Prevention, Treatment, and Control. Washington DC: National Academy Press.

McDonnell, F. 2005. Why so few choose physics: An alternative explanation for the leaky pipeline [Editorial]. *American Journal of Physics*, 73:583–86.

McFadden, D. 1993. A masculinizing effect on the auditory system of human females having male co-twins. *Proceedings of the National Academy of Sciences USA*, 90:11900–11904.

———. 1998. Sex differences in the auditory system. *Developmental Neuropsychology*, 14:261–98.

McFadden, D., J. C. Loehlin, S. M. Breedlove, et al. 2005. A reanalysis of five studies on sexual orientation and the relative length of the 2nd and 4th fingers (the 2D:4D ratio). *Archives of Sexual Behavior*, 34:341–56.

McFadden, D., and E. G. Pasanen. 1998. Comparison of the auditory systems of heterosexuals and homosexuals: Click-evoked otoacoustic emissions. *Proceedings of the National Academy of Sciences USA*, 95:2709–13.

Mealey, L. 1999. Evolutionary models of female intrasexual competition. *Behavioral and Brain Sciences*, 22:234.

Meaney, M. J., and J. Stewart. 1981. Neonatal androgens influence the social play of prepubescent rats. *Hormones and Behavior*, 15:197–213.

Medland, S. E., J. C. Loehlin, and N. G. Martin. 2008a. No effects of prenatal hormone transfer on digit ratio in a large sample of same- and opposite-sex dizygotic twins. *Personality and Individual Differences*, 44:1225–34.

Medland, S. E., J. C. Loehlin, G. Willemsen, et al. 2008b. Males do not reduce the fitness of their female co-twins in contemporary samples. *Twin Research and Human Genetics*, 11:481–87.

Mehl, M. R., S. Vazire, N. Ramírez-Esparza, et al. 2007. Are women really more talkative than men? *Science*, 317:82.

Merke, D. P., J. D. Fields, M. F. Keil, et al. 2003. Children with classic congenital adrenal hyperplasia have decreased amygdala volume: potential prenatal and postnatal hormonal effects. *Journal of Clinical Endocrinology and Metabolism*, 88:1760–65.

Meulenberg, P. M., and J. A. Hofman. 1991. Maternal testosterone and fetal sex. *Journal of Steroid Biochemistry and Molecular Biology*, 39:51–54.

Meyer-Bahlburg, H. F. 2005. Gender identity outcome in female-raised 46, XY persons with penile agenesis, cloacal exstrophy of the bladder, or penile ablation. *Archives of Sexual Behavior*, 34:423–38.

Miller, J. L., C. Macedonia, and B. C. Sonies. 2006. Sex differences in prenatal oral-motor function and development. *Developmental Medicine and Child Neurology*, 48:465–70.

Mills, D. L., K. Plunkett, C. Prat, and G. Schafer. 2005. Watching the infant brain learn words: effects of vocabulary size and experience. *Cognitive Development*, 20:19–31.

Mizuno, R. 2000. The male/female ratio of fetal deaths and births in Japan. *Lancet*, 356:738–39.

Moffatt, S. D., E. Hampson, and M. Hatzipantelis. 1998. Navigation in a "virtual" maze: Sex differences and correlation with psychometric measures of a spatial ability in humans. *Evolution and Human Behavior*, 17:73–87.

Molfese, D. L., and V. J. Molfese. 1979. Hemisphere and stimulus differences as reflected in the cortical responses of newborn infants to speech stimuli. *Developmental Psychology*, 15:505–11.

Molfese, D. L., and R. C. Radtke. 1982. Statistical and methodological issues in "Auditory evoked potentials and sex-related differences in brain development." *Brain and Language*, 16:338–41.

Monastersky, R. 2005. Primed for numbers? *Chronicle of Higher Education*, Vol. 51, No. 26 (March 4), http://chronicle.com/free/v51/i26/26a00102.htm.

Mondschein, E. R., K. E. Adolph, and C. S. Tamis-LeMonda. 2000. Gender bias in mothers' expectations about infant crawling. *Journal of Experimental Child Psychology*, 77:304–16.

Moody, M. S. 1997. Changes in scores on the mental rotations test during the menstrual cycle. *Perceptual and Motor Skills*, 84:955–61.

Moore, D. S., and S. P. Johnson. 2008. Mental rotation in human infants: A sex difference. *Psychological Science*, 19:1063–66.

Mordecai, K. L., L. H. Rubin, and P. M. Maki. 2008. Effects of menstrual cycle phase and oral contraceptive use on verbal memory. *Hormones and Behavior*, 54:286–93.

Morris, R. E. 2003. Where the boys *aren't*: Gender demographics in the L2 classroom. Paper presented at the 36th Annual Conference of the Tennessee Foreign Language Teaching Association, Nashville, TN (November 7).

Morrison, F. J., E. M. Griffith, and D. M. Alberts. 1997. Nature-nurture in the classroom: entrance age, school readiness, and learning in children. *Developmental Psychology*, 33:254–62.

Morrongiello, B. A., and T. Dawber. 2000. Mothers' responses to sons and daughters engaging in injury-risk behaviors on a playground: implications for sex differences in injury rates. *Journal of Experimental Child Psychology*, 76:89–103.

Morrongiello, B. A., C. Midgett, and K-L. Stanton. 2000. Gender biases in children's appraisals of injury risk and other children's risk-taking behaviors. *Journal of Experimental Child Psychology*, 77:317–36.

Moss, H. A., and K. S. Robson. 1968. Maternal influences in early social visual behavior. *Child Development*, 39:401–8.

Muehlenbein, M. P., and R. G. Bribiescas. 2005. Testosterone-mediated immune functions and male life histories. *American Journal of Human Biology*, 17:527–58.

Mueller, S. C., V. Temple, E. Oh, et al. 2008. Early androgen exposure modulates spatial cognition in congenital adrenal hyperplasia (CAH). *Psychoneuroendocrinology.* 33:973–80.

Muhle, R., S. V. Trentacoste, and I. Rapin. 2004. The genetics of autism. *Pediatrics*, 113:e472–86.

Nagamani, M., P. G. McDonough, J. O. Ellegood, and V. B. Mahesh. 1979. Maternal and amniotic fluid steroids throughout human pregnancy. *American Journal of Obstetrics and Gynecology*, 134:674–80.

Nagy, E., H. Kompagne, H. Orvos, and A. Pal. 2007. Gender-related differences in neonatal imitation. *Infant and Child Development*, 16:267–76.

Nagy, E., K. A. Loveland, H. Orvos, and P. Molnár. 2001. Gender-related physiologic differences in human neonates and the greater vulnerability of males to developmental brain disorders. *Journal of Gender-Specific Medicine*, 4:41–49.

National Education Association. 2003. *Status of the American Public School Teacher*, 2000–2001 (August), http://www.nea.org/home/2233.htm.

National Science Foundation. 2006. Women, minorities, and persons with disabilities in science and engineering. Table C-8: Bachelor's degrees awarded in engineering, by sex, race/ethnicity, and citizenship: 1990–2007, http://www.nsf.gov/statistics/wmpd/pdf/tabc-8.pdf.

Nelson, R. J. 2000. *An Introduction to Behavioral Endocrinology,* 2nd Ed. Sunderland MA: Sinauer.

Neuman, S. B. 1986. The home environment and fifth-grade students' leisure reading. *Elementary School Journal,* 86:335–43.

Newcombe, N. S., L. Mathason, and M. Terlecki. 2002. Maximization of spatial competence: More important than finding the cause of sex differences. In *Biology, Society and Behavior: The Development of Sex Differences in Cognition,* eds. A. McGillicuddy-De Lisi and R. De Lisi, 183–206. Westport CT: Ablex.

Newmark, M., P. Merlob, I. Bresloff, et al. 1997. Click evoked otoacoustic emissions: inter-aural and gender differences in newborns. *Journal of Basic and Clinical Physiology and Pharmacology,* 8:133–39.

Ng, W. H., Y. L. Chan, K. S. Au, et al. 2005. Morphometry of the corpus callosum in Chinese children: relationship with gender and academic performance. *Pediatric Radiology,* 35:565–71.

NICHD Early Child Care Research Network. 2005. *Child Development and Child Care: Results from the NICHD Study of Early Child Care and Youth Development.* New York: Guilford Press.

———. 2007. Age of entry to kindergarten and children's academic achievement and socioemotional development. *Early Education and Development,* 18:337–68.

NIMH Fact Sheet. 2004/2007. Autism spectrum disorders research at the NIMH. Department of Health and Human Services, http://www.nimh.nih.gov/publicat/autism.cfm.

O'Brien, M., and A. C. Huston. 1985. Activity level and sex-stereotyped toy choice in toddler boys and girls. *Journal of Genetic Psychology,* 146:527–33.

Okagaki, L., and P. A. Frensch. 1994. Effects of video game playing on measures of spatial performance: Gender effects in late adolescence. *Journal of Applied Developmental Psychology,* 15:33–58.

Okten, A., M. Kalyoncu, and N. Yaris. 2002. The ratio of second- and fourth-digit lengths and congenital adrenal hyperplasia due to 21-hydroxylase deficiency. *Early Human Development,* 70:47–54.

Olson, S. 2005. *Count Down: The Race for Beautiful Solutions at the International Mathematics Olympiad.* Boston: Houghton Mifflin.

Ontario Ministry of Education. 2004. *Me Read? No Way! A Practical Guide to Improving Boys' Literacy Skills,* http://www.edu.gov.on.ca/eng/document/brochure/meread/meread.pdf.

Organization for Economic Cooperation and Development (OECD). 2004. *Learning for Tomorrow's World: First Results from PISA 2003.* Paris: OECD, http://www.pisa.oecd.org/dataoecd/1/60/34002216.pdf.

———. 2005. *School Factors Related to Quality and Equity: Results from PISA 2000.* Paris: OECD, http://www.oecd.org/dataoecd/15/20/34668095.pdf.

Örnkloo, H., and C. von Hofsten. 2007. Fitting objects into holes: On the development of spatial cognition skills. *Developmental Psychology*, 43:404–16.

Overman, W. H. 2004. Sex differences in early childhood, adolescence, and adulthood on cognitive tasks that rely on orbital prefrontal cortex. *Brain and Cognition*, 55:134–47.

Overman, W. H., J. Bachevalier, E. Schuhmann, and P. Ryan. 1996. Cognitive gender differences in very young children parallel biologically based cognitive gender differences in monkeys. *Behavioral Neuroscience*, 110:673–84.

Ozel, S., J. Larue, and C. Molinaro. 2004. Relation between sport and spatial imagery: Comparison of three groups of participants. *Journal of Psychology*, 138:49–63.

Paradise, J. L., H. E. Rockette, D. K. Colborn, et al. 1997. Otitis media in 2253 Pittsburgh-area infants: prevalence and risk factors during the first two years of life. *Pediatrics*, 99:318–33.

Parliamentary Office of Science and Technology (UK). 2003. Sex selection. *Postnote*, No. 198 (July).

Pasterski, V., P. Hindmarsh, M. Geffner, et al. 2007. Increased aggression and activity level in 3- to 11-year-old girls with congenital adrenal hyperplasia (CAH). *Hormones and Behavior*, 52:368–74.

Pasterski, V. L., M. E. Geffner, C. Brain, et al. 2005. Prenatal hormones and postnatal socialization by parents as determinants of male-typical toy play in girls with congenital adrenal hyperplasia. *Child Development*, 76:264–78.

Pellegrini, A. D., and P. K. Smith. 1998. Physical activity play: the nature and function of a neglected aspect of play. *Child Development*, 69:577–98.

Pellis, S. M., and V. C. Pellis. 2007. Rough-and-tumble play and the development of the social brain. *Current Directions in Psychological Science*, 16:95–98.

Peplau, L. A., L. R. Spalding, T. D. Conley, and R. C. Veniegas. 1999. The development of sexual orientation in women. *Annual Review of Sex Research*, 10:70–100.

Perelman, R. H., M. Palta, R. Kirby, and P. M. Farrell. 1986. Discordance between male and female deaths due to the respiratory distress syndrome. *Pediatrics*, 78:238–42.

Pergament, E., P. B. Todydemir, and M. Fiddler. 2002. Sex ratio: a biological perspective of "Sex and the City." *Reproductive BioMedicine Online*, 5:43–46.

Perry, D. F., J. DiPietro, and K. Costigan. 1999. Are women carrying "basketballs" really having boys? Testing pregnancy folklore. *Birth*, 26:172–77.

Peters, M., P. Servos, and R. Day. 1990. Marked sex differences on a fine motor skill task disappear when finger size is used as a covariate. *Journal of Applied Psychology*, 75:87–90.

Peterson, R. S. 2004. Language skills are in need of emergency aid. *Sacramento Bee* (November 26).

Peterzell, D. H., J. S. Werner, and P. S. Kaplan. 1995. Individual differences in contrast sensitivity functions: longitudinal study of 4-, 6- and 8-month-old human infants. *Vision Research*, 35:961–79.

Pettit, G. S., A. Bakshi, K. A. Dodge, and J. D. Coie. 1990. The emergence of social dominance in young boys' play groups: developmental differences and behavioral correlates. *Developmental Psychology*, 26:1017–25.

Piburn, D. 2005. Just say NO, to no touch. *MenTeach.org* (October), http://www.menteach.org/node/970.

Piccinelli, M., and G. Wilkinson. 1999. Gender differences in depression. Critical review. *British Journal of Psychiatry*, 177:486–92.

Pickles, A., E. Starr, S. Kazak, et al. 2000. Variable expression of the autism broader phenotype: findings from extended pedigrees. *Journal of Child Psychology and Psychiatry*, 41:491–502.

Pinker, Steven. 1995. *The Language Instinct*. New York: HarperPerennial.

———. 2005a. The science of difference: sex ed. *New Republic* (February 14), 232:15–17.

———. 2005b. The science of gender and science. Pinker vs. Spelke: A debate. *Edge* (May 16), http://www.edge.org/3rd_culture/debate05/debate05_index.html.

———. 2007. A brief history of violence. *New Republic* (March 19), 236:18–21.

Pinker, Susan. 2008. *The Sexual Paradox: Men, Women and the Real Gender Gap*, New York: Scribner.

Pipher, M. 1995. *Reviving Ophelia: Saving the Selves of Adolescent Girls*. New York: Ballantine.

Plante, E., V. J. Schmithorst, S. K. Holland, and A. W. Byars. 2006. Sex differences in the activation of language cortex during childhood. *Neuropsychologia*, 44:1210–21.

Plewis, I. 1997. Inferences about teacher expectations from national assessment at key stage one. *British Journal of Educational Psychology*, 67:235–47.

Pollack, W. 1998. *Real Boys: Rescuing Our Sons from the Myths of Boyhood*. New York: Henry Holt.

Pomerleau, A., G. Malcuit, L. Turgeon, and L. Cossette. 1997. Effects of labelled gender on vocal communication of young women with 4-month-old infants. *International Journal of Psychology*, 32:65–72.

Poppick, S. 2005. Scientists refute gender claims: Professors say society is responsible for gender imbalance in sciences. *Yale Daily News* (January 31), http://www.yaledailynews.com.

Potegal, M., and J. Archer. 2004. Sex differences in childhood anger and aggression. *Child and Adolescent Psychiatric Clinics of North America*, 13:513–28.

Preis, S., L. Jancke, J. Schmitz-Hillebrecht, and H. Steinmetz. 1999. Child age and planum temporale asymmetry. *Brain and Cognition*, 40:441–52.

Proverbio, A. M., V. Brignone, S. Matarazzo, et al. 2006. Gender and parental status affect the visual cortical response to infant facial expression. *Neuropsychologia*, 44:2987–99.

Pruessner, J. C., D. L. Collins, M. Pruessner, and A. C. Evans. 2001. Age and gender predict volume decline in the anterior and posterior hippocampus in early adulthood. *Journal of Neuroscience*, 21:194–200.

Puts, D. A., M. A. McDaniel, C. L. Jordan, and S. M. Breedlove. 2008. Spatial ability and prenatal androgens: meta-analyses of congenital adrenal hyperplasia and digit ratio (2D:4D) studies. *Archives of Sexual Behavior*, 37:100–11.

Quas, J. A., M. Hong, A. Alkon, and W. T. Byce. 2000. Dissociations between psychobiological reactivity and emotional expression in children. *Developmental Psychobiology*, 37:153–75.

Quigley, C. A. 2002. The postnatal gonadotropin and sex steroid surge—insights from the androgen insensitivity syndrome [Editorial]. *Journal of Clinical Endocrinology and Metabolism*, 87:24–28.

Quinn, P. C., and L. S. Liben. 2008. A sex difference in mental rotation in young infants. *Psychological Science*, 19:1067–70.

Quinn, P. C., J. Yahr, A. Kuhn, et al. 2002. Representation of the gender of human faces by infants: A preference for female. *Perception*, 31:1109–21.

Quiñones-Jenab, V. 2006. Why are women from Venus and men from Mars when they abuse cocaine? *Brain Research*, 1126:200–203.

Raag, T., and C. L. Rackliff. 1998. Preschoolers' awareness of social expectations of gender: relationships to toy choice. *Sex Roles*, 38:685–700.

Rahman, Q., G. D. Wilson, and S. Abrahams. 2004. Sex, sexual orientation, and identification of positive and negative facial affect. *Brain and Cognition*, 54:179–85.

Rais-Bahrami, K., S. Nunez, M. E. Revenis, et al. 2004. Follow-up study of adolescents exposed to di(2-ethylhexyl) phthalate (DEHP) as neonates on extracorporeal membrane oxygenation (ECMO) support. *Environmental Health Perspective*, 112:1339–40.

Raivio, T., J. Toppari, M. Kaleva, et al. 2003. Serum androgen bioactivity in cryptorchid and noncryptorchid boys during the postnatal reproductive hormone surge. *Journal of Clinical Endocrinology and Metabolism*, 88:2597–99.

Raizada, R. D., T. L. Richards, A. Meltzoff, and P. K. Kuhl. 2008. Socioeconomic status predicts hemispheric specialisation of the left inferior frontal gyrus in young children. *Neuroimage*, 40:1392–401.

Rapp, S. R., M. A. Espeland, S. A. Shumaker, et al. 2003. Effect of estrogen plus progestin on global cognitive function in postmenopausal women: the Women's Health Initiative Memory Study, a randomized controlled trial. *JAMA*, 289:2663–72.

Räty, H., J. Vänskä, K. Kasanen, and R. Kärkkäinen. 2002. Parents' explanations of their child's performance in mathematics and reading: A replication and extension of Yee and Eccles. *Sex Roles*, 46:121–28.

Rauscher, F. H., G. L. Shaw, L. J. Levine, et al. 1997. Music training causes long-term enhancement of preschool children's spatial-temporal reasoning. *Neurological Research*, 19:2–8.

Reid, R., and K. R. Harris. 1993. Self-monitoring of attention versus self-monitoring of performance: Effects on attention and academic performance. *Exceptional Children*, 60:29–40.

Reiner, W. G., and J. P. Gearhart. 2004. Discordant sexual identity in some genetic males with cloacal exstrophy assigned to female sex at birth. *New England Journal of Medicine*, 350:333–41.

Reinisch, J. M., and S. A. Sanders. 1992. Prenatal hormonal contributions to sex differences in human cognitive and personality development. In *Handbook of Behavioral Neurobiology, Vol. 2: Sexual Differentiation*, eds. A. A. Gerall, H. Moltz, I. L. Ward, 221–43. New York: Plenum.

Resnick, S. M., I. I. Gottesman, and M. McGue. 1993. Sensation-seeking in opposite-sex twins? An effect of prenatal hormones? *Behavior Genetics*, 23:323–29.

Reyes, F. I., R. S. Boroditsky, J. S. Winter, and C. Faiman. 1974. Studies on human sexual development. II. Fetal and maternal serum gonadotropin and sex steroid concentrations. *Journal of Clinical Endocrinology and Metabolism*, 38:612–17.

Rhodes, G. 2006. The evolutionary psychology of facial beauty. *Annual Review of Psychology*, 57:199–226.

Ribeiro, F. M., and R. M. Carvallo. 2008. Tone-evoked ABR in full-term and preterm neonates with normal hearing. *International Journal of Audiology*, 47:21–29.

Richards, M.P.M., J. F. Bernal, and Y. Brackwill. 1976. Early behavioral differences: gender or circumcision? *Developmental Psychobiology*, 9:89–95.

Richter, L. M. 2006. Studying adolescence. *Science*, 312:1902–5.

Rimer, S. 2005. For women in sciences, the pace of progress in academia is slow. *New York Times* (April 15).

Rimm, S. 1999. *See Jane Win*. New York: Three Rivers Press.

Robert, M., and G. Héroux. 2004. Visuo-spatial play experience: forerunner of visuo-spatial achievement in preadolescent and adolescent boys and girls? *Infant and Child Development*, 13:49–78.

Roberts, J. E., and M. A. Bell. 2000. Sex differences on a mental rotation task: Variations in electroencephalogram hemispheric activation between children and college students. *Developmental Neuropsychology*, 17:199–223.

———. 2002. The effects of age and sex on mental rotation performance, verbal performance, and brain electrical activity. *Developmental Psychobiology*, 40:391–407.

Roberts, S. 2007. Young earners in a big city, a gap in women's favor. *New York Times* (August 3).

Robertson, J. A. 2002. Sex selection for gender variety by preimplantation genetic diagnosis. *Fertility and Sterility*, 78:463.

Robinson, N. M., R. D. Abbott, V. W. Berninger, and J. Busse. 1996. The structure of abilities in math-precocious young children: Gender similarities and differences. *Journal of Educational Psychology*, 88:341–52.

Robles de Medina, P. G., G.H.A. Visser, A. C. Huizink, et al. 2003. Fetal behaviour does not differ between boys and girls. *Early Human Development*, 73:17–26.

Rodgers, C. S., B. I. Fagot, and A. Winebarger. 1998. Gender-typed toy play in dizygotic twin pairs: a test of hormone transfer theory. *Sex Roles*, 39:173–84.

Rogan, W. J., and N. B. Ragan. 2003. Evidence of effects of environmental chemicals on the endocrine system in children. *Pediatrics*, 112:247–52.

Rolnick, A., and R. Grunewald. 2003. Early childhood development: Economic development with a high public return. *fedgazette* (December), Federal Reserve Bank of Minneapolis, http://minneapolisfed.org/pubs/fedgaz/03-03/earlychild .cfm.

Ronald, A., F. Happé, and R. Plomin. 2006. Genetic research into autism. *Science*, 311:952.

Rose, R. J., J. Kaprio, T. Winter, et al. 2002. Femininity and fertility in sisters with twin brothers: Prenatal androgenization? Cross-sex socialization? *Psychological Science*, 13:263–67.

Rosen, T. S., and D. Bateman. 2004. The role of gender in neonatology. In *Principles of Gender-Specific Medicine*, Vol. 1, ed. M. J. Legato, 3–11. San Diego: Elsevier Academic Press.

Rosenberg, L., and S. Park. 2002. Verbal and spatial functions across the menstrual cycle in healthy young women. *Psychoneuroendocrinology*, 27:835–41.

Rosenblitt, J. C., H. Soler, S. E. Johnson, and D. M. Quadagno. 2001. Sensation seeking and hormones in men and women: Exploring the link. *Hormones and Behavior*, 40:396–402.

Rosenbloom, S. 2005. Big girls don't cry. *New York Times* (October 13).

Rosenfeld, R. M. 1995. What to expect from medical treatment of otitis media. *Pediatric Infectious Disease Journal*, 14:731–37.

Ross, J. L., G. A. Stefanatos, H. Kushner, et al. 2002. Persistent cognitive deficits in adult women with Turner syndrome. *Neurology*, 58:218–25.

Ross, M. T., D. R. Bentley, and C. Tyler-Smith. 2006. The sequences of the human sex chromosomes. *Current Opinion in Genetics and Development*, 16:213–18.

Rowling, J. K. 2005. *Harry Potter and the Half-Blood Prince*. New York: Scholastic.

Rubin, J. Z., F. Provenzano, and Z. Luria. 1974. The eye of the beholder: parents' view on sex of newborns. *American Journal of Orthopsychiatry*, 44:512–19.

Ruskai, M. B. 1991. Are there innate cognitive gender differences? Some comments on the evidence in response to a letter from M. Levin. *American Journal of Physics*, 59:11–14.

Rust, J., S. Golombok, M. Hines, et al. 2000. The role of brothers and sisters in the gender development of preschool children. *Journal of Experimental Child Psychology*, 77:292–303.

Rutter, M., A. Caspi, D. Fergusson, et al. 2004. Sex differences in developmental reading disability: New findings from 4 epidemiological studies. *JAMA*, 291:2007–12.

Ryan, A. S., Z. Wenjun, and A. Acosta. 2002. Breastfeeding continues to increase into the new millennium. *Pediatrics*, 110:1103–9.

Sadker, M., and D. Sadker. 1995. *Failing at Fairness: How Our Schools Cheat Girls*. New York: Scribner.

Salomone, R. C. 2003. *Same, Different, Equal: Rethinking Single-Sex Schooling*. New Haven CT: Yale University Press.

Sandberg, D. E., H. F. Meyer-Bahlburg, A. A. Ehrhardt, and T. J. Yager. 1993. The prevalence of gender-atypical behavior in elementary school children. *Journal of the American Academy of Child and Adolescent Psychiatry,* 32:306–14.

Sanders, J. 1997. Teacher education and gender equity. *ERIC Digest,* ED408277, http://www.ericdigests.org/1998-1/gender.htm.

Sandstrom, N. J., and C. L. Williams. 2004. Spatial memory retention is enhanced by acute and continuous estradiol replacement. *Hormones and Behavior,* 45:128–35.

Sang-Hun, C. 2007. Where boys were kings, a shift toward baby girls. *New York Times* (December 23).

Sax, L. 2005. *Why Gender Matters.* New York: Doubleday.

Schemo, D. J. 2006. More small women's colleges opening doors to men. *New York Times* (September 21).

Schmidt, I. M., M. Chellakooty, A-M. Haavisto, et al. 2002. Gender difference in breast tissue size in infancy: correlation with serum estradiol. *Pediatric Research,* 52:682–86.

Schmidt, I. M., C. Mølgaard, K. M. Main, and K. F. Michaelsen. 2001. Effect of gender and lean body mass on kidney size in healthy 10-year-old children. *Pediatric Nephrology,* 16:366–70.

Schmitt, D. P., A. Realo, M. Voracek, and J. Allik. 2008. Why can't a man be more like a woman? Sex differences in Big Five personality traits across 55 cultures. *Journal of Personality and Social Psychology,* 94:168–82.

Schoeters, G., E. Den Hond, W. Dhooge, et al. 2008. Endocrine disruptors and abnormalities of pubertal development. *Basic and Clinical Pharmacology and Toxicology,* 102:168–75.

Schrader, S., and T. Ansley. 2006. Sex differences in the tendency to omit items on multiple-choice tests: 1980–2000. *Applied Measurement in Education* 19:41–65.

Schrauf, M., E. R. Wist, and W. H. Ehrenstein. 1999. Development of dynamic vision based on motion contrast. *Experimental Brain Research,* 124:469–73.

Schugurensky, S. 2002. Rosenthal and Jacobson publish *Pygmalion in the Classroom* (1968). *History of Education: Selected Moments of the 20th Century,* http://fcis.oise.utoronto.ca/~daniel_schugurensky/assignment1/1968rosenjacob.html.

Schulman, J. D., and D. S. Karabinus. 2005. Scientific aspects of preconception gender selection. *Reproductive BioMedicine Online,* 10 Suppl. 1:111–15.

Schumacher, J., P. Hoffmann, C. Schmäl, et al. 2007. Genetics of dyslexia: the evolving landscape. *Journal of Medical Genetics,* 44:289–97.

Schwab, J., H. E. Kulin, E. J. Susman, et al. 2001. The role of sex hormone replacement therapy on self-perceived competence in adolescents with delayed puberty. *Child Development,* 72:1439–50.

Seifritz, E., F. Esposito, J. G. Neuhoff, et al. 2003. Differential sex-independent amygdala response to infant crying and laughing in parents versus non-parents. *Biological Psychiatry,* 54:1367–75.

Serbin, L. A., J. M. Connor, C. J. Burchardt, and C. C. Citron. 1979. Effects of peer presence on sex-typing of children's play behavior. *Journal of Experimental Child Psychology,* 27:303–9.

Serbin, L. A., D. Poulin-Dubois, K. A. Colburne, et al. 2001. Gender stereotyping in infancy: visual preferences for and knowledge of gender-stereotyped toys in the second year. *International Journal of Behavioral Development*, 25:7–15.

Servin, A., G. Gohlin, and L. Berlin. 1999. Sex differences in 1-, 3-, and 5-year-olds' toy-choice in a structured play-session. *Scandinavian Journal of Psychology*, 40:43–48.

Servin, A., A. Nordenström, A. Larsson, and G. Bohlin. 2003. Prenatal androgens and gender-typed behavior: a study of girls with mild and severe forms of congenital adrenal hyperplasia. *Developmental Psychology*, 39:440–50.

Setchell, K. D., L. Zimmer-Nechemias, J. Cai, and J. E. Heubi. 1997. Exposure of infants to phyto-estrogens from soy-based infant formula. *Lancet*, 350:23–27.

Shapleske, J., S. L. Rossell, P.W.R. Woodruff, and A. S. David. 1999. The planum temporale: a systematic, quantitative review of its structural, functional and clinical significance. *Brain Research Reviews*, 29:26–49.

Sharp, L. M. 2004. Pink when you wanted blue (January 25). BabyUniversity.com, http://babyuniversity.com/pink-when-you-wanted-blue.

Sharpe, R. M., B. Martin, K. Morris, et al. 2002. Infant feeding with soy formula milk: effects on the testis and on blood testosterone levels in marmoset monkeys during the period of neonatal testicular activity. *Human Reproduction*, 17:1692–1703.

Shaywitz, B. A., S. E. Shaywitz, K. R. Pugh, et al. 1995. Sex differences in the functional organization of the brain for language. *Nature*, 373:607–9.

Shaywitz, S. E., B. A. Shaywitz, J. M. Fletcher, and M. D. Escobar. 1990. Prevalence of reading disability in boys and girls: results of the Connecticut Longitudinal Study. *JAMA*, 264:998–1002.

Sheiner, E., A. Levy, M. Katz, et al. 2004. Gender does matter in perinatal medicine. *Fetal Diagnosis and Therapy*, 19:366–69.

Sherwin, B. B. 2003. Estrogen and cognitive function in women. *Endocrine Reviews*, 24:133–51.

Shettles, L. B., and D. M. Rorvik. 2006. *How to Choose the Sex of Your Baby: Fully Revised and Updated*, 6th Ed. New York: Broadway Books.

Shucard, J. L., D. W. Shucard, K. R. Cummins, and J. J. Campos. 1981. Auditory evoked potentials and sex-related differences in brain development. *Brain and Language*, 13:91–102.

Shumaker, S. A., C. Legault, L. Kuller, et al. 2004. Conjugated equine estrogens and incidence of probable dementia and mild cognitive impairment in postmenopausal women: Women's Health Initiative Memory Study. *JAMA*, 291: 2947–58.

Siegel, L. S., and I. S. Smyth. 2005. Reflections on research on reading disability with special attention to gender issues. *Journal of Learning Disabilities*, 38:473–77.

Simmons, R. 2002. *Odd Girl Out: The Hidden Culture of Aggression in Girls*. New York: Harcourt.

Simner, M. L. 1971. Newborn's response to the cry of another infant. *Developmental Psychology*, 5:136–50.

Simon, R. W., and L. E. Nath. 2004. Gender and emotion in the United States: Do men and women differ in self-reports of feelings and expressive behavior? *American Journal of Sociology*, 109:1137–76.

Simos, P. G., J. M. Fletcher, E. Bergman, et al. 2002. Dyslexic-specific brain activation profile becomes normal following successful remedial training. *Neurology*, 58:1203–13.

Simpson, E., P. Chandler, E. Goulmy, et al. 1987. Separation of the genetic loci for the H-Y antigen and for testis determination on human Y chromosome. *Nature*, 326:876–78.

Sininger, Y. S., B. Cone-Wesson, and C. Abdala. 1998. Gender distinctions and lateral asymmetry in the low-level auditory brainstem response of the human neonate. *Hearing Research*, 126:58–66.

Slabbekoorn, D., S.H.M. van Goozen, J. Megens, et al. 1999. Activating effects of cross-sex hormones on cognitive functioning: a study of short-term and long-term hormone effects in transsexuals. *Psychoneuroendocrinology*, 24:423–47.

Smith, Y. A., E. Hong, and C. Presson. 2000. Normative and validation studies of the nine-hole peg test with children. *Perceptual and Motor Skills*, 90:823–43.

Smithers, A., and P. Robinson. 2006. *The Paradox of Single-Sex and Co-educational Schooling*. Buckingham, England: Carmichael Press, http://buckingham.ac.uk/education/research/ceer/pdfs/hmcsscd.pdf.

Sommer, I. E., A. Aleman, M. Somers, et al. 2008. Sex differences in handedness, asymmetry of the planum temporale and functional language lateralization. *Brain Research*, 1206:76–88.

Sommer, I.E.C., A. Aleman, A. Bouma, and R. S. Kahn. 2004. Do women really have more bilateral language representation than men? A meta-analysis of functional imaging studies. *Brain*, 127:1845–52.

Speakman, J. R. 2004. Obesity: the integrated roles of environment and genetics. *Journal of Nutrition*, 134:2090S-105S.

Spelke, E. S. 2005. Sex differences in intrinsic aptitude for mathematics and science? A critical review. *American Psychologist*, 60:950–58.

Spencer, S. J., C. M. Steele, and D. M. Quinn. 1999. Stereotype threat and women's math performance. *Journal of Experimental Social Psychology*, 35:4–28.

Spielhofer, T., T. Benton, and S. Schagen. 2004. A study of the effects of school size and single-sex education in English schools. *Research Papers in Education*, 19:133–59.

Stern, M., and K. H. Karraker. 1989. Sex stereotyping of infants: a review of gender labeling studies. *Sex Roles*, 20:501–22.

Stipek, D. 2002. At what age should children enter kindergarten? A question for policy makers and parents. *Society for Research in Child Development*, Social Policy Report, Vol. 16, No. 2.

Stoel, R. D., E.J.C. De Geus, and D. I. Boomsma. 2006. Genetic analysis of sensation seeking with an extended twin model. *Behavior Genetics*, 36:229–37.

Stone, J., C. I. Lynch, M. Sjomeling, and J. M. Darley. 1999. Stereotype threat effects on black and white athletic performance. *Journal of Personality and Social Psychology,* 77:1213–27.

Storm, H., and M. Jenkins. 2002. *Go Girl! Raising Healthy, Confident and Successful Girls Through Sports.* Naperville IL: Sourcebooks.

Strauss, V. 2005. Educators differ on why boys lag in reading. *Washington Post* (March 15).

Strom, B. L., R. Schinnar, E. E. Ziegler, et al. 2001. Exposure to soy-based formula in infancy and endocrinological and reproductive outcomes in young adulthood. *JAMA,* 286:807–14.

Strube, M. J. 1981. Meta-analysis and cross-cultural comparison: Sex differences in child competitiveness. *Journal of Cross-Cultural Psychology,* 12:3–20.

Stuart, A., and E. Y. Yang. 2001. Gender effects in auditory brainstem responses to air- and bone-conducted clicks in neonates. *Journal of Communication Disorders,* 34:229–39.

Subrahmanyam, K., R. E. Kraut, P. M. Greenfield, and E. F. Gross. 2000. The impact of home computer use on children's activities and development. *The Future of Children: Children and Computer Technology,* Vol. 10, No. 2 (Fall/Winter), http://www.futureofchildren.org.

Susman, E. J., J. W. Finkelstein, V. M. Chinchilli, et al. 1998. The effect of sex hormone replacement therapy on behavior problems and moods in adolescents with delayed puberty. *Journal of Pediatrics,* 133:521–25.

Swaab, D. F., W.C.J. Chung, F.P.M. Kruijver, et al. 2002. Sexual differentiation of the human hypothalamus. In *Pediatric Gender Assignment: A Critical Reappraisal,* eds. S. A. Zderic, D. A. Canning, M. C. Carr, and H. M. Snyder, 75–105. New York: Kluwer Academic/Plenum.

Swaab, D. F., and M. A. Hofman. 1988. Sexual differentiation of the human hypothalamus: Ontogeny of the sexually dimorphic nucleus of the preoptic area. *Developmental Brain Research,* 44:314–18.

Swan, S. H., R. L. Kruse, F. Liu, et al. 2003. Semen quality in relation to biomarkers of pesticide exposure. *Environmental Health Perspectives,* 111:1478–84.

Swan, S. H., K. M. Main, F. Liu, et al. 2005. Decrease in anogenital distance among male infants with prenatal phthalate exposure. *Environmental Health Perspectives,* 113:1056–61.

Symonds, C. S., P. Gallagher, J. M. Thompson, and A. H. Young. 2004. Effects of the menstrual cycle on mood, neurocognitive and neuroendocrine function in healthy premenopausal women. *Psychological Medicine,* 34:93–102.

Tamimi, R. M., P. Lagiou, L. A. Mucci, et al. 2003. Average energy intake among pregnant women carrying a boy compared with a girl. *British Medical Journal,* 326:1245–46.

Tanner, J. M. 1990. *Fetus into Man: Physical Growth from Conception to Maturity* (Revised and enlarged edition). Cambridge MA: Harvard University Press.

Temple, E. 2002. Brain mechanisms in normal and dyslexic readers. *Current Opinion in Neurobiology*, 12:178–83.

Temple, E., G. K. Deutsch, R. A. Poldrack, et al. 2003. Neural deficits in children with dyslexia ameliorated by behavioral remediation: Evidence from functional MRI. *Proceedings of the National Academy of Sciences USA*, 100:2860–65.

Terlecki, M. S., and N. S. Newcombe. 2005. How important is the digital divide? The relation of computer and videogame usage to gender differences in mental rotation ability. *Sex Roles*, 53:433–41.

Thomas, H., W. Jamison, and D. D. Hummel. 1973. Observation is insufficient for discovering that the surface of still water is invariantly horizontal. *Science*, 181:173–74.

Thompson, T., and C. Ungerleider. 2004. Single sex schooling: Final report. Canadian Centre for Knowledge Mobilisation, http://www.cckm.ca/pdf/SSS%20Final%20Report.pdf.

Thomsen, T., K. Hugdahl, L. Erlsand, et al. 2000. Functional magnetic resonance imaging (fMRI) study of sex differences in a mental rotation task. *Medical Science Monitor*, 6:1186–96.

Thordstein, M., N. Lofgren, A. Flisberg, et al. 2006. Sex differences in electrocortical activity in human neonates. *Neuroreport*, 17:1165–68.

Thorne, B. 1993. *Gender Play: Girls and Boys in School*. Piscataway NJ: Rutgers University Press.

Tyre, P. 2005. Boy brains, girl brains: Are separate classrooms the best way to teach kids? *Newsweek* (September 19).

———. 2008. *The Trouble with Boys: A Surprising Report Card on Our Sons, Their Problems at School, and What Parents and Educators Must Do*. New York: Crown.

U.S. Department of Education, Office of Planning, Evaluation and Policy Development. 2005. *Single-Sex Versus Coeducational Schooling: A Systematic Review*, http://www.ed.gov/rschstat/eval/other/single-sex/single-sex.pdf.

U.S. Department of Education. 2007. Total fall enrollment in degree-granting institutions, by attendance status, sex of student, and control of institution: Selected years, 1947 through 2005 (Digest of Education Statistics, Table 179), National Center for Education Statistics, http://nces.ed.gov/programs/digest/d07/tables/dt07_179.asp.

U.S. Department of Labor. 2002. Women in high-tech jobs. *Facts on Working Women*, No. 02–01 (July), http://www.dol.gov/wb/factsheets/hitech02.htm.

U.S. Government Accounting Agency. 2003. *Women's Earnings: Work Patterns Partially Explain Difference between Men's and Women's Earnings*. Report to Congressional Requesters, GAO-04-35 (October), http://www.gao.gov/new.items/d0435.pdf.

UNICEF. 2001. A league table of child deaths by injury type in rich nations. *Innocenti Report Card*, Issue 2 (February), http://www.unicef-icdc.org/publications/pdf/repcard2e.pdf.

University of California Pediatric Environmental Health Specialty Unit, University of Kansas. 2008. Health care provider guide to safer plastics (July), http://www.coeh.uci.edu/PEHSU/BPA_HealthCareProviders.pdf.

Vadlamudi, L., R. Hatton, K. Byth, et al. 2006. Volumetric analysis of a specific language region—the planum temporale. *Journal of Clinical Neuroscience,* 13:206–13.

Van Anders, S. M., P. A. Vernon, and C. J. Wilbur. 2006. Finger-length ratios show evidence of prenatal hormone-transfer between opposite-sex twins. *Hormones and Behavior,* 49:315–19.

van Beijsterveldt, C. E., J. J. Hudziak, and D. I. Boomsma. 2006. Genetic and environmental influences on cross-gender behavior and relation to behavior problems: a study of Dutch twins at ages 7 and 10 years. *Archives of Sexual Behavior,* 35:647–58.

van de Beek, C., S. H. van Goozen, J. K. Buitelaar, and P. T. Cohen-Kettenis. 2009. Prenatal sex hormones (maternal and amniotic fluid) and gender-related play behavior in 13-month-old infants. *Archives of Sexual Behavior,* 38:6–15.

Van Goozen, S. H., P. T. Cohen-Kettenis, L. J. Gooren, et al. 1995. Gender differences in behaviour: activating effects of cross-sex hormones. *Psychoneuroendocrinology,* 20:343–63.

Van Tilburg, M.A.L., M. L. Unterberg, and J.J.M. Vingerhoets. 2002. Crying during adolescence: The role of gender, menarche, and empathy. *British Journal of Developmental Psychology,* 20:77–87.

Vandenbergh, J. G. 2003. Prenatal hormone exposure and sexual variation. *American Scientist,* 91:218–25.

Vanderkam, L. 2005. What math gender gap? *USA Today* (April 11).

Vanston, C. M., and N. V. Watson. 2005. Selective and persistent effect of foetal sex on cognition in pregnant women. *Neuroreport,* 16:779–82.

Vasey, P. L., and J. G. Pfaus. 2005. A sexually dimorphic hypothalamic nucleus in a macaque species with frequent female-female mounting and same-sex sexual partner preference. *Behavioral Brain Research,* 157:265–72.

Vasta, R., J. A. Knott, and C. E. Gaze. 1996. Can spatial training erase the gender differences on the water-level task? *Psychology of Women Quarterly,* 20:549–67.

Vellutino, F. R., and D. M. Scanlon. 2001. Emergent literacy skills, early instruction, and individual differences as determinants of difficulties in learning to read: The case for early intervention. In *Handbook of Early Literacy Research,* eds. S. B. Neuman and D. K. Dickinson, 295–321. New York: Guilford.

Verner, D. 2005. What factors influence world literacy? Is Africa different? World Bank Policy Research Working Paper 3496 (January).

Vilain, E. 2000. Genetics of sexual development. *Annual Review of Sex Research,* 11:1–25.

Vito, C. C., and T. O. Fox. 1981. Androgen and estrogen receptors in embryonic and neonatal rat brain. *Brain Research,* 254:97–110.

vom Saal, F. S., and F. H. Bronson. 1980. Sexual characteristics of adult female mice are correlated with their blood testosterone levels during prenatal development. *Science*, 208:597–99.

Voracek, M., and S. G. Dressler. 2007. Digit ratio (2D:4D) in twins: Heritability estimates and evidence for a masculinized trait expression in women from opposite-sex pairs. *Psychological Reports*, 100:115–26.

Voyer, D., A. Postma, B. Brake, and J. Imperato-McGinley. 2007. Gender differences in object location memory: a meta-analysis. *Psychonomic Bulletin and Review*, 14:23–38.

Voyer, D., S. Voyer, and M. P. Bryden. 1995. Magnitude of sex differences in spatial abilities: A meta-analysis and consideration of critical variables. *Psychological Bulletin*, 117:250–70.

Wada, J. A., R. Clarke, and A. Hamm. 1975. Cerebral hemispheric asymmetry in humans: Cortical speech zones in 100 adult and 100 infant brains. *Archives of Neurology*, 32:239–46.

Wade, N. 2006. A peek into the remarkable mind behind the genetic code. Review of *Francis Crick, Discoverer of the Genetic Code*, by Matt Ridley. *New York Times* (July 11).

Wager, T. D., K. L. Phan, I. Liberzon, and S. F. Taylor. 2003. Valence, gender, and lateralization of functional brain anatomy in emotion: a meta-analysis of findings from neuroimaging. *Neuroimage*, 19:513–31.

Walker, D., C. Greenwood, B. Hart, and J. Carta. 1994. Prediction of school outcomes based on early language production and socioeconomic factors. *Child Development*, 65:606–21.

Wallen, K. 1996. Nature needs nurture: the interaction of hormonal and social influences on the development of behavioral sex differences in rhesus monkeys. *Hormones and Behavior*, 30:364–78.

———. 2005. Hormonal influences on sexually differentiated behavior in nonhuman primates. *Frontiers in Neuroendocrinology*, 26:7–26.

Wallon, G. 2005. Aptitude or attitude? *EMBO Reports*, 6:400–402.

Wang, Z., L. J. Young, G. J. De Vries, and T. R. Insel. 1998. Voles and vasopressin: a review of molecular, cellular, and behavioral studies of pair bonding and paternal behaviors. *Progress in Brain Research*, 119:483–99.

Watson, J. S., L. A. Hayes, L. Dorman, and P. Vietze. 1980. Infant sex differences in operant fixation with visual and auditory reinforcement. *Infant Behavior and Development*, 3:107–14.

Weinraub, M., L. P. Clemens, A. Sockloff, et al. 1984. The development of sex role stereotypes in the third year: Relationships to gender labeling, gender identity, sex-typed toy preference, and family characteristics. *Child Development*, 55:1493–503.

Weiss, E., C. M. Siedentopf, A. Hofer, et al. 2003. Sex differences in brain activation pattern during a visuospatial cognitive task: a functional magnetic resonance imaging study in healthy volunteers. *Neuroscience Letters*, 344:169–72.

Wells, J.C.K. 2000. Natural selection and sex differences in morbidity and mortality in early life. *Journal of Theoretical Biology*, 202:65–76.

West, J., A. Meek, and D. Hurst. 2000. Children who enter kindergarten late or repeat kindergarten: Their characteristics and later school performance. U.S. Department of Education (NCES 2000–039).

Whitehurst, G. J., F. L. Falco, C. J. Lonigan, et al. 1988. Accelerating language development through picture book reading. *Developmental Psychology*, 24:552–59.

Whitehurst, G. J., and C. J. Lonigan. 2001. Emergent literacy: Development from pre-readers to readers. In *Handbook of Early Literacy Research*, eds. S. Neuman and D. Dickinson 11–29. New York: Guilford Press.

Whiting, B. B., and C. P. Edwards. 1992. *Children of Different Worlds: The Formation of Social Behavior.* Cambridge, MA: Harvard University Press.

Whitley, B. E., A. B. Nelson, and C. J. Jones. 1999. Gender differences in cheating attitudes and classroom cheating behavior: A meta-analysis. *Sex Roles*, 41:657–80.

Wichstrøm, L. 1999. The emergence of gender difference in depressed mood during adolescence: The role of intensified gender socialization. *Developmental Psychology*, 35:232–45.

Wiegand, S. 1998. Interview with Melanie Wood, Olympiad team member. *AWM Newsletter*, Vol. 28, No. 4 (July–August), http://www.awm-math.org/newsletter/199807/wood.html.

Wilcox, A. J., C. R. Weinberg, and D. D. Baird. 1995. Timing of sexual intercourse in relation to ovulation: Effects on the probability of conception, survival of the pregnancy, and sex of the baby. *New England Journal of Medicine*, 333:1517–21.

Williams, C. L., and W. H. Meck. 1991. The organizational effects of gonadal steroids on sexually dimorphic spatial ability. *Psychoneuroendocrinology*, 16:155–76.

Wingfield, J. C. 2005. A continuing saga: The role of testosterone in aggression. *Hormones and Behavior*, 48:253–55.

Winter, J.S.D., I. A. Hughes, F. I. Reyes, and C. Faiman. 1976. Pituitary-gonadal relations in infancy: II. Patterns of serum gonadal steroid concentrations in man from birth to two years of age. *Journal of Clinical Endocrinology and Metabolism*, 42:679–86.

Wiseman, R. 2002. *Queen Bees and Wannabees: Helping Your Daughter Survive Cliques, Gossip, Boyfriends, and Other Realities of Adolescence.* New York: Crown.

Witelson, S. F., I. I. Glezer, and D. L. Kigar. 1995. Women have greater density of neurons in posterior temporal cortex. *Journal of Neuroscience*, 15:3418–28.

Witelson, S. F., and W. Pallie. 1973. Left hemisphere specialization for language in the newborn. *Brain*, 96:641–46.

Wordes, M., and M. Nunez. 2002. Our vulnerable teenagers: Their victimization, its consequences, and directions for prevention and intervention. National Council on Crime and Delinquency, http://www.ncvc.org/ncvc/AGP.Net/Components/documentViewer/Download.aspxnz?DocumentID=32558.

Work/life & Women's Initiative Executive Committee. 2004. A decade of changes in the accounting profession: Workforce trends and human capital practices. American Institute of Certified Public Accountants, http://www.aicpa.org/download/career/wofi/ResearchPaper_v5.pdf.

Wynn, T. G., F. D. Tierson, and C. T. Palmer. 1996. Evolution of sex differences in spatial cognition. *Yearbook of Physical Anthropology*, 39:11–42.

Yankova, M., S. A. Hart, and C. S. Woolley. 2001. Estrogen increases synaptic connectivity between single presynaptic inputs and multiple postsynaptic CA1 pyramidal cells: a serial electron-microscopic study. *Proceedings of the National Academy of Sciences USA*, 98:3525–30.

Yawman, D., C. R. Howard, P. Auinger, et al. 2006. Pain relief for neonatal circumcision: a follow-up of residency training practices. *Ambulatory Pediatrics*, 6:210–14.

Yurgelun-Todd, D. A., W.D.S. Killgore, and C. B. Cintron. 2003. Cognitive correlates of medial temporal lobe development across adolescence: a magnetic resonance imaging study. *Perceptual and Motor Skills*, 96:3–17.

Zaichowsky, L. D., L. B. Zaichowsky, and T. J. Martinek. 1980. *Growth and Development: The Child and Physical Activity*. St. Louis: Mosby.

Zhang, Y., X. Jin, X. Shen, et al. 2008. Correlates of early language development in Chinese children. *International Journal of Behavioral Development*, 32:145–51.

Zimmer, B. 2006. Tracking snowclones is hard. Let's go shopping! *Language Log* (March 2), http://itre.cis.upenn.edu/~myl/languagelog/archives/002892.html.

Zimmerman, F., D. Christakis, and A. Meltzoff. 2007. Associations between media viewing and language development in children under age 2 years. *Journal of Pediatrics*, 151:364–68.

Zucker, K. J., S. J. Bradley, G. Oliver, et al. 1996. Psychosexual development of women with congenital adrenal hyperplasia. *Hormones and Behavior*, 30:300–18.

Index

variances in body mass index (BMI)
as attributable to, 320
See also DNA
genital tubercle, 26, 27
geography, 229, 230
geometry, 142, 208, 221, 230, 231, 235, 242
gesturing, 69–70, 192
Giedd, Jay, 289, 290
gifted children, 16, 146, 182, 214
Gilligan, Carol, 293
glass ceiling, 207, 211, 252, 274, 301, 343
glia, 6
Gneezy, Uri, 278
gonadarche, 288
gonadotropins, 87
grades (academic performance), 177, 179, 208, 237, 281, 297, 309
Greece, 169
Green, Richard, 114, 115, 334
grip strength, 64
gross motor skills, 65, 94, 97–98, 131, 133, 140–41, 144, 155
guitarists, 214
guns (toy), 127–29, 132, 220
Gurian, Michael, 9, 10, 14, 126, 173, 305, 320

Hagger, Corinne, 89
Halpern, Diane, 228
Hampson, Elizabeth, 190
hands-on learning, 163, 164–65, 202, 247–48
hands-on-play, 133
handwriting, 147–48, 165–66, 171, 184, 204–5
Haviland, Jeanette, 75, 76
head size, 57–58, 82
hearing. *See* auditory abilities
Hedges, Larry, 184
height, 12, 57
hemophilia, 21, 24
Henderson, Brenda, 41
heritability, of mental traits, 7, 193, 193n

high-school dropout rates, 4, 18, 300, 301
Hines, Melissa, 108, 157
hippocampus, 8, 31, 191, 221, 224–25, 253, 320
Hippocrates, 49
homicide, 264
homosexuality, 39–40, 43, 112n, 114–15, 287, 334. *See also* bisexuality; lesbians; sexual orientation
Hong Kong, 236
hormone replacement therapy (HRT), 191, 228, 308n, 341
hormones, 4, 6, 7, 16, 302, 308n
congenital adrenal hyperplasia (CAH) and, 36–37
cortisol (stress hormone), 36n, 74
depression and, 257–58
DHEA (dehydroepiandrosterone), 287–88
interest of girls in babies and, 129–30
language/verbal skills and, 189–91, 201
math/science differences and hormones after puberty, 226–28
mini-puberty in infancy and, 86–90
morning sickness and, 50
in newborns/infants, 55–56, 268
oxytocin, 263, 269, 305, 320
plastics and, 101
progesterone, 19, 190, 228
prolactin, 263, 269
puberty/sexuality and, 287–88, 290–93
social experience and, 262–63
spatial skills and, 225–28
toddlers/preschool age and, 104–5
treatment of Turner's syndrome with, 29
vasopressin, 263, 269
visual abilities and, 219–20
See also anti-Mullerian hormone (AMH); estrogen; testosterone
How to Choose the Sex of Your Baby (Shettles), 23